FOR DUMMIES™
COMPUTER
BOOK SERIES
FROM IDG

Modems For Dummies

D0817830

Cheat Sheet

If you're too busy even to read *Modems for Dummies*, here are some essentials.

Get Their Number

When You See This	It Has Something to Do with This	Turn to Glossary or Chapter
8,N,1 or 7,E,1	data word format, port settings, parameters	Chapters 3, 4
V.42, V.32*bis*	modem's capabilities	Chapter 4; Appendix A
14.4, 9600, 2400	speed of modem	Chapter 4; Appendix A
VT102	terminal emulation	Chapters 3, 4
80, 40, 24	BBS display	Chapter 4
F1	modem software Help key	Chapter 4, Appendix B
*70	disables call waiting	Chapter 4
&626%##@))_ _ _	garbage, line-noise	Chapter 15
COM1, COM2, IRQ4	com port settings	Chapters 3, 4
ATX4,Q0&K3	modem initialization string	Chapter 4

DOS Software Shortcuts

To Do This	Try Typing This (Your Software May Vary)
Open dialing directory	Alt-D
Download/upload file	PgDn/PgUp
Hang up	Alt-H
Change to terminal mode	Esc
Exit	Alt-X

Generic Software Settings to Try First:

If You're Calling	Set This	To This
CompuServe	Port Settings	7,E,1, 2400bps
	Terminal Type	TTY or VT102
	Duplex, Echo	Full
BBS	Port Settings	8,N,1, 2400bps
	Terminal Type	ANSI BBS
	Duplex, Echo	Full
GEnie	Port Settings	8,N,1, 2400bps
	Terminal Type	VT102
	Duplex, Echo	Half

. . . For Dummies: #1 Computer Book Series for Beginners

COMPUTER BOOK SERIES FROM IDG

Modems For Dummies

Cheat Sheet

What Stuff's On Your COM Ports?

Gadget	Port It's On	Interrupts
	COM1	IRQ 4
	COM2	IRQ 3
	COM3	IRQ 4
	COM4	IRQ 3

Write down the locations of your modem, mouse, scanner, serial printer, and any other serial devices you own. Don't try to use two gadgets occupying squares of the same "color" at the same time, because they'll try to do something ghastly, like share interrupts.

Tech-Support Numbers for Major On-Line Services

On-Line Service	"Voice" Phone Number
America Online	1-800-827-6364
CompuServe	1-800-848-8990
DELPHI	1-800-695-4005
Dow Jones	1-800-522-3567
GEnie	1-800-638-9636
MCI Mail	1-800-444-6245
Prodigy	1-800-284-5933
ImagiNation	1-800-SIERRA-1

Tone Phone Dialing Shortcuts

To Do This	Try Putting This Before the Phone Number
Disable call waiting	*70
Get an outside line and pause	9,

The words go by too fast!

Press These Simultaneously	To Make Service Do This
Ctrl-S	Pause text flow
Ctrl-Q	Resume text flow (after Ctrl-S)
Ctrl-O (Ctrl key and letter O)	Stop text flow to your computer without quitting current area
Ctrl-C or Break key	Quit current activity or area

IDG BOOKS

. . . For Dummies: #1 Computer Book Series for Beginners

™

References for the Rest of Us

COMPUTER BOOK SERIES FROM IDG

Are you intimidated and confused by computers? Do you find that traditional manuals are overloaded with technical details you'll never use? Do your friends and family always call you to fix simple problems on their PCs? Then the *"...For Dummies"*™ computer book series from IDG is for you.

"...For Dummies" books are written for those frustrated computer users who know they aren't really dumb but find that PC hardware, software, and indeed the unique vocabulary of computing make them feel helpless. *"...For Dummies"* books use a lighthearted approach, a down-to-earth style, and even cartoons and humorous icons to diffuse computer novices' fears and build their confidence. Lighthearted but not lightweight, these books are a perfect survival guide to anyone forced to use a computer.

> *"I like my copy so much I told friends; now they bought copies."*
> **Irene C., Orwell, Ohio**

> *"Quick, concise, nontechnical, and humorous."*
> **Jay A., Elburn, IL**

> *"Thanks, I needed this book. Now I can sleep at night."*
> **Robin F., British Columbia, Canada**

Already, hundreds of thousands of satisfied readers agree. They have made *"...For Dummies"* books the #1 introductory level computer book series and have written asking for more. So if you're looking for the most fun and easy way to learn about computers look to *"...For Dummies"* books to give you a helping hand.

IDG BOOKS

MODEMS FOR DUMMIES

by Tina Rathbone

IDG BOOKS

IDG Books Worldwide, Inc.
An International Data Group Company

San Mateo, California ✦ Indianapolis, Indiana ✦ Boston, Massachusetts

Modems For Dummies

Published by
IDG Books Worldwide, Inc.
An International Data Group Company
155 Bovet Road, Suite 310
San Mateo, CA 94402

Library of Congress Catalog Card No.: 93-61072

ISBN 1-56884-001-2

Printed in the United States of America

10 9 8 7 6 5 4 3 2 1

Distributed in the United States by IDG Books Worldwide, Inc.

Distributed in Canada by Macmillan of Canada, a Division of Canada Publishing Corporation; by Computer and Technical Books in Miami, Florida, for South America and the Caribbean; by Longman Singapore in Singapore, Malaysia, Thailand, and Korea; by Toppan Co. Ltd. in Japan; by Asia Computerworld in Hong Kong; by Woodslane Pty. Ltd. in Australia and New Zealand; and by Transword Publishers Ltd. in the U.K. and Europe.

For information on where to purchase IDG Books outside the U.S., contact Christina Turner at 415-312-0633.

For information on translations, contact Marc Jeffrey Mikulich, Foreign Rights Manager, at IDG Books Worldwide; FAX NUMBER 415-358-1260.

For sales inquiries and special prices for bulk quantities, write to the address above or call IDG Books Worldwide at 415-312-0650.

 is a trademark of IDG Books Worldwide, Inc.

About the Author

Tina Rathbone fell into writing about computers and software after Dan Gookin hired her as his assistant editor at San Diego's *ComputerEdge* magazine. After most of the weirdos left the magazine, she headed for *Supercomputing Review,* where she put together a monthly magazine full of neat-o graphics showing really big computers doing complex stuff like charting wind flow over airplane wings.

She purchased her first computer in 1987: an odd PC-compatible from San Diego's own Kaypro Computers. Right away she knew she needed a modem. Back then, there were no books — friendly or otherwise — on choosing or using modems. It took a second trip to the computer store, but she finally found some cables that matched up and made it on-line.

Today, Tina writes books about buying, fixing, and using personal computers. She enjoys prowling software stores and on-line services, finding programs that let people put those computers they bought to ever-weirder uses.

Tina lives with her terrific husband Andy and their cat, Laptop, in Ocean Beach, a San Diego beach community generally forgotten by time. When you're in San Francisco's Chinatown, Tina recommends heading for the Pearl City Dim Sum house and ordering Chicken Claws in Broth.

About IDG Books Worldwide

Welcome to the world of IDG Books Worldwide.

IDG Books Worldwide, Inc., is a division of International Data Group, the world's largest publisher of computer-related information and the leading global provider of information services on information technology. IDG publishes over 194 computer publications in 62 countries. Forty million people read one or more IDG publications each month.

If you use personal computers, IDG Books is committed to publishing quality books that meet your needs. We rely on our extensive network of publications, including such leading periodicals as *Macworld*, *InfoWorld*, *PC World*, *Computerworld*, *Publish*, *Network World*, and *SunWorld*, to help us make informed and timely decisions in creating useful computer books that meet your needs.

Every IDG book strives to bring extra value and skill-building instruction to the reader. Our books are written by experts, with the backing of IDG periodicals, and with careful thought devoted to issues such as audience, interior design, use of icons, and illustrations. Our editorial staff is a careful mix of high-tech journalists and experienced book people. Our close contact with the makers of computer products helps ensure accuracy and thorough coverage. Our heavy use of personal computers at every step in production means we can deliver books in the most timely manner.

We are delivering books of high quality at competitive prices on topics customers want. At IDG, we believe in quality, and we have been delivering quality for over 25 years. You'll find no better book on a subject than an IDG book.

John Kilcullen
President and C.E.O.
IDG Books Worldwide, Inc.

IDG Books Worldwide, Inc. is a division of International Data Group. The officers are Patrick J. McGovern, Founder and Board Chairman; Walter Boyd, President. International Data Group's publications include: **ARGENTINA's** Computerworld Argentina, InfoWorld Argentina; **ASIA's** Computerworld Hong Kong, PC World Hong Kong, Computerworld Southeast Asia, PC World Singapore, Computerworld Malaysia, PC World Malaysia; **AUSTRALIA's** Computerworld Australia, Australian PC World, Australian Macworld, Network World, Reseller, IDG Sources; **AUSTRIA's** Computerwelt Oesterreich, PC Test; **BRAZIL's** Computerworld, Mundo IBM, Mundo Unix, PC World, Publish; **BULGARIA's** Computerworld Bulgaria, Ediworld, PC & Mac World Bulgaria; **CANADA's** Direct Access, Graduate Computerworld, InfoCanada, Network World Canada; **CHILE's** Computerworld, Informatica; **COLOMBIA's** Computerworld Columbia; **CZECH REPUBLIC's** Computerworld, Elektronika, PC World; **DENMARK's** CAD/CAM WORLD, Communications World, Computerworld Danmark, LOTUS World, Macintosh Produktkatalog, Macworld Danmark, PC World Danmark, PC World Produktguide, Windows World; **EQUADOR's** PC World; **EGYPT's** Computerworld (CW) Middle East, PC World Middle East; **FINLAND's** MikroPC, Tietoviikko, Tietoverkko; **FRANCE's** Distributique, GOLDEN MAC, InfoPC, Languages & Systems, Le Guide du Monde Informatique, Le Monde Informatique, Telecoms & Reseaux; **GERMANY's** Computerwoche, Computerwoche Focus, Computerwoche Extra, Computerwoche Karriere, Information Management, Macwelt, Netzwelt, PC Welt, PC Woche, Publish, Unit; **HUNGARY's** Alaplap, Computerworld SZT, PC World; **INDIA's** Computers & Communications; **ISRAEL's** Computerworld Israel, PC World Israel; **ITALY's** Computerworld Italia, Lotus Magazine, Macworld Italia, Networking Italia, PC World Italia; **JAPAN's** Computerworld Japan, Macworld Japan, SunWorld Japan, Windows World; **KENYA's** East African Computer News; **KOREA's** Computerworld Korea, Macworld Korea, PC World Korea; **MEXICO's** Compu Edicion, Compu Manufactura, Computacion/Punto de Venta, Computerworld Mexico, MacWorld, Mundo Unix, PC World, Windows; **THE NETHERLAND'S** Computer! Totaal, LAN Magazine, MacWorld; **NEW ZEALAND's** Computer Listings, Computerworld New Zealand, New Zealand PC World; **NIGERIA's** PC World Africa; **NORWAY's** Computerworld Norge, C/World, Lotusworld Norge, Macworld Norge, Networld, PC World Ekspress, PC World Norge, PC World's Product Guide, Publish World, Student Data, Unix World, Windowsworld, IDG Direct Response; **PANAMA's** PC World; **PERU's** Computerworld Peru, PC World; **PEOPLES REPUBLIC OF CHINA's** China Computerworld, PC World China, Electronics International, China Network World; **IDG HIGH TECH BEIJING's** New Product World; **IDG SHENZHEN's** Computer News Digest; **PHILLIPPINES'** Computerworld, PC World; **POLAND's** Computerworld Poland, PC World/Komputer; **PORTUGAL's** Cerebro/PC World, Correio Informatico/Computerworld, MacIn; **ROMANIA's** PC World; **RUSSIA's** Computerworld-Moscow, Mir-PC, Sety; **SLOVENIA's** Monitor Magazine; **SOUTH AFRICA's** Computing S.A.; **SPAIN's** Amiga World, Computerworld Espana, Communicaciones World, Macworld Espana, NeXTWORLD, PC World Espana, Publish, Sunworld; **SWEDEN's** Attack, ComputerSweden, Corporate Computing, Lokala Natverk/LAN, Lotus World, MAC&PC, Macworld, Mikrodatorn, PC World, Publishing & Design (CAP), Datalngenjoren, Maxi Data, Windows World; **SWITZERLAND's** Computerworld Schweiz, Macworld Schweiz, PC & Workstation; **TAIWAN's** Computerworld Taiwan, Global Computer Express, PC World Taiwan; **THAILAND's** Thai Computerworld; **TURKEY's** Computerworld Monitor, Macworld Turkiye, PC World Turkiye; **UNITED KINGDOM's** Lotus Magazine, Macworld, Sunworld; **UNITED STATES'** AmigaWorld, Cable in the Classroom, CD Review, CIO, Computerworld, Desktop Video World, DOS Resource Guide, Electronic News, Federal Computer Week, Federal Integrator, GamePro, IDG Books, InfoWorld, InfoWorld Direct, Laser Event, Macworld, Multimedia World, Network World, NeXTWORLD, PC Games, PC Letter, PC World Publish, Sumeria, SunWorld, SWATPro, Video Event; **VENEZUELA's** Computerworld Venezuela, MicroComputerworld Venezuela; **VIETNAM's** PC World Vietnam

Dedication

To my family and all my on-line buddies, old and new.

Acknowledgments

This book owes a great big Thank You! to many helpful people:

Darren Albert, Amy Arnold, Tony Augsburger, Ian Barnard, Erik Basil, Rick Beardsley, Mary Bednarek, Kevin Behrens, Desirree Biggs, Gale Blackburn, Sandy Blackthorn, Valery Bourke, Jan Bowers, Mary Breidenbach, Munira Brooks, Trudy Brown, Barbara Byro, Angie Ciarloni, Kristen Cocks, Chris Collins, Boston Computer Society, Steve Crippen, Ben Cunningham, Janna Custer, Erik Dafforn, Leigh Davis, Ron Dippold, Brian Ek, Richard Ernst , Susan Estrada, Brad Fikes, Rob Fouer, Sam Fousias, Bob Francis, Cleveland Freenet, Allen Garrett, Harold Goldus, Sherry Gomoll, Dan and Sandy Gookin, Lynette Graham, Ghalib Habib, Carolyn Halliday, Jim Harrer, Rick Hemming, Tina Hildago, Steve Hine, Beth Jenkins, David Johnston, Sara Kavanagh, John Kilcullen, David Kishler, Tom Krogh, Allen Lahosky, Laurie Lance, Dennis Laufenberger, Jeff Leibowitz, Linda C. Lindley, Dave Loetham, Bob and Tracey Mahoney, Nico Mak, Barbara Maxwell, Patricia McCafferty, Theresa McGeary, Bill McKiernan, Kevin McLaughlin, Sandy Meeker, Mark Miller, Pushpendra Mohta, Drew Moore, Greg Ogarrio, Bob Olliver, Marta Partington, Cindy L. Phipps, Tom Powers, Bob Pritchard, Andy Rathbone, Tricia Reynolds, Jack Rickard, Greg Robertson, Mike Robertson, Alexis Rosen, Pete Royston, Doris Runyan, Margaret Ryan, Nick Sargologos, Lisa Senkevich, David Shargel, Beth Slick, Jeff Smith, Norris Parker Smith, Shelly Sofer, David and Terrie Solomon, Kerry Stanfield, Nancy Stout, Phil Talsky, Marc Teitler, Doug Wade, Matt Wagner, Russell Wagner, Chris Walcott, Milford Webster, John Williams, Rusty Williams, Lutz Winkler, Kathleen Yerby, Josef Zankowicz and many, many others. Thanks!

(The publisher would like to give special thanks to Patrick J. McGovern, without whom this book would not have been possible.)

Credits

Publisher
David Solomon

Managing Editor
Mary Bednarek

Acquisitions Editor
Janna Custer

Production Manager
Beth Jenkins

Senior Editors
Sandy Blackthorn
Diane Graves Steele

Production Coordinator
Cindy L. Phipps

Acquisitions Assistant
Megg Bonar

Editorial Assistant
Patricia R. Reynolds

Project Editor
Gregory R. Robertson
H. Leigh Davis

Editors
Kristin Cocks
Erik Dafforn
Marta J. Partington

Technical Reviewer
Beth Slick

Production Staff
Tony Augsburger
Mary Breidenbach
Valery Bourke
Chris Collins
Sherry Gomoll
Drew R. Moore
Gina Scott

Proofreader
Sandy Grieshop

Indexer
Anne Leach

Book Design
University Graphics

Say What You Think!

Listen up, all you readers of IDG's international bestsellers: the one — the only — absolutely world-famous ...*For Dummies* books! It's time for you to take advantage of a new, direct pipeline to the authors and editors of IDG Books Worldwide.

In between putting the finishing touches on the next round of ...*For Dummies* books, the authors and editors of IDG Books Worldwide like to sit around and mull over what their readers have to say. And we know that you readers always say what you think.

So here's your chance. We'd really like your input for future printings and editions of this book — and ideas for future ...*For Dummies* titles as well. Tell us what you liked (and didn't like) about this book. How about the chapters you found most useful — or most funny? And since we know you're not a bit shy, what about the chapters you think can be improved?

Just to show you how much we appreciate your input, we'll add you to our Dummies Database/Fan Club and keep you up to date on the latest ...*For Dummies* books, news, cartoons, calendars, and more!

Please send your name, address, and phone number, as well as your comments, questions, and suggestions, to our very own ...*For Dummies* coordinator at the following address:

...For Dummies Coordinator
IDG Books Worldwide
3250 North Post Road, Suite 140
Indianapolis, IN 46226

(Yes, Virginia, there really is a ...*For Dummies* coordinator. We are not making this up.)

Please mention the name of this book in your comments.

Thanks for your input!

IDG BOOKS

x

Contents at a Glance

Cartoons at a Glance
By Rich Tennant

About the Taglines

Taglines

What are the little sayings at the bottom of each page? Why, they're taglines.

Taglines are the bumper stickers of the on-line world. You can add one to the bottom of each e-mail message you write whenever you're feeling jovial. Taglines can be humorous "Thoughts for the Day" or as practical as advertising your company name and phone number. They range from racy to quizzical to just plain dumb.

Some of the livelier modem software even automates *tagging* for you, choosing taglines at random, adding them to your messages, and helping you "steal" the choicest taglines from messages you read.

You'll find taglines sprinkled throughout *Modems For Dummies*, both to give you the flavor of life on-line and to provide you with plenty of good ones to stick into your own messages.

*** Been there, done that, got the T-shirt. ***

Tagline Caution: From time to time, you'll see prank taglines in the messages of others. These contain computer instructions designed to fool you, the on-line beginner. These naughty taglines may *look* authoritative — but shun them anyway. They're bogus and they'll cause you great grief if you follow their advice. Below is a sampling of taglines to ignore:

- ✔ <Ctrl><Alt><Del> to read the next message
- ✔ ONLINE? Good! Hit <Alt-H> to take the I.Q. Test
- ✔ To steal this tagline, press <Ctrl><Alt><Del> now

The first and third ones will reboot your computer. And number two will probably take you off-line.

Clever. Not! Don't worry; most of the people you'll meet on-line really do have a life. Just ignore the computery-looking taglines and you'll be fine. Of course, you won't find any of these renegade taglines in *this* book.

Table of Contents

Part II: Modems Need Software (to Do Any Real Damage)...55

Part IV: Modem on the Blink (and Not Just Its Little Lights) 339

Chpater 13: Common Modem Mysteries 341

Introduction

*M*aybe you've seen modem geeks before, shuffling through the computer sections of bookstores and mumbling in quivery tones about *external protocols* to the nearest clerk. They're the ones who have gone over the edge ... the frail fringe who find social interaction, entertainment, and even employment through the *modem* gizmos plugged into their computers.

Hey, that's great, but ... yecchh! The word *telecommunications* still brings a shiver in your social circles. An outlandish thought takes hold: *You're* reading a book on modems; could it happen to you?

You stifle the urge to check your reflection for green ear hairs.

Relax. You're not into computers in a hardcore way. And you're a smart person, not a dummy. It's just that computers in general — and modems in particular — make you *feel* like a dummy sometimes.

Now, suddenly you need to use a modem. It could be a new productivity scheme they've dreamed up at work. Or maybe your kids are clamoring to get on one of the "on-line" services ... probably after one too many viewings of the Prodigy commercial, where the whole family — even the dog — gathers in the living room and happily squints at a computer screen, pestering some other family for brownie recipes.

Whatever the reason, you've decided to pick up a modem and use it. Fortunately, you've also picked up this book — to get you plugged in, turned on, and connected.

About This Book

This is a reference book. You can pick it up, consult it for help, and put it down again. I won't exhort you to read it cover to cover or insist that you let each chapter "build" on previous ones. (No sane person wants to know *all there is to know* about modems — at least not in one sitting.)

Most of all, I won't try to convince you that a modem is a "gateway to the world" or the magic key to a new lifestyle. You already have a life. You just need a modem reference.

You'll find plain, self-contained chapters on subjects like:

- ✔ Finding out what modems do — and *don't* do
- ✔ Plugging in your modem
- ✔ Setting up software to work with your modem
- ✔ Sending and receiving e-mail
- ✔ Joining CompuServe, America Online, or the Internet, for starters
- ✔ Knowing what to do when it doesn't work

Using This Book

Early modems were merely confusing. Today's super-duper models boast many more advances — and much more confusion — than their elders. Ditto for modem software. (Devious plot? No. Computer stuff naturally sprouts complexities, the way a shark pops out new teeth.)

Unfortunately, dozens of terms and abbreviations cling to each enhancement like barnacles on a whale. When the confusion starts to accumulate around a specific task or term, pad over to the Index or the Table of Contents.

(When you really zonk out, just browse randomly, counting the number of "x"s in a particular chapter or seeing how many times you spot the word "aardvark." Modems will do that to a person. Come back and join us anytime you're ready.)

When you seek a more general overview, check the Table of Contents for the appropriate chapter titles and page numbers.

You can keep the book open to that section while trying stuff out with your computer and modem, if it will help. By all means, keep reading in a given section if the subject seems interesting and you somehow feel driven to learn more.

If you have to type something, the text to type will look like this:

```
C:\> A:INSTALL THIS &^&%^$)** SOFTWARE
```

In this example, you type **A:INSTALL THIS &^&%^$)** SOFTWARE** after the C:\> and then press the Enter key on your keyboard. You'll be shown and usually told what to type, as in the preceding example.

A bird in the hand is a big mistake.

"But Mine's a Brand X!"

This book isn't about DOS or a popular software package, like Windows, for example, where you and I could go through *the same program* together, defining specific commands and finding out how to do stuff. Nooooo.

Instead, this book is about modems, which happen to work with a large assortment of communications software. The modem or software you bought may be different from the ones in the chapters. The service you're dialing may look different, too. Yet you'll still be able to follow along. All these things have more similarities than differences, especially when it comes to the bare-bones basics.

When a modem task involves specific DOS, Mac, or Windows commands, these are covered here, too. For general, jargon-free information about DOS, Macs, and Windows, you can refer to *DOS For Dummies*, *Macs For Dummies*, and *Windows For Dummies*, this book's venerable ancestors.

More on You

No, *not* moron. Nor even dummy. Actually, you're *neither one*. You're a very bright person, or you wouldn't be reading this book. But you have a modem. And trying to use the thing is making you feel like a dummy.

Just for once, you want to experience an effortless connection, get the information you need, hang up, and get back to that Tetris game. You want to see the office computer guru's bewilderment when he hears the screech of your modem instead of your screech for help!

Whether your questions focus on buying a modem or using one, this book is your key to reserving your guru for only the most critical help sessions. (Like when a friend spills Tang on your important floppy ...)

Just Skip It!

No computer gizmo can touch a modem when it comes to weird acronyms. For starters, *modem* itself is a contraction. Then there's the multisyllabic stuff they didn't bother abbreviating. (They must've run out of letter combinations.)

Why ask why? It's probably because when you're using a modem, you're betting on the successful interaction of a computer, a modem, software, and phone lines — at *each end* of the connection. (The abbreviations speed up conversations to tech support.)

To cut to the quick and solve your problem, this book discusses terms in clearly labeled, understandable nuggets. Any acronyms are defined. Some are explained. If I go into detail on some insanely obtuse aspect of modems, forgive me — I'm sort of a nerd. But these detours are clearly labeled with the ... *For Dummies* equivalent of those "Danger, Wet Floor" signs you see in the supermarket after they've mopped up the spilled milk.

When the road ahead appears muddy and you need to don your waders, you'll see the computer geek symbol. You can happily dodge whatever Technical Stuff comes next. Reading these sections will only expose you to the Muck of Geek, and bring you one step closer to the perilous line between normal human and *modem enthusiast*.

What's in This Book

This book is organized into five parts that cover general topics like modems, communications software, places to call, and so on. Parts consist of two or more chapters. These divvy up further into sections, which are clearly marked with headings, subheadings, and the Tips, Remembers, and other little *icons* (pictures) described at the end of this Introduction.

Each chapter ends with a special Section of Tens, containing ten (more or less) things to remember; ten bizarre BBSs to call; ten cool things to do on CompuServe; and so on. Everything relates somehow to the chapter's subject matter.

Bring on the parts.

Part I: Meet the Modem

Despite it being the perfect name for a snack food, *modem* is a contraction of *modulator/demodulator*, a device that lets you dial up other computers.

You'll discover reasons to do that. And you'll also be relieved to learn that most of the fears associated with modems are groundless. If you're dragging your modem out of the closet for another try, here's where you find out which type you have and what you can do with it. Not sure whether you own such a beast? You'll learn the tell-tale signs of a lurking modem. After you have a modem in hand, this part shows how to plug it in and make it work.

Part II: Modems Need Software (to Do Any Real Damage)

Here's where you find out you need to buy still more stuff before you can call someplace. Modem software is a large, glib family of programs that seem complex but actually *save* you from *sending direct commands to your modem*, in its own language. Shudder.

You'll find out how to install your software and set it up to take full advantage of your modem's capabilities.

Part III: Making the Modem Do Something

Where can you go and what can you do with a modem? It's going to take a while to exhaust all the suggestions stacked up in this part.

Here's where you find out about e-mail, file transfers, and other on-line staples. The top ten (or so) commercial on-line services display their wares here. And you find out just what this BBS (bulletin board system) stuff is all about. Into the Internet? It gets a chapter here, too.

Part IV: Modem on the Blink (and Not Just Its Little Lights)

Help! It's broken! (Probably not.) This part helps you tell for sure. Here's a question/answer tour through the thicket of settings and adjustments that are sometimes necessary to get everything working together. You'll find a field guide to tech support staffers and other soothing balms for when your modem jangles more than your phone line.

Part V: Bonus Part of Even More Tens

So many tips and secrets in modem land; we needed an extra Part of Tens just to fit everything in. Here you'll find more lists of ten: dangerous modem practices to avoid; the meaning of those little lights on external modems; translations for the weird, "C-U-L8TR" acronyms found sprinkled inside many on-line messages.

Icons Used in This Book

Dedicated modem users are different from the rest of us. When you encounter ... er, *interesting* behavior for the first time, here's where you can find out what it all means.

After you're engaged in a modem connection, Macintosh computers and IBM (and IBM-compatible) computers behave in a surprisingly similar fashion. Where Macs and PCs part ways, you'll see this icon.

This icon flags an important point to keep in mind.

Read these sections at your own risk. (Learning this much about modems can be hazardous to your social life.)

If there's an easier way, you'll find it nestled next to this icon.

It's best to walk slowly and drink plenty of water in the vicinity of this icon.

What's Next?

If the thought of modems gives you the shakes, Chapter 1 will shake your fears. Otherwise, plunge ahead to whatever pressing modem questions you may have.

This book is the resource that lets you log on with a minimum of fuss, weed through all the extraneous matter, and find whatever drove you on-line in the first place.

`Committee: a group that keeps minutes and wastes hours.`

Part I
Meet the Modem

The 5th Wave   By Rich Tennant

"NO, SIR, THIS ISN'T A DATING SERVICE. THEY INTRODUCE PEOPLE THROUGH A COMPUTER SO THEY CAN TALK TO EACH OTHER IN PERSON. WE INTRODUCE PEOPLE IN PERSON SO THEY CAN TALK TO EACH OTHER THROUGH A COMPUTER."

In this part...

Years ago, modems looked like a telephone handset cradle with a penchant for Sara Lee. Modem software? Most users wrote their own.

How-to books? Forget it. Trial and error drove early users to distraction ... or eventual mastery of the beasts. A geeky few pursued modems into the byways of obsession. Journalists followed, tossing back pages about hackers, cyberpunks, and phone freaks (phreaks). Normal people started wondering what all the fuss was about — lumping "modems" into the category of "dangerous playthings of the nerdy and obscure."

It's different now. Modems come in every shape and size. Software abounds. Nerds still *master* their modems. But increasing numbers of normal people buy and use the little devices.

Trouble is, after the first few tries fail, most of those people toss 'em in the closet. This part of the book retrieves your long-lost modem and gets you connected.

Chapter 1
The Modem (Is Not a Monster)

A modem (pronounced *mow-dum*), that techie-looking little box on your desk or the hidden electronic circuitry in your computer, is not the monster many people fear it to be. The modem is just another computer accessory. It lets your computer talk to other computers the way a phone lets you talk to other people.

"Why would someone allow a computer to have a social life?" you're wondering. "One computer's enough of a headache." That's true. And sometimes modems can be as much or more of a pain than computers.

Are they worth it? You bet. After a modem connects your computer to another PC through the phone line — and you're *on-line* — you can exchange files with other computer users. Type messages back and forth with old (and new) friends. Play games. Even ask computer gurus you meet on-line for help with your deepest, most secret, inner computer problems.

The key word here is *people*. A modem isn't a bridge between computers; it's a bridge between the humans sitting in front of them.

Don't worry — you can still do computery things with a modem. You can roam vast information warehouses, or *databases*, on giant university computers, looking for factoids for your reports. You can find the latest hot games and other programs on-line, and copy them to your computer. Or you can spend entire evenings — and great wads of cash — shopping in one of the many electronic "malls." Omigod!

Are you a Klingon or is that a turtle on your head?

You'll soon see that you don't have any good reasons to fear modems. This chapter tells you what modems are and what they do — easing you into some modem basics. Any indigestible material here will be burped back up in later parts. (Particularly gristly pieces are chewed more thoroughly in their own chapters.)

Meet the Modem

A modem is a keen-o device that lets you link up with other computers over the phone lines. It looks like either a run-of-the-mill answering machine or an electronic circuit board, depending on whether it's *external* or *internal*. An external modem is a separate device you hook up to your computer; an internal modem is a circuit board you place inside your computer. More info on both types is in Chapter 2.

Because a modem combines the power of your computer with the reach of the phone system, it gives you "global access to data and resources." (That's modem-nerd gush for "poking around on other peoples' computers, meeting people, or grabbing software for your computer.")

What's in it for me?

A modem is the quickest way for you to get a document from computer A to computer B without typing it all over again and using a spelling checker. Sending a file by modem — in minutes instead of overnight — beats Federal Express ... and saves you a good chunk of change. And a modem lets you grab a file while you're on the road — off your base computer — or even run programs at home, when you're not.

Lots of other modem tricks await you in Chapter 5 and beyond — after you jump over the hurdles of hooking it up (Chapter 2) and setting up the modem's software (Chapters 3 and 4).

Feel free to skip this introductory chapter at any time and go right for the meatier parts. Besides, you can always return here to browse while you're waiting for your modem to outlast the eternal busy signal on that popular electronic bulletin board system (*BBS*). For an overview of BBSs and other types of places on-line, refer to "You can call an electronic bulletin board" in this chapter. All you ever wanted to know about BBSs hides out in Chapter 11.

Can we talk?

The miracle of modems lets computers that normally snub each other "talk." If you want to share your Macintosh document with an IBM/PC clone user, for example, just save it in your word processor as a *text*, or *ASCII*, file. Send it by modem, and, magically, your Mac text is readable on your friend's alien clone. (You won't have much luck sharing *programs* between different computer families, though.)

Using the same "text" trick, modems can shuttle some types of data between two similar computers running different, incompatible programs. After you save it as text, you can spurt that WordPerfect file to the clone across town that only uses WordSwell, for example. To find out more stuff about text files, refer to Chapter 4.

Modemspeak and why it exists

Having a modem expands your computer's reach, sort of like the way cable expands your TV's reach. Yet no one feels the uncontrollable urge to dissect the inner workings of a cable box. Imagine how few customers would sign up for cable TV if they had to sit through a dull lecture first, or if they had to inhale lists of acronyms and abbreviations.

Bizarre terms pile up around modems faster than toothpaste flecks on a bathroom mirror. Don't let the mumbo-jumbo bog you down. With modems, as with anything else, you need know only enough to make it work for you. Making it work means getting *on-line*, or connected to another computer, successfully; doing what you want after you're on-line; finding what you need there; and hanging up (or *logging off*, in modemspeak).

Glitchbusters

After you're on-line, you'll discover an amazing phenomenon: Most modem users enjoy helping "newbies." Maybe they remember how it felt to be a rank beginner or something. Often, you can cajole these on-line gurus into solving general computer problems — as well as your modem enigmas.

Why are these total strangers so helpful? I don't know. But show your appreciation by thanking everyone who offers suggestions, listening carefully, and not asking the same question twice (of the same person, that is). Also, know that truly challenging queries require bribes of beer, pizza, or homemade German chocolate cake.

How can I use it when I can't use it?

After printers, modems are among the first computer add-ons people buy. Unlike printers, however, half the modems sold are never used. When people sense the dark nature of modems and communications software, they give up and throw the whole mess in the closet for "later."

That's because working with modems involves a frightful number of unknowns. You have to worry about all these things working right:

- ✔ Your computer
- ✔ Your modem
- ✔ Your software
- ✔ The phone line
- ✔ The computer at the other end
- ✔ The software on the computer at the other end

Even the *weather* counts. A connection can't go through unless all the elements are coexisting in peace and harmony. What's worse, each of these factors spawns its own lexicon of confusing terms and techno-geekisms.

So why does anybody bother with modems at all? It's because they let you and your computer do so many cool new things.

What's Using a Modem Like?

Whether it's a picture, a spreadsheet, or your letter, it's all *data* to your computer. Data is stored in units called *bits*. Well, any bits that can be stored on a computer can be sent and received by a modem.

Suppose that you want to print your résumé on your friend's hot new laser printer. You don't have time to drive across town to your friend's house. Sending it over the phone line is the only efficient way.

First, you select the data you want to send, or *upload*. You fire up your modem software, and it dials the phone number of your friend's modem. After they're connected, the two modems agree on the procedure for sending and receiving your résumé (the data).

You tap your keyboard, and your modem software starts the data flow from your computer into your modem, which turns the data into sound. The data shoots out over the phone line to your friend's house across town, as shown in Figure 1-1.

Figure 1-1:
By turning data into sound and back again, modems move files and stuff from one computer to another over ordinary phone lines.

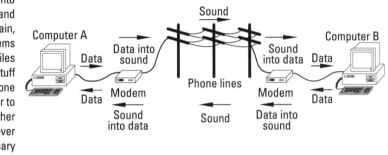

Your friend's phone line carries your data into his modem, which catches it, listens, and turns the data sounds back into plain old data that's understandable to the computer. During the whole process, the modems check for errors — in case a bad phone connection kept them from "talking" and "listening" correctly.

This process takes only minutes. Now your friend can exit the modem software, print your résumé, and use the expensive paper.

Skippable drivel on how modems work

Computer nerds named *modem* after the droll contraction of *mo*dulator/*dem*odulator. *Modulate* means changing those bits of data into sound to make it "fit" through the phone lines — like our voices do. And *demodulate* means changing sounds back into data.

Mo'dem Things Modems Can Do

The fun with modems isn't in *how* they work (except among the crowd who wear out their pocket protectors). It's what you can *do* with a modem that counts. All you need is your modem and some *communications software*. And a place to call, of course. After everything's hooked up and working correctly, you can:

- ✔ Call an on-line service and check the travel weather (and share pictures of your Chihuahua Brutus with total strangers while you're there), get sports info and stock prices, send faxes (even if you don't have a fax machine), and get help!

- ✔ Check your electronic mailbox at work for messages from modem-wielding coworkers

- ✔ Join CompuServe's Showbiz Forum and ask movie critic Roger Ebert how much popcorn he nibbles in an average year

- ✔ Call an electronic bulletin board system and send and receive files almost anywhere

- ✔ Skip the line at the fax machine — if you have a fax modem

- ✔ Run a spreadsheet on the computer at work, from the computer at home

- ✔ Play "host" to another modem caller to transmit files, or just for fun

- ✔ "Attend" live conferences where billionaire software magnates try to convince you that their products aren't overpriced or where you can exchange ideas with people in your field just by typing back and forth

- ✔ Make your friend buy a modem so that you can play modem games

- ✔ Fashion a new "face" for your on-line persona and play a few rounds of blackjack on ImagiNation

Doing Something Useful

Talking about modems is boring. It's much better to see one in action. It's like in MGM's *Wizard of Oz*, when the tornado forces Dorothy and Toto to dive-bomb the Wicked Witch's sister. Until then, the movie has unfolded in dull black and white. But one step over the farmhouse threshold into Oz and — ooh, aah, it's Technicolor Time.

Just talking about modems is like Dorothy in Kansas. Watching what a modem can do is like Dorothy in Oz (although Toto would skip the modem and opt for a palmtop organizer and a wireless link to MCI Mail, instead).

The following glimpses into Oz aren't in color, despite my pleadings with the publishers. (But it's OK to color them in when no one's looking.)

You can call an on-line service

Modems are a great tool to get people talking. Naturally, one of the most popular topics on-line is where else to call with a modem. Basically, you call three types of places: local or regional electronic bulletin boards, or *BBSs*; the *Internet*, a network of computer networks; and commercial *on-line services*. Some of the best places you'll see mentioned are these huge, exciting on-line services.

The commercial on-line services differ from the Internet and from the smaller bulletin board systems run by your local computer nerd — both in size and in the enormous scope of things you can do there. Most of the on-line services offer you the capability to make airline reservations, for example.

On-line services vary, but one thing they have in common is that they're run by large companies on large computers that let many people, sometimes tens of thousands, do stuff on-line at the same time.

Around the clock, people meet and discuss their interests. They go off by themselves to read the latest news from the Associated Press. They make travel reservations. Or they find new software and *download* it (copy it to their computers). They play "real-time" games, attend "real-time" discussion groups, and even develop "real-time" romantic interests. *Real-time* simply means typing stuff to someone and having them type right back at you. (Some of this stuff takes a *real* lot of time, too.)

Prodigy and CompuServe are two of the most popular on-line services, but many others compete for your dollar. The main ones get their own chapters later in this book. Flip to Chapter 5 to get an overview of on-line services and what they can offer you.

Getting your shareware of software

On-line services draw users in two ways. The biggest draw is having huge software libraries packed with goodies that users can *download*. A good on-line system carries file areas bulging with games, pictures, sounds, word processors, spreadsheet programs, and modem software — plus just about any other kind of software that exists.

The one kind of program you *won't* find on-line is commercial software. That's because it's illegal to upload, download, or even copy your programs for friends. When you see a word processor available for downloading, it's because it's being distributed as *shareware*. The program's author wants folks to share it on-line, in the hope of attracting more users.

The concept behind shareware is that you don't pay until you've used the program for a while and decide you like it. Then you send the required donation to the address posted in the prominent "guilt message" that pops up every time you start the program. (Registering the program does away with this annoying "send money" message.) Many excellent shareware programs are much cheaper than their commercial counterparts — and every bit as good (if not better). Shareware's a try-before-you-buy marketing scheme that has proven enormously successful for some program authors.

And the second-biggest draw for an on-line service? Plenty of chat areas, conferences, and other ways for people to meet and talk.

Poking around on Prodigy

One of the most well-known on-line services is called *Prodigy*. The first thing you see on Prodigy is a Highlights menu that suggests nifty areas to visit. Notice how the screen in Figure 1-2 lists what's new since the last time you called. (And displays an ad for AT&T, among others — Prodigy is the only major on-line service to toss commercials at you.) Prodigy gets its unique look from special Prodigy software you buy and install on your computer.

Prodigy lines up tools for getting around the service — right along the bottom of your screen. JUMP, the best tool, instantly beams you to any other Prodigy area. Figure 1-3 shows what happens when you type **JUMP EXPLORE**.

Zeroing in further, you decide to click the Travel box (or press 6). Figure 1-4 shows the screen you get. All the travel services on Prodigy pop up to dazzle you with dreams of traveling to far-away places.

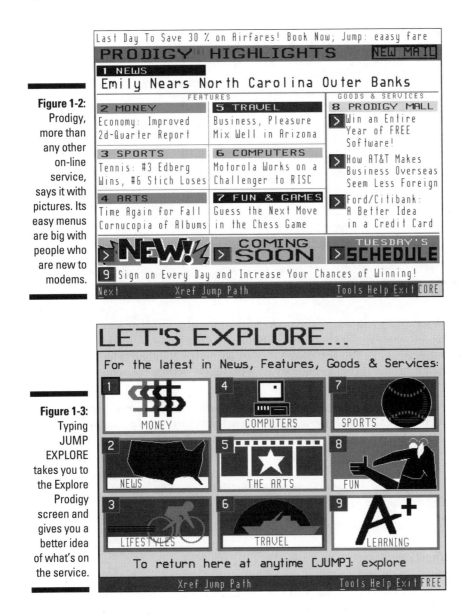

Figure 1-2: Prodigy, more than any other on-line service, says it with pictures. Its easy menus are big with people who are new to modems.

Figure 1-3: Typing JUMP EXPLORE takes you to the Explore Prodigy screen and gives you a better idea of what's on the service.

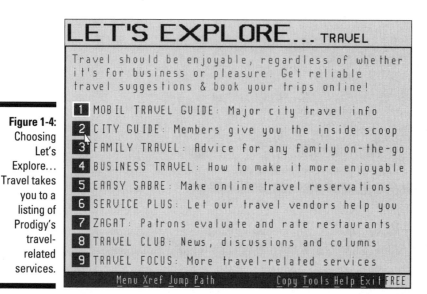

Figure 1-4:
Choosing
Let's
Explore...
Travel takes
you to a
listing of
Prodigy's
travel-
related
services.

✔ Prodigy is the biggest or second-biggest on-line service. (It depends on whether you count family members sharing the same account as separate members — as Prodigy does.) This megaservice, jointly run by IBM and Sears, has an urgent yet homey air. For help in getting connected to Prodigy, jump to Chapter 6.

✔ Nerds call Prodigy's picture screens a *graphical user interface*. Long-time members called it *slo-o-o-w*. These days, it can be downright zippy if you have a fast enough modem. Head to Appendix A to find out more about speedy modems versus the other kind.

✔ Prodigy has a soft spot for kids. There's Grolier's Academic American Encyclopedia, a weekly Background on the News, and National Geographic's Nova area, for starters.

✔ In Prodigy's TV ad, the computer sat too close to the fireplace.

Cruising CompuServe

CompuServe's the King Daddy of the on-line services — in age and reach (and probably membership). If you can imagine it, you can find it somewhere on CompuServe — whether "it" happens to be an obscure company's annual report, clues for the hottest new adventure game, or a recipe for guacamole pizza.

The welcome screen in Figure 1-5 shows how CompuServe looks when you're using a special, friendlier interface than CompuServe's normal look. It's called CompuServe Information Manager for Macintosh, or *MacCIM* for short. (Compu-Serve's Windows versions look just like this, so nobody should feel left out.)

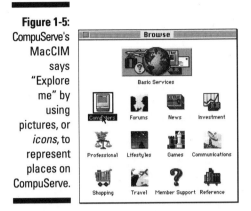

Figure 1-5:
CompuServe's
MacCIM
says
"Explore
me" by
using
pictures, or
icons, to
represent
places on
CompuServe.

Clicking Computers jolts MacCIM into dialing CompuServe and zooming you over to Computing Support (see Figure 1-6).

Figure 1-6:
Because
you clicked
the
Computers
icon to start,
MacCIM
dumps you
right into
Computing
Support.

From there, a new menu awaits your clicking pleasure. The highlighted item, Software Forums, is a good place to start for some answers to that nagging margins problem you're having in Microsoft Word, for example.

MacCIM slaps a friendly face on CompuServe; versions for Windows and DOS enliven sessions on PCs. But many first time callers don't yet have a special CompuServe Information Manager Program. Figure 1-7 shows how CompuServe looks when you're using a plain-vanilla (with no sweetener) communications program. Notice how the places to go show up as numbered items.

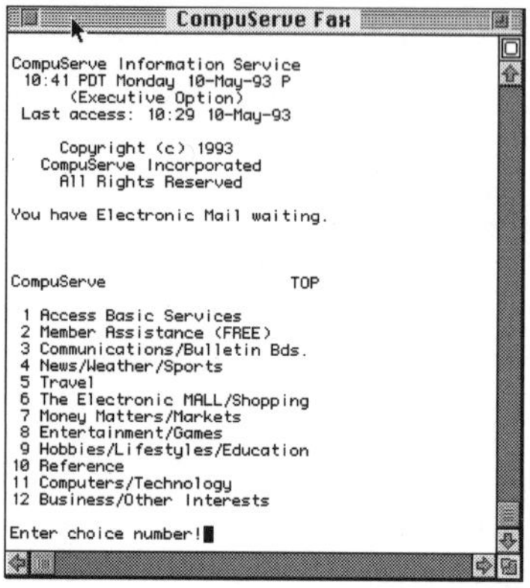

Figure 1-7:
CompuServe's
opening
screen looks
like this if
you're
calling with
a general-
purpose
communi-
cations
program.

```
╔═════════════ CompuServe Fax ═══════════════╗
CompuServe Information Service
10:41 PDT Monday 10-May-93 P
      (Executive Option)
Last access: 10:29 10-May-93

    Copyright (c) 1993
    CompuServe Incorporated
    All Rights Reserved

You have Electronic Mail waiting.


CompuServe                    TOP

 1 Access Basic Services
 2 Member Assistance (FREE)
 3 Communications/Bulletin Bds.
 4 News/Weather/Sports
 5 Travel
 6 The Electronic MALL/Shopping
 7 Money Matters/Markets
 8 Entertainment/Games
 9 Hobbies/Lifestyles/Education
10 Reference
11 Computers/Technology
12 Business/Other Interests

Enter choice number!█
```

Not quite as glamorous as MacCIM, eh? But some old hands prefer the texty directness of numbered lists.

✔ CompuServe started its $8.95 per month basic fee fairly recently. Before that, it charged almost that much per *hour*. (Some of CompuServe's "extended" services, like the forums, where people go to share special interests, still tally an hourly charge.)

That's why you may see it called Compu$erve by old timers who started modeming back in the days before all the on-line services, including Prodigy, charged hourly fees for their "plus" services.

Charge over to Chapter 7 for more details on rates and membership.

✔ CompuServe is so vast that you really need a guide. Otherwise, you'll never find that area your boss told you to call. Chapter 7 also points you in the direction of the CompuServe Directory Almanac and *CompuServe Magazine*, for starters.

✔ Every major computer company hosts at least one special area, called a *forum*, on CompuServe. You'll see the Lotus forum, the Hayes forum ... even tiny MegaMondo Modems, Inc., probably has a forum.

You can find support for a company's product on its forum, with no fear of suffering through hours of easy listening music while you're on hold. Discussions in computer forums give a sense of camaraderie among members who are likely to be struggling with the same problems as you.

You can call an electronic bulletin board

After shelling out the cash for a modem, some folks aren't ready to start shelling out even more bucks for an on-line service. So they start by calling the free places: a local bulletin board system, dubbed *BBS* by the committee in charge of computer initials.

A BBS is just a computer — usually a PC or a Macintosh — sitting in somebody's garage or den. Sometimes businesses run BBSs to give callers technical support, similarly to the support-oriented forums on CompuServe.

People can call up the BBS with their own computers and swap messages with other callers (see Figure 1-8). Some of these boards let you *download* programs: bring them into your own computer over the phone lines.

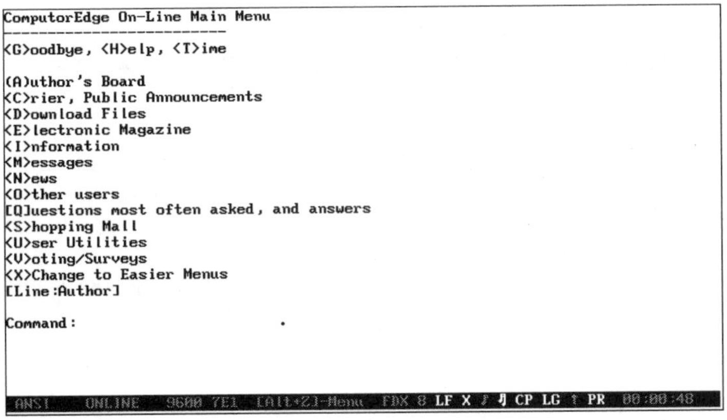

Figure 1-8:
A BBS
screen lists
areas for
swapping
messages,
downloading
files, and
discussing
current
events,
among other
diversions.

```
ComputorEdge On-Line Main Menu
-----------------------------
<G>oodbye, <H>elp, <I>ime

<A>uthor's Board
<C>rier, Public Announcements
<D>ownload Files
<E>lectronic Magazine
<I>nformation
<M>essages
<N>ews
<O>ther users
[Q]uestions most often asked, and answers
<S>hopping Mall
<U>ser Utilities
<V>oting/Surveys
<X>Change to Easier Menus
[Line :Author]

Command :

ANSI    ONLINE   9600 7E1  [Alt+Z]-Menu  FDX 8 LF X ♪ ♫ CP LG ↑ PR  00:00:48
```

Why would someone buy a PC and stick it in the garage for everyone to play with? Who knows? They range from eccentric computer nerd hobbyists to hip teenagers, all making their mark in the electronic age.

Suppose that you decide to view some messages in the Modems and BBSing section. Figure 1-9 shows an experienced computer user responding to a novice's question. Although the wizened guru sent this message, notice how he quotes the novice's original message in the first few lines. The question and answer appear together, yet it's easy to tell them apart. This makes it easier for *lurkers* who never post messages of their own to eavesdrop and perhaps learn something.

> ✔ Ask your computer dealer for the numbers of some local boards. After you've called up one board, you can discover the numbers for dozens, hundreds, perhaps thousands of other boards.

> ✔ Local boards are a good way to weed out bad computer dealers. You'll find plenty of people calling up and posting news of shady dealings or exceptional service.

Figure 1-9:
Nailing a
computer
problem can
be as simple
as calling a
local BBS
and posting
your
question.

```
Msg# :25998 *Modems & BBSing*
04/02/93 07:59:22
From: BOB FRANCIS
   To: MICAH MAY (Rcvd)
Subj: REPLY TO MSG# 25950 (MODEM)
MM>when i hook up to a bbs a lot of times it gets an extra line for each line s
MM>that it looks real screwed up can anyone help me with this problem also does
MM>anyone know how i talk to other people i tried everything and nothing works
MM>please help
MM>
Some BBS's require YOUR comm software to add "linefeeds" (such as here);
while others do it for you.  Obviously, the BBS's you're getting the
extra lines on don't.  Check your comm software's manual to learn how to
toggle this on-and-off.  It's usually some simple setting that can be
set individually for each BBS you call.


<->, <C>ubby, <R>eply, <A>gain, <N>ext, or <S>top?
ANSI    ONLINE  38400 8N1  [Alt+Z]-Menu   FDX 8 LF X ♪ ♫ CP LG ↑ PR   00:13:54
```

When you log on to a local board, you're usually required to type your name, address, and phone number. *But don't ever type your credit card number,* no matter who asks. Some unscrupulous folks prey on unsuspecting callers.

Save your credit card numbers for times you're ordering goods and services on the national services, like CompuServe or Prodigy.

Modem and Fax: What a Concept!

Modems have been around for years. Fax machines, too. Slowly, "I'll fax it" began replacing "I'll mail it" in corporate America. Then some Einstein noticed that modems and fax machines basically work the same way.

Voilà, the fax modem.

Today, almost any new modem can send and receive faxes. Like anything else, fax modems have their pros and cons:

✓ **Pro:** Faxes save paper. There's no printing something before faxing it. And you can view, then delete, most of the mundane faxes you get. No need to print 'em unless you want to.

✓ **Pro:** No cursing at the curly, evil fax paper.

✓ **Pro:** You can swamp dozens of people by "broadcasting" the same fax. Just give the fax software a list of folks to call and press a button.

✓ **Pro:** You can set up your fax software to send long-distance faxes after hours when the rates go down.

✔ **Pro:** Faxes sent by a fax modem look much better than documents you print and then end up accidentally smearing pastrami grease on while you fax them.

✔ **Con:** You can't fax someone the latest Callahan comic unless you somehow shove it into your computer first. This involves a *scanner* gizmo, plus all sorts of awkward software and words like "digitize."

✔ **Con:** Received faxes take up a lot of space on your hard drive. And, if your hard drive's too full, the fax won't show up at all.

✔ **Con:** Instead of hearing office gossip while in line at the fax machine, you'll hear only the efficient squeal of your fax modem. Bo-ring!

Look, I Just Want to Call the Office Computer

If you already know exactly what you need your modem to do, look it up in the table of contents and scoot over to that chapter. The rest of you can find other ideas in Chapter 5.

Reasons Why You Probably Already Have a Modem

Whether it was your idea or someone else's, you probably already have a modem.

✔ If your computer came with a modem but you've never been able to use it, shuffle forward to the installation tips in Chapter 2. You'll learn how to install and navigate tricky software currents in Chapters 3 and 4. After you get it set up and you want someplace to call, graze in Chapter 5 for a sample platter.

✔ If you bought a modem to connect with the office but gave up, dust it off and try again. If it's hooked up and ready to go, head over to Chapter 4.

✔ Did you like that TV commercial for Prodigy? Well, you won't have to *be* a prodigy to set up Prodigy if you turn to Chapter 6.

✔ You bought a modem just because it sounded like a neat thing to have. If you're brimming with enthusiasm, put it on your night table so you'll see it when you wake up in the morning. Then saunter coolly over to Chapter 2. After you hook up the modem, tweak your software's settings in Chapter 3.

```
The food here is terrible, and the portions are too small.
```

Modem Ahoy! (Finding Your Modem)

Not long ago, modems came in two shapes (see Figure 1-10). The external ones (right) sat on your desk and looked like squished, black Kleenex tissue boxes. The internal ones (left) looked more like a circuit board. (They *were* circuit boards.)

Figure 1-10:
For years, modems looked like one of these.

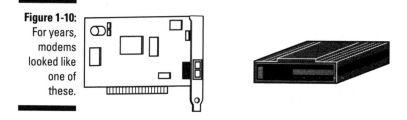

These days, it's entirely possible to have a modem and not even know it. Fashionable external modems like the ones in Figure 1-11 come in boxes that look more like cube puzzles. The vertical ones look like high-tech building vents or like something Mr. Spock would talk into.

Figure 1-11:
Today's racy designs obscure the modem's true identity.

If one of these sits on your desk, challenging you to "name that object," this could be the breakthrough you've been waiting for. Internal modems still look like a circuit board. (They still *are* circuit boards.) But because they hide inside your computer's case, it's doubly hard to tell whether you have one.

If you suspect a modem is lurking nearby, start your sleuthing here:

  ✔ Look for a *phone jack*: that little hole the plastic thingie on the end of a phone line plugs into. In fact, some modems have two jacks. Scrutinize the back of your PC's case. If one of those vertical slots houses something that looks like Figure 1-12, you have an internal modem.

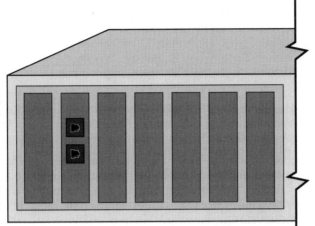

Figure 1-12:
If one of these things peeks out the back of your computer, you have an internal modem.

✔ Look for a phone line. I once worked in a building where the phone line crept down the hall toward my office and then snaked up the wall outside my door, plunging toward my desk and spiraling endlessly onto a nest of wires and cables. At the end of that line sat my modem.

If a similar mess sits at your feet, a modem may be anchoring that line. Trace the phone line to one of these four places:

1. The line leads to a boxy gizmo with letters along the front. This means you have an external modem. Compare it with Figure 1-10 and 1-11, just in case.

2. It leads to the jack in back of your computer, which looks like the one in Figure 1-12. If so, you have an internal modem.

3. The line leads nowhere. You don't see a boxy thing on your desktop, and there's no slot/jack gizmo on your PC's backside. It appears that you don't have a modem yet. On the bright side, when you *do* get a modem, having a ready phone line makes it easier to set up.

4. At the end of the line sits a noisy gray crate with "HIGH VOLTAGE" stickers plastered all over it. Go find the men in neon orange overalls and tell them to come and carefully take it away.

✔ Find the receipt from when you bought your computer. Try to decipher the salesperson's handwriting. The scribbled "medmo" or "mnmdm" may be a modem. (You still have to play "find the modem," though.)

✔ Look for modem software. On a Mac, it's usually in a folder called COMM (although Macs let you get as wordy as you want and you may see a folder labeled "Communications Program" or "Software for Calling CompuServe," or something even wordier). No luck? Try searching for folders with names like Microphone, ZTerm, White Knight, VersaTerm, or Smartcom. (These weird-sounding names identify Mac modem programs.)

✔ Internal modems are rare birds in the Macintosh world, but they cause a similar looking phone jack to peek out the back of your Mac. Rules for spotting external models are the same as for PCs: look for a boxy thing or cubistic sculpture attached to lots of cords.

✔ PCs with modems usually harbor a directory name like \COMM. If you use a DOS menu program, scan the file listing for words like Procomm, Smartcom, Qmodem, Microphone, Telemate, Telix, or Crosstalk. Or just file names with "talk," "tele," "comm," or the beloved "term" in them. No? Consider cornering someone who looks like she knows such things and asking bluntly whether your computer has a modem.

✔ Look in your closet. You never know what's lurking in there. Just don't get sidetracked and start reading your high school yearbooks and stuff.

Whew. Did you find one? If you found a modem and it's plugged in and set up, read Chapter 2 and then test it. Chapter 2 is also where you'll find out how to hook up your modem if you still need to do that.

Terms for term software

Modem surfers love jargon. For proof, look at the zillions of names they substitute for the simple-but-elegant *modem software*. You'll see *communications, comm, terminal, term,* and even *telecom* software. In the course of your modem career, you'll see even worse terms pop up. Luckily for you, this book sticks pretty much to *communications software* or *modem software*. (It isolates bizarre terms with a Technical Stuff icon so that you don't have to bother with them if you don't feel like it.)

Ten Phantom Modem Phobias

Among computer components, the prize for worst PR can safely go to modems.

These simple computer/phone devices have a knack for seizing the imaginations of otherwise rational beings. After it's embedded in the brain's fertile byways, the modem takes hold. Inevitably, irrevocably, this harmless device transforms into an unloved, pitiless monster ... the perverse spawn of the phone company's infinite reach and the computer's relentless drive for world domination.

Balderdash! (Don't you love that word?)

PCs and modems aren't really engaged in an evil campaign to rub out humankind and revel, unabated, for all eternity, in a sandbox of silicon dust.

While we're here, you can delete these other ridiculous notions from your brain cells:

Fear of viruses

Viruses don't just sit around in the lobby of Virus Central, awaiting word of a new modem to infiltrate. A few simple precautions rub out any worry of viral infection; some tips flow freely from Appendix C.

The modem will call the Pentagon and order the army to wear plaid shorts

A modem would insist on Jockeys. And no, your modem can't call anywhere without you giving the order first.

I'll look dumb in front of contemptuous youths

So what's new? Besides, modem-wielding youths will soon respect your adventuresome spirit. They may even end up being a great source of on-line tips.

The modem will call the IRS and tell them how much I really donated to charity last year

Don't worry, the modem naturally avoids governmental involvement. That's because if the government were awake, it'd be charging you extra for the privilege of having something as cool as a modem hooked up to your computer and phone line. Shhh.

My inborn fear of the phone company's capability to ruin me financially

There's some truth to this one, unfortunately. An on-line service's hourly rates can add up. Such dangers even accompany BBSing, for example, when you're downloading a mega-huge file and you're parked way outside your local toll-free zone.

Fortunately for you, this book's filled with tips for saving money while you're on-line. For example, most modem software displays a little clock that shows you every minute of connect time; you can use that clock to keep your budget in line while you're on-line. And besides, lots of fun places to call await you — right within Zone 1. (Unless you live in some pristine rural area; in which case, what are you doing in front of a computer when you could be fishing in some Alpine lake?)

It'll launch a SCUD by accident

Yep. The movie *Wargames* definitely gave modems a lot to live down. A young computer whiz (redundant, I know) played by Matthew Broderick thinks he's calling a game company's BBS and chooses a game named "Global Thermo-nuclear Warfare." Only it's *not* a game company.

Nope. That world-domination scenario is old hat. Modern modems definitely prefer redistributing the national debt, à la *Sneakers*.

My modem will call the Larry King Show and run for president

It won't do that. But if it did, it could probably come up with some better lines than the ones spouted by most candidates.

My mother told me never to talk to strangers

Since when did you start listening to your mother? Besides, she's probably already on-line, complaining to strangers that you never send her any e-mail.

Leaving one attached to my computer leaves me open to random invasion by another modem

This actually seems plausible. After all, if any phone can call your phone, why can't any modem call your modem? Trust me, you don't know enough at this point to get into this much trouble.

The modem will open a super-savers account at the credit union — just to get the freebie microwave cookbook

Everyone knows modems prefer fast food to cooking at home.

Chapter 2

Hooking Up Your Modem (and Making It Work Right)

* * *

In This Chapter

▶ Letting somebody else do it

▶ Plugging in an external modem

▶ Hooking up an internal modem

▶ Attaching a phone to your modem

▶ Testing your modem

▶ Ten reasons to be glad you installed an internal or external modem

* * *

*O*kay, you bought, borrowed, or found what seems to be a modem. Before this strange device can work with your computer, though, you need to introduce them.

"Computer, modem. Modem, computer."

Actually, there's a little more to it than that. This chapter explores the grisly details of adding a modem to your computer.

Remember when I said I'd flag any technoglop with the *Dummies* equivalent to those "Danger, Wet Floor" signs? Well, look at the top of this page — it's a Technical Stuff icon warning you away from this *entire chapter.* Keep reading — if you dare.

Unless you woke up this morning and said, "Yippee! Today I get to install my modem!" I strongly suggest that you skip this chapter. Go find a nerdly friend or pay someone to install your modem for you.

That goes double for internal modems.

This chapter gives you something to hand to your nerd pal so that it looks like you're being helpful. Or you can call out the installation steps to him or her — if it makes you feel better. (Stop if your friendly helper tells you to, though.)

A good friend *keeps* the surplus zucchini!

I Don't Have a Modem Yet

This chapter assumes that you have a modem and want to use it as more than a rather ugly paperweight. Actually, an *internal modem* (the kind that fits inside your computer) *would* make a cool paperweight if you could avoid catching your sweaters on the jagged parts.

If you haven't bought a modem yet, go right to Appendix A. Do not pass Go; do not collect $200. (You can get a pretty nice modem if you *spend* $200, though.)

If you choose an *external modem*, Appendix A reminds you to return here before you go shopping and read the sections on serial cables. By the way, you may as well buy a cable while you're at the computer store, because you'll need one before you can plug in your new toy.

Innie or Outie?

Like navels, modems come in internal and external models. If you're not sure which one you're contemplating, flip back to Chapter 1 for a refresher course.

What About Installing My Fax Modem?

Fax modems are just like regular modems, except that they can send and receive faxes. The difference comes in using the fax software. For installation tips, keep reading this chapter.

Macs Are Different, Aren't They?

Yes. And they like it that way. As far as modems go, connecting one to your Mac is the same plug-and-play operation it is for PC users. (Actually, it's a lot easier, because the place you plug it in is marked by a little picture of a phone. But [*whisper*] let's not discourage the others.)

Before You Move Your Desk ...

Both internal and external modems come with a phone cord. Usually it's long enough to reach from your wall to the modem. If not, you can buy an extra phone cord and a connector, called an RJ-11 coupler, that fits between the two cords. This is regular phone cord — the same kind you trip over until you spring for a cordless phone.

Hooking Up an External Modem (IQ: 100)

Handy Tools: your modem's manual, strong thumbs, a screwdriver, and a flashlight

What Else You'll Need:

- ✔ **Phone cable:** This is a flat wire that looks like the one in Figure 2-1. It runs between your modem and the phone jack in the wall. Most modems come with one. If yours didn't, Radio Shack's full of 'em, although you may be forced to call it an "RJ-11 cable." (Whisper it, though.) While you're at it, buy a nice long one.

Figure 2-1:
A typical modem phone cable.

- ✔ **Power adapter:** This plugs into the wall to give your modem electricity. External modems *always* come with one, similar to the one shown in Figure 2-2. If you lost the adapter, call the modem company and order a new one. (Your boom box's AC adapter will not do — and neither will the one from your friend's old modem. Modems use different sizes and styles of power adapters.)

Figure 2-2:
Your basic modem power adapter.

Most portable modems for laptops use batteries, although they usually come with an AC adapter for home or office use. A few portable modems don't use batteries or adapters; they suck power straight from the computer. Check your manual to see which one you have.

✔ **Serial cable:** Also called an RS-232 cable, this is what connects the back of your modem to a special plug on the back of your computer. Most modems don't come with one. Before you can buy one, you need to scrutinize the sockets on your PC's serial port and on the modem and then buy a cable that fits just right. Check out Table 2-1.

✔ When you go out to buy the serial cable, you need to make sure that you get one that fits your modem's port and your PC's serial port. Here's what to look for when shopping.

✔ Buy your serial cable from a nice dealer who'll cheerfully exchange it if you buy the wrong size.

✔ The best cables come with large thumbscrews at each end.

✔ Weirdly, connectors come with their own genders. Plugs, pins, or prongs make them "male." Holes, sockets, and such count as "female."

✔ If you're having problems finding a cable that fits, you can buy a special *9- to 25-pin adapter*, like the one in Figure 2-3. It alters the cable to suit your PC's whims and fancies (and sockets). Make sure that you get the adapter that has the holes and pins that fit your setup; they come both ways.

✔ You can plug into any PC that comes along if you buy a multiheaded serial cable, like the one in Figure 2-4. It's an ugly sight to behold, but it'll work just about anywhere. If you can find one, that is. Try the mail-order places in the back of computer magazines.

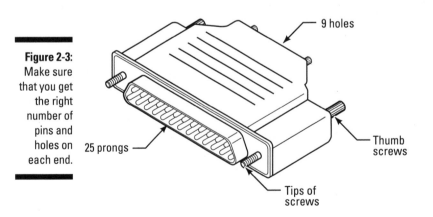

Figure 2-3: Make sure that you get the right number of pins and holes on each end.

9 holes

25 prongs

Thumb screws

Tips of screws

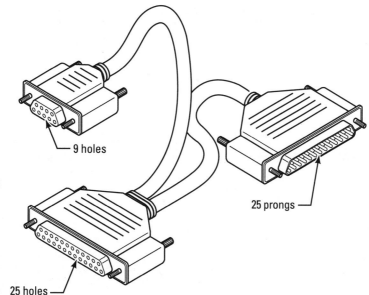

Figure 2-4:
A multi-
headed
serial cable
takes on all
comers.

9 holes

25 prongs

25 holes

✔ Mac users have it easy: A Mac serial cable has a round end (nerds, who shun "easy," call it a *9-pin DIN connector*). This plugs into the Mac's serial port (labeled with a little phone picture). The rectangular end plugs into the back of the modem, as with a PC.

Ahhh, simplicity itself.

✔ Your best bet is to draw a picture of the connectors on your modem and on your serial port. Dash down to the computer store with the picture safely stowed in your pocket — along with $15-20 bucks for the cable. Ouch!

✔ Any external modem can work with a Macintosh computer, but Macs need a special serial cable. Specify a "Macintosh hardware-handshaking" cable to the teenaged clerk behind the computer store's counter. This cable will whiz through any modem technology the nerds dream up — and is essential for use with something called *hardware flow control*, which is covered in Chapter 4.

✔ **Serial port:** Most PCs have two, lurking in one of your PC's slots. You need to find one that doesn't already have something plugged into it. Refer to Table 2-1 for the most common port lineup.

Table 2-1	What Serial Cable Should I Buy?
Modem's Socket	*Computer's Port*

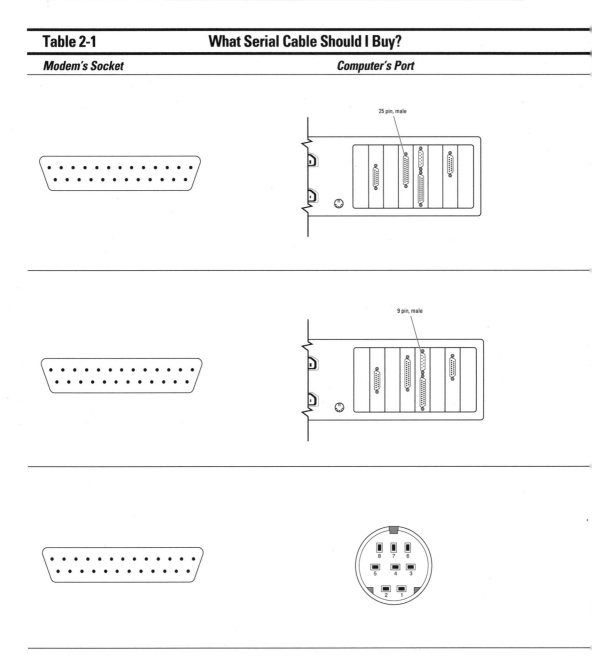

Cable You Need	Description
	Your serial cable needs connectors with 25 prongs at one end and 25 holes at the other. *Nerd's translation:* A DB-25P to DB-25S RS-232 cable.
	Your cable should have 25 prongs at one end, 9 holes on the other. *Nerd's translation:* A DB-25P to DB-9S RS-232 cable.
	You need a Mac serial cable. *Nerd's translation:* Mac RS-232 hardware-handshaking cable.

To keep your milk sweet, leave it in the cow.

✔ **Phone wall jack:** This is where you plug in your normal, everyday phone. Most homes in the twentieth century come with one; make sure that you have a newer, *modular* jack like the one in Figure 2-5. People who collect phone books for a hobby call this an *RJ-11* jack, in case you have to actually buy one. Radio Shack sells an adapter for those folks living in Victorian houses who've never updated from their old four-prong phone jack. Offices without modular jacks will have to install one, or else have the network geek figure it out.

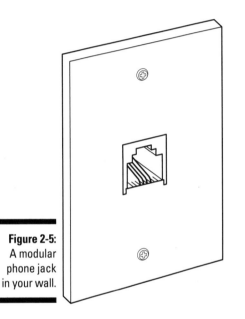

Figure 2-5:
A modular
phone jack
in your wall.

Follow these steps to hook up your modem and computer:

1. **Save your work, exit the programs you're using, and turn off your PC's power.**

   Unplug it, too. Turn off and unplug your monitor and anything else that's attached to your computer. Label the cables so that you know what goes where when you put everything back together.

2. **Find the modem's power switch; make sure that it's off.**

3. **Find the modem's power adapter.**

   The small end that looks like a nose-hair remover goes in the modem. Plug the other end into a power strip or standard wall socket.

Visual cues

Icons show up in books like this one; just look at the little "tip" picture to the upper left. You see them on elevators and on just about everything else these days. Some external modems even wear friendly icons to show where everything plugs in. The modem in Figure 2-6 models little pictures of a phone jack, a phone, a computer, and a power adapter.

Gently remove your modem from its box. Ogle its underbelly for any pictures or labels. (Soon we'll devolve back to counting little bison icons on cave walls.)

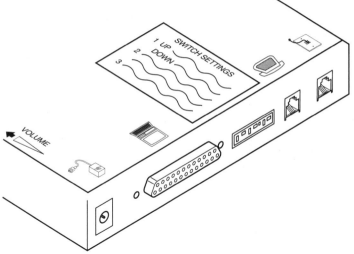

Figure 2-6: Friendly modems like this US Robotics Sportster give you hints about what plugs in where.

Yikes! Power spikes!!

Plug your computer stuff (computer, monitor, modem, and anything else that you use with your computer) into a power strip with *surge suppression.* That protects your computer from sudden electrical mood swings. Alas, there's no way to protect it from *your* mood swings.

Also, be sure to unplug your computer's power strip at the first signs of an electrical storm. Be sure to unplug your modem from the wall, too. *Spikes* suit your computer even less than dreadlocks.

4. **Find an empty serial port on the PC.**

Look at the back of your PC, where all the weird-looking plugs are. Some plugs like the ones in Table 2-1 should peek back at you.

If this step brings forth a puzzled look or a quaint, unprintable colloquialism from your nerd pal, dog-ear Chapter 13 and shove it at your friend, along with a generous portion of fish tacos. Yummy. Not in California? Cod-flavored Doritos should substitute nicely.

5. **Grab the serial cable with one hand and the modem with the other.**

Aligning the pins and holes, gently jiggle one end of the cable into the back of the modem. Jiggle the other end into your PC's empty serial port.

6. **Turn the serial cable's thumbscrews with strong, confident thumbs.**

You can use some other fingers if need be. Some cables have little screws that need a little screwdriver to tighten them.

7. **Save this step for when you're not expecting any phone calls: Unplug your phone's cable from the wall jack.**

Take your modem's phone line and plug one end into the wall jack. Plug the other end into the modem jack that says "telco" or "line." If there's no label, try the left jack. (Isn't Telco the Ronco subsidiary that makes the new CompuNerd Dehydrator?)

The other jack is where you plug in your telephone. It may come as a shock after the obscure "telco," but this is the one labeled "phone."

8. **Plug everything back in.**

Don't forget the monitor.

TIP

Huge plug syndrome

Power adapters typically suffer from plug-elephantiasis. These huge, boxy plugs take up more than their share of power strip. If your modem plug covers up too many outlets, buy a short extension cord and plug it into the strip. The Pluggus Gigantus can cover up your new extension cord instead.

If you don't feel like buying YAG (yet another gizmo), sometimes you can plug them in at the very end outlet; that way, the big part hangs over the edge of the surge protector and all is well.

Surging in the phone lines

You can buy a phone-line surge suppressor to ward off yet another electrical hazard: power surges zapping the phone line. You plug your phone line into the gizmo and then plug it into your wall jack.

This is a Good Thing for people who live in stormy areas. One good lightning jolt may ruin your whole day, not to mention an expensive modem and possibly the (even-far-more-expensive) PC attached to it. Some stores stock power strips with built-in phone jacks for one-stop surge suppression.

9. You're all finished.

Take time for a good sneeze or two after crawling around and inhaling all the dust bunnies piled up around your wall jack. Actually, this is a good time to dust the back of your computer, because dust can get sucked inside and make it run hotter. (Use a vacuum cleaner — with a hose attachment, not in its upright position. And turn your keyboard upside-down and vacuum up all *that* crud too, while you're at it.) Then head for "Testing the Waters" at the end of this chapter.

10. Turn on the power switches for your computer, monitor, and modem.

The Fine Print About Internal Modems

Congratulations! You've found your way all the way down here. I guess that means you want to install your internal modem. Before you start, be aware of or do these things:

✔ Installing an internal modem is bound to expose you to all sorts of icky technical terms and practices.

✔ You don't need a power cord or an electrical outlet for your internal modem. You *do* need to make sure that your computer's *power supply* is powerful enough to handle the newcomer. (Most of you won't have to worry about this. But those of you with very old computers and very new, powerful internal modems need to ask your dealer or nerdy pal to check your computer's current power supply. The power supply is the computer part that feeds electricity to your computer.)

✔ Internal modems come on little cards that plug into something called an *expansion slot* inside your PC's case. Your modem needs a slot all to itself. If your PC is fairly new and you don't have many gizmos attached yet, proceed with confidence. You probably have a free slot.

✔ *Expansion slots* are the parking spaces for computer ports inside your PC.

If the back of your PC bristles with cords, cables, and connectors, pop the case off and look for a spare slot. Otherwise, it's back to the store for an external modem.

✔ Your PC's slots come in different sizes: small (called *8-bit*), medium (*16-bit*), and large (*32-bit*, rare in the PCs that belong to most people). Most internal modems come on the smaller, *8-bit cards*, but a few bulk up on fax capabilities and other stuff and grow to the larger, 16-bit size. Make sure that the modem card fits your available slot size.

COM1, COM2, What Should I Do?

Your internal modem talks to your computer through something called a *serial* port. (You'll also see it called a *COM*, for "communications," port.) Other parts of your computer talk through COM ports, too, especially a serial mouse.

A *port* is a place on your computer that you use to connect another piece of computer equipment to your computer. Ports come in two varieties. You use *serial* ports to connect your modem and serial mouse to your computer. You use *parallel*, or *LPT*, ports to connect a printer to your computer. Internal modems provide their own serial port, but because of interrupt conflicts and other potential difficulties, most new users don't go beyond the traditional COM1 and COM2. Chapter 13 covers ports in greater detail.

A serial mouse usually claims the computer's first communications port, COM1. You need to look at the part of your internal modem's manual that tells you how to set special *switches* or *jumpers* to make the modem use COM2. *Hint:* Most internal modems come preset to COM2. (How do they know?) Refer to Chapter 13 to dig up the technical details on jumpers, switches, and internal modem assignments.

If a mouse and some other serial gizmo hog up COM ports 1 and 2, you'll need to set your internal modem to COM3 or COM4. It's rare that you'll have to do this, however. Refer to the Chapter 13 section "Now I Gotta Learn about COM Ports" to see how.

✔ Feel free to install the internal modem yourself, but be willing to take your chances. Who knows, maybe you'll get lucky and it'll work on the first shot. If not, though, Chapter 13 is waiting for you.

✔ Unless it's very old, or really bizarre, your modem has two phone jacks on it. Why? One is for the phone line that runs to the wall. The other is for your telephone. (You had to unplug it from the wall to plug in the modem, remember? Thoughtful of them, eh?)

Installing Your Internal Modem (IQ: 120)

Handy Tools: your modem's manual, a medium Phillips head screwdriver, and a stubborn streak. Be sure *not* to use a magnetic screwdriver.

What Else You'll Need:

✔ **Phone cable:** The flat cable that runs between your phone jack in the wall and your modem. Most modems include one. Refer to Figure 2-1 for a visual refresher.

✔ **Phone wall jack:** Your ordinary phone is already plugged into a wall jack. The closer the jack is to your PC, the better. Refer to Figure 2-5 for a look-see.

✔ **Expansion slot:** An empty slot inside your PC's case, which is where you insert the modem card.

Expansion slots are literally slots into which you stick modem cards and other computer add-ons.

✔ **Unoccupied desk space:** You'll need room to spread out for this one.

To install your internal modem, follow these steps:

1. **Save your work; exit any programs, including Windows.**

   Turn off your PC's power. Unplug it. Turn off and unplug printers, monitors, and any other stuff attached to your computer. To be safe, label the cables so that you know what goes where when you put everything back together.

2. **Unscrew the PC's cover screws.**

   There are usually five, around the outer edges, similar to Figure 2-7. Remove the cover. This sounds scary, but it's actually easy, especially on the newer computers, where the cover lifts off. If you have an older computer with a cover that slides off, you need to make plenty of room on your desk.

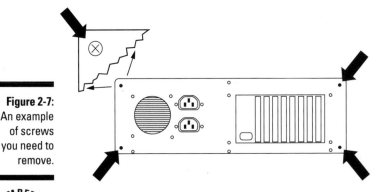

Figure 2-7:
An example
of screws
you need to
remove.

REMEMBER

WARNING!

Be sure to set aside the screws for your PC's cover in a handy, safe place. You don't want to lose one and have to traipse all over town looking for one that matches it.

When sliding off your computer's cover, proceed slowly and carefully. You don't want to knock loose any cables that connect your hard drive to the main board.

3. Tap smartly on the unpainted metal surface on your PC's case.

This grounds you of static electricity charges that could ruin your PC's sensitive parts.

4. Gaze respectfully at the inside of your PC.

Find the expansion slots. Make double sure that there's an unused slot for the modem card. Figure 2-8 shows a typical PC's expansion bus with an empty slot.

If you don't have a free slot after all, you'll have to do one of the following:

a.) Return your internal modem for an external model.

b.) Take everything to the dealer and let a technician do it (highly recommended).

c.) Remove one of the cards — say, a scanner or something you don't use very often — from your PC to free up a slot (recommended only as the last possible alternative, and by all means get a nerd friend to perform this drastic measure).

5. Grip the screwdriver, smile, and remove the metal slot cover from the empty slot you've selected.

Figure 2-9 shows how.

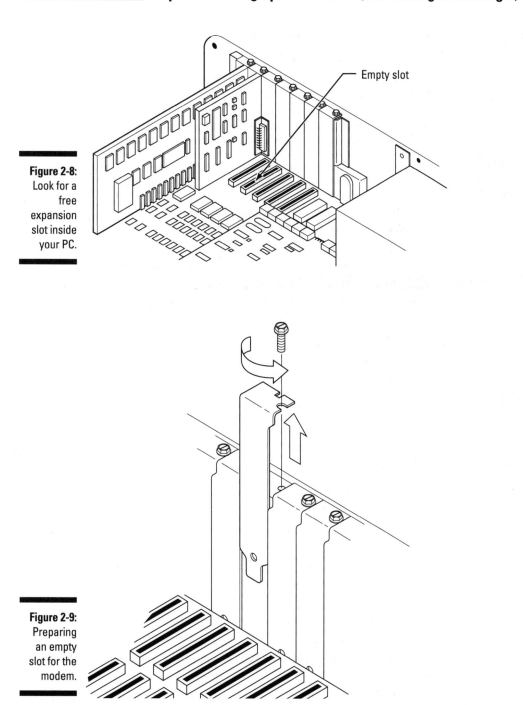

Empty slot

Figure 2-8:
Look for a
free
expansion
slot inside
your PC.

Figure 2-9:
Preparing
an empty
slot for the
modem.

The problem with the gene pool is that there's no lifeguard.

Save the screw in a cup or someplace where it won't roll away. You'll need the screw to put in the modem card.

If you drop the screw in the PC's case, carefully extract it with tweezers before you continue. If a screw's out of reach, don't shake your computer in the hopes of dislodging it! And don't even *think* about using a magnet. Instead, wrap some tape "sticky-side out" around a pencil's end and dab that around those tricky crevices. Loose screws and computers don't mix.

Save the slot cover for when you remove a card from your PC. Covering the empty slot keeps your PC's innards from getting so dusty.

6. **Handling the modem card carefully by its edges, position its mounting bracket and phone jacks so that they face the computer's rear.**

   Make sure that you put the shiny side that looks sort of like a comb facing down toward the slot. This *connector* fits into the slot in Step 7.

7. **Lower the card into an expansion slot; when it's positioned directly inside the slot, give it a firm but gentle downward push.**

   You should feel it "snap" into place. Got it? Then screw it down with the screw you saved earlier.

   If your modem card doesn't fit into the slot easily, don't force it. The card may be designed for a different style of computer slot. Happily, most modem cards fit most personal computers. If your slots and card don't match, read the buying tips in Appendix A and head back to the store for the right kind of modem.

   Count any remaining free slots. Write this number down for later. It'll come in handy the next time you want to add some gizmo to your computer.

8. **Plug in the computer, along with your printer, monitor, and anything else you unplugged.**

9. **You should now see two telephone jacks on the back of your computer.**

   Plug the telephone cable into the modem jack that's labeled "line" or "telco," like the bottom one in Figure 2-10.

   No label? Check the modem's manual to see which jack is for the phone line. Then plug the phone cable's other end into the wall jack.

10. **Plug your telephone's phone line into the other jack on the modem, sometimes labeled "phone."**

    Theoretically, you won't need to check your manual for this one; there's only one jack left.

    If it's a rare, older modem with only one jack, you'll have to play "plug and dial" between modem and phone each time, over at the wall jack.

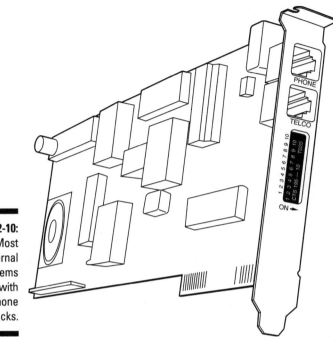

Figure 2-10:
Most
internal
modems
come with
two phone
jacks.

11. **Carefully, without touching anything inside, plug in your computer and monitor.**

    Turn them on. Now go to the section "Testing the Waters." When it works, head to Step 12. You're almost finished.

    If it doesn't work, you need to turn off your computer and monitor and unplug everything, because you need to set those funny jumpers or switches to make your modem talk to a different COM port on your computer. Refer again to your modem manual, because modems are like snowflakes: No two are exactly alike. (You probably have to remove the modem card to do this, so I didn't want you to put the case cover back on just yet.) Now head to the Chapter 13 section "Now I Gotta Learn about COM Ports."

12. **Pat yourself on your back, but don't forget to put the screwdriver down first.**

    You're finished. Put the computer's cover back on and screw it down again. Plug everything back in and turn on the PC and its peripherals.

Modem Troublemakers

The two biggest troublemakers when installing modems are:

1. Plugging the phone line into the wrong jack on the back of your modem

2. Choosing the wrong COM port

- •If you think you've messed up on the modem jack, it won't hurt to try swapping the two phone cords. If that doesn't fix it, then you may have the wrong COM port. That bit of madness is covered in Chapter 13.

Testing the Waters

Not sure whether your modem works? You're certainly not alone. Here's a quick test to see whether your external modem is turned on and plugged into the right places. (You installed an internal modem? Your checklist is in the next section.)

Mac users can't test their modems from outside their modem software. Head over to Chapter 3 to learn how to test your modem's readiness for Mac duty.

Testing your external modem

First, do your best to turn it on. Most external modems make you flip a little switch in the back or toggle a rocker switch along the side. Also, make sure that the phone cord is plugged into the modem and into a working phone jack in the wall.

✔ Some little lights along your modem's front should light up.

✔ One, generally marked **MR**, for "Modem Ready," should stay on. Nerds call this the *power-source light*, in case you're forced to tell someone on the phone what lights you see.

✔ If your modem's hooked up just right and your software's installed and everybody's getting along great right from the get-go, three or four lights on your external modem should remain on.

✔ In an ideal world, all modem brands would have the same little letters under their lights. They don't. Modem lights, their little letters, and what it all means to you come to light in Chapter 17, "Ten External Modem Lights and What They Mean."

In DOS, at your C:\> prompt thing, type the following:

```
C:\> ECHO AT > COM1
```

Then, when you press Enter, watch the modem's lights and see whether they flicker. Did they do anything? It happens real fast!

What's a *DOS prompt*? It's the signal DOS gives that it's ready for your input — generally, for you to type a DOS command or the word that starts a program.

Throughout the book, the DOS prompt looks like this:

```
C:\>
```

The prompt on your computer may look more like this:

```
C>
```

- ✔ If the system you use returns you to a menu or some other program when you exit an application, you need to ask your computer guru how you can get to the DOS prompt. (While the guru's doing that, have him or her test your modem — heh, heh.)

- ✔ DOS prompts can be changed to suit your preferences by using the PROMPT command.

- ✔ The prompt used in this book is handy, because it shows you the current drive (indicated by the letter) you're *logged to*, or *using*, in normal humanspeak. It also shows you the current directory you're in (indicated by any directory name after the \).

- ✔ For information about DOS, DOS prompts, drives, and directories, pick up *DOS For Dummies*. To alter your DOS prompt, ask your computer guru to add this line to your system's AUTOEXEC.BAT file:

```
PROMPT $p$g
```

If you saw a flicker, your modem is hooked up to a magical place called COM1. Better write that COM1 thing down, because your modem software is very interested in that informational tidbit.

Or maybe you saw this:

```
Write fault error writing device COM1

Abort, Retry, Ignore, Fail?
```

If so, type **A** for abort. Your modem is not hooked up to COM1; however, your system is equipped with COM1 (if it wasn't, you wouldn't see any error message at all).

If you didn't see anything — no light flash, not even a flicker — that, too, means it's not hooked up to COM1. It also means that your system does not have a serial port on COM1.

If the COM1 thing didn't work, try typing this at the `C:\>` prompt:

```
C:\> ECHO AT > COM2
```

Did the modem's lights flash this time? Yes? Then you've found the password. Write down COM2 for further reference. Oh, and if you saw that scary `Write fault error writing device` stuff, just press A to tell the computer to knock it off. That message is a *lot* less scary than it looks. In fact, it's handy; it means that you have a COM2 — where you may already have a serial mouse (or some other device) plugged in, or it's free for future serial devices.

Testing your internal modem

Here's how to test an internal modem, in case you have one.

You don't have to worry about turning on the power to your internal modem: it feeds off the power in your PC, so it's on when you turn on your computer. However, make sure that the phone cord is plugged into the modem's Line socket and into a working phone jack in your wall.

In uppercase letters, type this at the `C:\>` prompt:

```
C:\> ECHO ATA > COM2
```

That's the word **ECHO**, a space, the letters **ATA**, the greater-than symbol, >, and the "word" for your PC's COM port. (That's usually COM2 for internal modems, but try COM1 if COM2 doesn't give any results.)

You should hear a dial tone. If not, head to Chapter 13 for help in troubleshooting.

To turn off that obnoxious dial tone, just press Enter and, at the next `C:\>` prompt, type:

```
C:\> ECHO ATH COM2
```

Substitute your COM port "word" if it's not COM2.

If none of this stuff worked right, then you can find out how to change some switches and try again. You'd best head for the "modem mechanic's manual," which is Chapter 13. Don't be too alarmed, though. Your modem *still* may work, even though it didn't pass the first test.

If you're still in a troubleshooting mood, keep on going: Your modem may be hooked up to two more places, COM3 or COM4. Go ahead; try those two places, just like you did with COM1 and COM2. If you still get the error message, just press A to abort.

Modem phone lines

Did you know that the average hard-core modem user's house has at least two phone lines installed? (Three phone lines, if there's a teenager living there.)

Why do they bother? That way, the nerds reason, both the phone and modem can be working at the same time. The nerd can call a friend who also has two phone lines, and they can type and talk at the same time.

Plus, the two-line modem users don't worry about someone else picking up the extension phone while they're downloading a huge file or receiving an important fax.

Best of all, people like this can toss out bad modem jokes like "I've got modem phone lines than you."

Heads Up!

It's not over 'til the modem sings, so to speak. And a modem's delighted screeching upon connecting with a kindred device cannot be described; it must be experienced.

Before you can witness this rapturous interlude, you need to write down the COM port you assigned to your modem. You need that number in the next chapter so that you can tell it to your modem's software. Then you and your modem will be making some, er ... beautiful music together.

Ten Reasons to Be Glad You Installed an Internal Modem

As when you stop banging your head against a brick wall, it feels good when you have finished installing your internal modem. Here are (almost ten) more things you can be thankful for.

It doesn't hog up your desk space

You still have room for that cool black halogen desk lamp you've been eyeing.

Plenty of room left on your power strip

Now you can plug in your lava lamp. (It's a good substitute for those little lights on the external modems.)

You saved a wee bit o' cash

Internal models cost a bit less than their less subtle brethren. But because you're not a nerd, you won't be spending your change on computer stuff ... yet.

Installing an internal modem makes you feel buff

Installing an internal modem is really quite an accomplishment. For fun, be sure to bring it up often at cocktail parties; you'll get the nerdier guests to start talking in numbers. Everyone else will just think you're a bore.

No need to buy an expensive cable

You just saved yourself $15-25 on a decent serial cable. You also saved yourself the sight of that unruly cable tangle in the back of your PC.

Hand, *n.* A singular instrument worn at the end of a human.

No one in their right mind would try to steal an internal modem

Who would voluntarily go through what you just went through just to steal an icky computer gizmo? Of course, they could just pick up the whole computer and climb back out the window, but hey, you can't worry about *everything*, can you?

Ten (Almost) Reasons to Be Glad You Installed an External Modem

Aren't you glad you've got all those cables and stuff plugged in? Doesn't the new modem look cool on your desk? Here's some other stuff to be happy about:

Those cute little lights

... give you the feeling of peeping inside and watching the action, rather than just staring at a closed door and wondering whether anybody's *really* behind there. Besides, those little lights let you know whether your modem is doing something ... unlike an internal modem, where you're always wondering whether the other guy hung up on you about five minutes ago ...

You can use it with any computer, anywhere

You can unplug it and use it with your computer at work, or with Jeremy's down the street, or with that laptop you're thinking about buying. Plus, it's easier to sell when you get a newer, faster one. You can even sell it to the guy with the four-year-old Amiga brand computer.

The speaker's volume control knob is easier to reach

You don't feel as helpless when you hear that screaming dial tone, the beeps, and the harsh, throw-up sound PCs make when they start to talk to each other.

If you ever get really carried away and start staying up late to download games and stuff, you won't keep your housemates awake and wondering whether you've totally lost it.

It doesn't hog up an expansion slot inside your PC

Don't worry; no matter how you try to avoid it, that slot will fill up sooner or later. Best to prolong it for as long as possible, though.

It looks impressive on your desktop

Especially with the speaker blaring and the little lights flashing a lot.

Part II
Modems Need Software
(to Do Any *Real* Damage)

In this part...

*W*here would comedy be without the phone company? Ma Bell and her Baby Bells — even AT&T — never fail to draw big laughs. Until the month's phone bill arrives, that is.

The phone companies stumble occasionally, but overall they do a great job. Pick up the phone, and — ruling out natural disasters — you can count on hearing a dial tone. Dial a number, and you'll usually get through to your party (or hear a nasal, disembodied voice telling you the number's been changed.)

Modems, unlike the telephone, need *modem software* before they'll work. And you need to learn some commands for your software — before your modem does more than belch and squeal and make your dog's ears rotate in alarm. This part tears the shrink wrap off modem software, telling you how to install it, what it does, and how to tweak its myriad settings.

Chapter 3
Your Basic Guide to Installing Software

. . . .

In This Chapter

▶ Installing your modem software

▶ Deciding where to put it

▶ Answering the software's pesky questions

. . . .

*W*ho cares about installing software? The trick is to get someone else to do it while plying them with humorous pet stories and soothing cakes.

Barring that, give this chapter a once-over before delivering those disks into your floppy drive's clutches.

Feel free to skip this chapter if you feel comfortable with software installation. Only the most basic pesky modem software questions are covered here. Chapter 4's awaiting you with more answers to pesky modem questions.

Two Types of Modem Software

Modem software comes in two primary types:

1. *General-purpose modem programs.* With names like ProComm or ZTerm, they're the computer version of Ernestine the operator. They have dialing directories you can set up to call on-line services, bulletin board systems, and lots of other places.

2. *Service-specific software.* This is the stuff on the disk inside Prodigy's Start-Up Kit or in the packet America Online sends when you call the toll-free number. You can't call Prodigy, America Online, or ImagiNation without first installing the service's software. After you set up the software, that's the *only* place it calls.

A good hot dog feeds the hand that bites it.

> ✔ This chapter concentrates on general-purpose modem programs.
>
> ✔ Service-specific software is easier to install; because it calls only one place, it doesn't ask so many questions. Those programs get their due in the chapters describing their services.
>
> ✔ A few other programs are like extremely focused general-purpose programs. They can do specialized modem tasks: fax software sends and receives faxes, of course. Remote control software — you guessed it! — lets you remotely control another computer. And there's some truly ghastly stuff called dedicated terminal emulation software. Nobody's really sure what it does except that it's like general-purpose software only nerdier. You meet these programs in Chapter 4.

Should you even bother installing the software that came with your modem?

You may want to call the office computer or a friend's BBS with your modem. Or you may join one of the on-line services requiring specialized software instead. Perhaps you have a fax modem and you just want to send faxes. Each one of these tasks requires its own, specialized program.

The modem companies can't guess what you'll be doing with your modem, so they usually leave out the software altogether. A few brave modems do come with software.

> ✔ Most people use the software that came with the modem, at least at first. If you install yours now, you can test it in Chapter 4.
>
> ✔ If the software's called BitComm, don't bother. It's buggy.
>
> ✔ If your modem came with fax software and you don't need to fax right now, install it later. You'll see more stuff on faxing in Chapter 4.
>
> ✔ General-purpose programs let you call lots of places, but if you're just planning to call Prodigy or America Online, turn to the appropriate chapter instead. You can install your modem's other software some rainy afternoon when there's nothing on TV and you get the urge to call somewhere else.
>
> ✔ If your modem didn't come with any software, don't feel ripped off (it probably wouldn't have been any good, anyhow). That's common, especially with external modems. Read through the following section, "Software Sleuthing." If you don't hit pay dirt, you can head to Appendix B for a list of must-have software features.

✔ Modem gurus tell me that most questions from new users — after the modem is installed correctly — come from trying to use the crummy free software.

Software sleuthing

You may already have modem software and not know it. Before you rush out to the software store, check these places:

✔ **Your computer's hard drive.** In their eagerness to close a sale, many computer dealers sweeten the deal with a basic selection of software. Call the computer store and ask whether they tossed in some "comm" software. Or you can poke around on your hard drive for *directories* (PCs) or *folders* (Macs) called Term, Comm, Modem, or similarly obscure words.

✔ **Your software shelf.** Dealers often throw in these jack-of-all-trades (and truly master of none) programs called *integrated software*. Look on your shelf for a big box labeled something like Microsoft Works, ClarisWorks (for Macs), Lotus Works, WordPerfect Works ... you sense a pattern here.

The "Works" programs are basically "bundles" containing several software *modules*: typically, a word processor for writing, a spreadsheet for calculating, a database for sorting stuff, and most important (for us), a communications program so that you can get on-line with a modem. These programs appeal to people who don't want to shop for each program they need separately.

✔ **Windows.** Microsoft Windows comes with a stripped-down but perfectly functional communications program with the ominous name Terminal. If you have Windows, look in Program Manager for an icon that shows a phone sitting atop a little box with lights next to a computer screen. That's Terminal.

Chinese parsley lasts longer in the fridge when you keep it "roots down" in a jar full of water.

My comm software can beat up your comm software

Modem zealots are a picky bunch. They're known for harboring deep, personal feelings about the number of *protocols* or *terminal emulators* packed inside their modem software.

Because the matter of which brand is best brings on many heated debates (known on-line as *flame wars*), modem companies take a "laissez faire" stance and let the fanatics, er, *users* pick their own software.

As modems creep into mainstream use, however, you'll see more software bundled with the gadgets. Some of the crankier oldtimers delight in ridiculing this "free" software. They sniff that it's only good for calling your first BBS and downloading some *real* modem software (the kind *they* use). *Hint:* The zealots will go back to ignoring you after you thank them nicely for their feedback. And their advice may bear the stamp of solid experience.

Nyahh, Nyahh: My Software's Installed Already

If so, reward your guru handsomely, then skip to "Testing Your Modem and Software" in Chapter 4 to make sure that everything's working okay. If you run up against problems, browse there for set-up tips.

Dog-ear this software chapter for later reference, though, even if your test call worked out great. Like a chiropractic patient, modem software is always begging for a tweak here and an adjustment there.

Dissecting Your Basic Software Box

Communications software strives for a sexy aura. Some publishers bundle colorful stickers with the program so that you can plaster the company's logo on your file cabinets. Others promise an automatic entry in their vacation drawing when you send in your registration card. Here's what else you're likely to find under the shrink wrap:

- ✔ **Floppy Disks:** Communications programs typically come on 3.5-inch, *high-density* disks, compatible with the high-density floppy drives commonly found on newer computers. If your PC needs different sized or density disks, you'll know right away: either the disk won't fit or it won't work in your drive.

- ✔ **User's Guide:** The user's guide usually contains installation and set-up information if no Quick Reference card is included; it attempts to explain the program's commands and how to use them.

- ✔ **Quick Reference Card:** Cherish this; it contains the essentials for installing, using, and exiting the program.

Trying to establish voice contact — please yell into keyboard.

✔ **Advanced Manual:** Bristling with chapters on Script Programming Languages, it's hairy, but leave it lying around the house anyway. Makes great casual reading material for when your nerdy friends drop by.

✔ **License Agreement:** Like storm clouds, these assume myriad forms. Sometimes they're printed on the envelope housing the disks, sometimes they pop up in the manual. Basically, you're "agreeing" that the program is to be used like a book, that is, "read" by one person on one computer at a time; and that you won't "softlift" the program by copying it for friends. If the software acts weird or wrecks your computer, the agreement also says you can't blame the publisher.

The tiny type on the License Agreement spells out about 90 other contingencies. No one actually reads these, because people rip them into shreds trying to get at the disks.

✔ **Read-Me Sheet:** Last-minute features and updates to the program that never made it into the manual go here. (Updates that never made it into the *program* go into magazine advertisements.)

Read it now, even if all the lingo doesn't seep into your brain cells quite yet; when troubles arise, you'll have the satisfaction of saying, "I know I've seen that somewhere before."

✔ **Registration Card:** Sending this in qualifies you for technical support plus a lifetime of computery junk mail.

These disks won't fit!

In case of a disk/drive mismatch, call the software's tech-support number and have the appropriate disk size or density sent to you. (They may levy a small charge for the privilege.)

If you're in too much of a hurry to wait the requisite week to ten days, ask a chum whose computer has both drive sizes to copy them onto disks your PC *can* read. Bribes of tiny colored marshmallows come in handy here.

One size really *does* fit all when it comes to Mac disks and drives. Some older Macs may not be able to handle the often-standard high-density disks, however; and the older Mac may not have enough memory to run the program properly.

✔ To verify that a program will work on your Mac, look on the program's box to see whether your Mac model is listed under the Compatibility blurb.

✔ That's not a bad tip for PC users to keep in mind, either: Check for your PC on the box, too. Some programs require Windows; others require extended or expanded memory.

✔ The boxes always leave out the most crucial detail: how much hard disk space the program will hog up. You may want to ask the software store's clerk for a rough estimate.

Shareware: Disk ain't no manual!

Some of the best communications software is that shareware stuff you met in Chapter 1. Well, the most popular way to get shareware is to use your modem to download it from a BBS or on-line service. You're probably thinking, "Catch 22: How can I download something when I don't have a program yet?" And you're right!

✔ Excellent commercial communications programs like Qmodem Pro, ProComm Plus, and White Knight all started out as excellent shareware. The shareware versions are alive and well (and still excellent). White Knight's shareware version goes by the mysterious moniker Red Ryder; these two are Macintosh classics.

✔ You can send in the coupons in the back of this book to get your hands on one of these programs.

✔ If you're in a hurry but you don't have your modem up and running yet, users' groups and friends are your best bet for finding a program quick 'n' dirty — and in person. (It's okay to pass shareware around on floppies — that's the whole idea.)

✔ Shareware doesn't come in a big box with foam pads and manuals and stickers and stuff. That's why it costs a fraction of the commercial versions.

✔ Have a friend who knows about shareware show you how to view the READ.ME file and access the software. (If you just can't wait, turn to Appendix C's section "Unpacking the New Arrival," because the program is bound to be *compressed*, or squished, to fit on fewer disks.)

✔ Don't worry: There's a manual for the shareware program. It's usually a text file squeezed onto the disk — along with the program files, a file named READ.ME or something similar, plus the README.1ST file, the NO, READ ME.1ST! file, and the MOTHER ALWAYS LIKED YOU BEST file.

Spelling your name slightly differently when filling out the registration card makes it easier to track the source of your junk mail.

Hayes Smartcomm for Windows takes the prize for most stuff crammed into a software box. A dip into a recent version yielded five manuals, eight disks (four of each size), several pamphlets for trial memberships to on-line services, a warranty sheet in five languages, a Quick Reference card, a registration card, and Three French Hens.

The first urge with shareware is to print the manual with your printer. This can be a long, laborious waste of paper; especially when some funky formatting code makes it skip every other page. (This can happen.) First, make sure that you really use and like the program.

Besides, *registering*, or paying for, the shareware — which you're supposed to do when you start using it regularly — usually results in the author sending you a nice printed manual.

Installing Your Store-Bought Modem Program

Modems don't do much without software, but modem software won't do *anything* unless you install it on your computer. The following steps tell you how:

1. **Open the *User's Guide* or *Manual* to its installation steps.**

   Review this carefully to see whether it's written in any current incarnation of English. If you can understand the steps it gives you, follow them. (Read the "Deciding Where to Put It" section here first, though.) Otherwise, you can follow the following steps.

2. **Turn on your external modem's power switch; also, turn on your computer if it's not on already.**

   If you have an internal modem, it's already on (silicon parasites, internal modems feed off your computer's power supply).

   This helps the smart installation programs guess what serial port you've assigned your modem, which saves you guesswork.

3. **Remove the floppy disks from the envelope and find the one marked Disk A, Disk 1, Setup, or Installation.**

   Hold it label-side up by its label end and stick it into drive A.

   If the disks came in two sizes, using the smaller ones can save on floppy-shuffling (they hold more). Don't forget to close the latch if you're using your 5.25-inch drive.

4. **Type the following at the DOS prompt:**

   ```
   C:\> A:
   ```

   Type **B:** instead of **A:** if you stuck Disk 1 in your B drive. Press Enter.

   Now you're *logged to drive A*, as the nerds (and manual) say.

You don't usually see that type of behavior in a major appliance.

5. **Next, look in your manual for the name of the installation program; typically, it's INSTALL.**

Type that:

```
A:\>INSTALL
```

After typing **INSTALL**, press Enter.

If the name is SETUP, type that instead (still press Enter afterwards):

```
A:\>SETUP
```

Looking up the program's installation command in the *User's Guide* can pay off. Two very popular DOS programs, for example, ask you to type two completely different commands: ProComm Plus has you type **PCINSTAL** and Qmodem Pro asks you to type **QINSTALL**.

Windows installation programs make you type the same stuff, but from within Windows. From the Program Manager, click the **F**ile menu at the top left, choose **R**un, and type the following in the **C**ommand Line box:

```
A:INSTALL
```

or

```
A:SETUP
```

Substitute **B:** for **A:** if you inserted Disk 1 into drive B. The program should whisk you along from here. Press Enter after typing one of the preceding commands.

It's currently fashionable for software to come with a miniprogram called INSTALL or SETUP, which whizzes you through the whole installation routine. Using this program is a good idea for two reasons:

✔ The program won't install correctly without it.

✔ It "guesses" about many of the pesky modem settings — making this whole process a little easier.

Follow the steps and read everything you see on the screen carefully.

Mac programs are easy to install: Just insert the floppy into the disk drive and drag the program's folder, which appears on-screen, onto your hard disk's icon.

✔ If an installation program exists, clicking the main program folder should bring it to life.

✔ Pay close attention to what the installation program is doing.

✔ If the program tells you to insert other disks that came in the box, welcome to floppy swapping.

Filling out the registration card is a wonderful thing to do while you're waiting for the installation process to finish.

Deciding Where to Put It

Your modem program needs its own place to live on your computer's hard drive. DOS and Windows call a program's home a *subdirectory*. Mac users stow their programs in *folders*. This section tells how to create a subdirectory or folder for your modem program. If you've installed software before and you feel comfortable with this whole shebang, go ahead and skip to the next section.

For more detailed information on organizing your hard disk with subdirectories and folders, refer to *DOS For Dummies* or *Macs For Dummies*, part of the proud *...For Dummies* family brought to you by IDG Books Worldwide, Inc.

What's a subdirectory or folder?

Think of a hard drive as a giant file cabinet and a subdirectory as a file folder inside the cabinet. (Macintosh computers dive right in with the "folder" metaphor.)

✔ Just as you can stuff paperwork from several projects into the same file folder, you can install several programs in the same subdirectory or folder.

✔ It's really best to separate your programs by creating a subdirectory for each one — just as you'd probably prefer to keep papers on various projects inside separate file folders. It's easier to find stuff and do work in your programs that way.

✔ You name directories and folders as you create them.

Making a subdirectory or folder during an installation routine

You saw earlier in this chapter how lots of modem programs come with an installation routine. As part of that routine, you're usually asked to name a subdirectory or folder where you'll put the program.

✔ Some of the bossier modem programs have installation routines that suggest a subdirectory or folder name. It's perfectly okay to go with that instead of thinking up a name yourself (if you're sure that you don't already have a subdirectory or folder with that name). When the program's screen pops up with a subdirectory name, you can generally press Enter to okay it.

✔ If for some reason you want to make up a different subdirectory or folder name, you can usually type over the suggested name in the "pathname" box or wherever the installation routine provides for this.

✔ Mac users can be as wordy as they want, naming their folder COMMUNI-CATIONS SOFTWARE, MICROPHONE, WHITE KNIGHT, or whatever they want.

✔ DOS and Windows users must confine their creative urges, keeping the subdirectory name to eight letters or fewer. For example, a logical, eight-letter subdirectory name for a DOS fax program called FAXOMITE might be FAXOMITE. Easy, huh?

✔ If the subdirectory or folder doesn't already exist on your hard drive (and it probably doesn't), your program's installation routine generally tells you so. Then it asks you whether it should go ahead and create the new subdirectory. (Your software may use different wording, of course; just read the installation routine's messages carefully and tell it to go ahead and make the subdirectory or folder.)

"Manually" creating a subdirectory or folder

A few programs shun the automated approach provided by installation routines. Instead, these programs enforce a more "hands-on" installation (with _you_ doing all the work of creating the subdirectory or folder and copying all the files to it).

Creating a subdirectory for a DOS modem program

If you have a DOS modem program, follow these steps to create a subdirectory for it:

1. **Make a new subdirectory.**

   From DOS, you create subdirectories by using a DOS command called MD (short for Make Directory). From your DOS prompt (which may look different than the ones in these examples), you type the following to make a subdirectory called MODEM:

   ```
   C:\> MD MODEM
   ```

   This command is MD, followed by a space and the name of the directory you want to create. In this case, DOS creates a subdirectory called MODEM.

2. **Copy the program's files to the new subdirectory.**

   It's time to copy the modem program's files from the disk in your floppy drive to the new directory on your hard disk. You do that with the DOS command called COPY:

   ```
   C:\> COPY A:*.* C:\MODEM
   ```

   That's the DOS command COPY, a space, the floppy drive letter A, a colon, an asterisk, a period, another asterisk, a space, the letter C, a colon, a backslash, and the name of the newly created subdirectory — in this case, MODEM.

   (Substitute **B:** for **A:**, if you stuck the disk into floppy drive B.)

   Now press Enter. You should hear a whirring of disk drives as the files scurry over to your new MODEM subdirectory.

Creating a folder for a Macintosh modem program

Finding a home for your Mac modem program is as easy as creating a new folder, naming it, and copying the program files to the new folder.

1. **Choose New Folder from the File menu.**

   From the Mac's Desktop, you create a new folder by choosing New Folder from the File menu. A new folder (you were expecting chopped liver?) appears, timidly named "untitled folder."

2. **Name the new folder.**

   The folder name is already highlighted, so to change it from "untitled folder" to the folder name MODEM, for example, you just type the word **MODEM**.

3. Copy the modem program files to the new folder.

Now you want to copy the program from its little disk to your new folder. Insert the program's disk in your Mac's floppy drive. A folder should appear, representing the modem program. You use the mouse pointer to drag this folder over to your new folder. (If you see the word `Install` instead of the program name, the program has an automated installation routine with steps you can follow.)

For help with basic Mac procedures like dragging, pointing, and clicking, refer to *Macs For Dummies*.

Quick Answers to Pesky Installation Questions

Although most software has an installation program that minimizes much of the guesswork, you'll still have to throw out some answers to get it to work. The software's questions will be pretty much the same, whether you're installing the program for DOS, Windows, or a Macintosh.

- ✔ Answers you give now can be changed anytime in the future from within the program. You're not carving anything into stone, so don't worry.

- ✔ If your program didn't ask you any questions, proceed to Chapter 4 to set it up manually.

- ✔ If your program is Mac or DOS shareware, it may not have an installation program. Copy its files onto your hard drive. You'll have to proceed to Chapter 4 and hunt down the set-up questions yourself. Appendix C talks more about installing shareware programs on your computer.

What's your name, rank, and serial number?

Most programs start by asking you to fill out your name, title, and company — that sort of thing. Feel free to be creative with your "company name" if it's for home use. But remember, you'll have to look at it every day, so don't pick something really, really irksome.

Programs trying for a "hard-liner" impression ask you to type the serial number from the manual here, too.

How should I know what files I want to install?!

ProComm Plus for Windows first asks you to decide what files to install from its installation "menu." This confusing start is rooted in good intentions: By leaving out some of the nonessential (to you) files, you can save hard disk space. Several other modem programs follow suit.

Go with the Base Installation and Features Demo files at first; you can add others at any time just by telling the program.

Keep an eye on that serial number

As long as you're dealing with the serial number, print it in big block letters on the front page of your manual, along with the number for tech support (if you can find it). When calling the software company for technical support, you'll almost always be asked for your serial number as your password or as some other essential part of getting through.

It wants to know my "connection"!

The program is asking you what type of connection you're attempting. Table 3-1 shows jargon it'll ask you and what it really means. You'll probably want to choose an option similar to the one that is boldfaced in the table: Serial or Modem.

Table 3-1	Modem Software Asks about Connection Types
Setup Program Jargon	*Means This Type of Connection*
Serial, Modem	A modem to your computer
Direct, PC to PC	A computer to your computer
NASI/NACS, INT14	A network computer to your computer

If there's a modem connected to your computer and that's how you'll be using the communications software, it's safe to pick "serial" or "modem," or "serial, modem" ... whatever looks closest in your particular program.

Trying to connect two computers? Head for the section "Connection type" in Chapter 4. Finally, if you're trying to connect to a network, your office network guru should be doing this.

Lemme Outta Here!

If you need to exit anytime during the installation process, one of these key combinations should do it:

- ✔ Press Ctrl and Break simultaneously.

- ✔ Press Esc.

- ✔ Mac and Windows programs offer you a Cancel button to click instead. Microphone's Mac and Windows products have a prominent Exit key.

- ✔ When all else fails and you're serious about splitting the grisly scene, you can press your computer's Reset button.

- ✔ It's best to avoid exiting the installation program in the middle; you'll have to install it another day anyway, and besides, the program may have already made some unfinished changes to your computer's settings. Also, it may be hazardous to your computer's health.

Remain COM, or, What COM Port Should I Use?

What's all the big fuss about COM ports? Well, a COM port is like a doorway for information flowing into and out of your computer. That's why it's important that two gadgets don't try to use the same COM port: Information gets stuck, like Laurel and Hardy trying to go out the same door at the same time.

You Mac people can happily ignore this section (reading it will only give you an empathy headache).

DOS computers can handle up to four COM ports: Each one is assigned its own number — COM1, COM2, COM3, or COM4 — so you can tell which one's which.

- ✔ If your modem is turned on, your installation program may run a small search on your computer and return, triumphantly announcing what COM port your modem is on.

- ✔ If this doesn't seem to be happening, it's up to you to tell the program what COM port you assigned to your modem when you installed it.

✔ Those of you who installed your modem in Chapter 2 wrote down the COM port you assigned it. If you haven't done so already, turn to the Quick Reference Card in the inside front cover of this book and jot down your modem's COM port on the chart there.

✔ If you're still not sure, go for the default COM port selection the program tosses at you, usually COM1. If you have an internal modem, choose COM2 instead. There are only two more choices in case COM1 and COM2 don't seem to be working: Try COM3 and COM4 in turn. Nothing gets hurt when you try.

✔ If it's not the right COM port, you'll know, because the communications program will tell you it can't "find" COM1 or whichever port you selected.

✔ Chapter 4 is where you'll test the software and modem together; for now, just try the suggestions here.

More excruciating flap about COM ports

The bogus thing about serial ports is that even though PCs running DOS Version 3.3 or later can handle up to four of them, only two COM ports can be used at the same time.

Table 3-2 shows how COM1/COM3 share "interrupt" IRQ4, and COM2/COM4 share IRQ3. Simply expressed, they can't *interrupt* the PC for attention at the same time.

Table 3-2	Two Ports with the Same "Gray Pattern" Can't Be in Use at the Same Time
COM Port:	**Interrupt:**
COM1 //////////	IRQ 4 //////////
COM2 +++++++++++	IRQ 3 +++++++++++
COM3 //////////	IRQ 4 //////////
COM4 +++++++++++	IRQ 3 +++++++++++

See how you can use a mouse with your modem software if the modem is on COM1 and the mouse is on COM2? If your mouse is on COM3, however, neither the mouse nor the modem can gain the PC's attention properly.

Why these clumsy conflicts on something as advanced as a computer? It's because the people who planned the first PCs lacked foresight; they didn't think anyone would need that many gadgets communicating to and from the PC through its serial ports. One of history's more glaring "fudges."

Port settings and changing your modem's port assignment receive more coverage in Chapter 13, "Common Modem Mysteries."

What does this baud and bps stuff mean?

These disgusting technoidisms boil down to one thing: Your software is asking you to choose your modem's fastest speed. If you don't know your modem's speed, choose 2400. As with any of these settings, this is something you can fine-tune in Chapter 4 or anytime you like.

It's tossing around terms like character format, data word format, parity, stop bits, and data bits (oh, my!)

This may cross over the line into Technical Stuff You Don't Want to Know, but your modem software needs this information to work properly. So brace yourself.

Your program wants to know what *data word format* you'll use when calling other computers. This is one thing both the terminal (you) and the host (remote) computers must agree on.

The most typical data word format is **8,N,1**. Most on-line services and all bulletin boards you'll call use this setting. Some, like CompuServe, use the second-most-common setting: **7,E,1**.

- ✔ Data bits (the first number): usually 8, more rarely 7
- ✔ Parity (the letter): typically None (N), more rarely Even (E)
- ✔ Stop bits (the second number): always 1
- ✔ Other settings: ignore them for now

Setting your terminal emulation (ugh ugh?)

Yes, it's English; choose ANSI for now. When you start calling more on-line services, you can switch to the more capable obscure terminal types if your screen looks strange.

Let's pretend ...

Strange relics like *terminal emulation* reek of the ghosts of computing past. Back then, very large, corporate, mainframe computers (the kind with punch cards warning not to "fold, spindle, or mutilate") worked by means of human operators who sat punching numbers and stuff into *terminals*. The keyboards wore several exotic keys that performed arcane Big Iron functions, sort of like the Ctrl-Break combination or the Scroll Lock key on today's personal computer keyboards — the ones you've never used and aren't likely to try out anytime soon.

PCs, Macs, and other types of computers, when pressed, can still communicate with bulky mainframes. (CompuServe is on a mainframe, among other services — although you don't need fancy terminal emulation to talk to CompuServe.) It helps when you have a smart communications program that gives you several ways to "emulate" terminals, though.

That's what that terminal emulation stuff does: It lets you choose different ways to talk to those different computers. The best modem programs even let you "remap" your keyboard — choosing specialty functions by using your normal PC keyboard — making it easier for you to jump through its hoops.

It wants to know my modem type

If you see your particular modem listed on the program's little menu, choose it. Otherwise, it's safe to go with the generic "Hayes" selection that best matches your modem's speed.

The bolder modem programs may then proceed to test and fine-tune your modem's initialization commands and other stuff.

What's my outside-line access number?

If you're setting up the program for home use, you can usually get by fine by leaving this area blank.

Some offices have fancy phones that require you to dial a 9 before you can get an outside line; otherwise, you'll just be reaching Joe in production down the hall, or Janet in accounting. If you can't make regular "voice" calls before dialing 9 or other special numbers first, go ahead and type those numbers now where the program tells you to. Not sure what numbers reach an outside line? Ask someone who's worked there longer than you.

`Dinner not ready: (A)bort (R)etry (P)izza`

If you type **9** or another access number, adding one comma (**9,**) causes it to pause, ensuring that it has time to reach the outside line.

Should I take the tour?

Most programs end the set-up routine by asking whether you want to take a tour. It's a good idea, if only to make you more familiar with the program's look and feel.

Shutdown alert!

Some Windows installation routines end by asking you to exit and restart Windows, because the modem software cannot work if you don't — these routines even provide a little button you can click to do so. First, before you exit Windows, be sure to exit all other Windows programs that may be running.

Ten Common Questions and Answers about Modem Software

Modem programs generally raise more questions than they answer. Here's a sampling to get you started:

Q: I copied the program files that came on my modem software's little disks over to my hard drive, but the program doesn't want to do anything.

A: Some software lets you install it by copying files from the floppy over to the hard disk. But most communications programs are too picky for that. They make you run an installation program. This unpacks their files in the right order. Sometimes it asks you whether it can tweak your computer's start-up files.

You just have to sit back and let the modem program do what it wants. The program's files are packed onto the floppies with special compression software and won't work if you just try to copy them to your hard disk.

Q: I tried to borrow (okay, steal) my brother's faster modem. But I already set up my modem's software to work with my modem and it's a different brand. Can I get a different modem to work with my software?

A: Yes. It's easy to set up your communications software to work with other modems. Just run the software's internal set-up program (refer to its manual if you need a refresher). Then hunt through its menu for the new brand of modem. The software itself will change the settings, so you can avoid answering all those questions again.

Q: The manual keeps mentioning something called a Terminal and a Host! Do I have these? What do these words mean?

A: These words are leftovers from the days of big mainframe *host* computers that only communicated with the outside world by means of a smaller (but brainless) *terminal* connected to them. Today's modem software still insists on calling the dialing computer (you, usually) the *terminal* and the computer that's receiving the call the *host*. Someday you might get a friend to dial up your computer and then you get to play host.

Q: It's all set up, but it still doesn't work!

A: You may have to fine-tune some settings in the next chapter. See you there.

Q: Why do the little lights on my modem blink when I first start my software?

A: Because your modem is hooked up and working, you lucky dog! If those lights don't flash, something's probably wrong with the COM port you've chosen.

Q: I just want to call Prodigy. Do I need to install any other software?

A: Nope. Just install Prodigy's software, and you'll be fine. You'll be fine, that is, until you want to call CompuServe, or GEnie, or your cousin Jeff's new bulletin board that's filled with family photos.

Q: My friend uses a Mac with his modem, and he's crazy about this one BBS, Mac's Magoo. Can I call it with my PC-type modem software?

A: Yes. Macs and PCs can call the same BBSs, but if your friend's fave BBS focuses completely on Mac files and stuff, you may not find too much to interest you! BBSs can be system-specific or completely neutral, depending on who runs them. Turn to Chapter 11 for more of the low-down on BBSs.

Q: Can I have more than one communications program on my hard drive at the same time?

A: Sure, you can have as many as you want. Just don't try to use two of them *at the same time.* (Some "multitasking" software, like Windows or DesqView, lets you run more than one program at a time.) You don't want the two programs knocking at the same COM port's door, however.

Q: Will a DOS modem program work under Windows?

A: Sure! But probably not as well or as easily as a Windows version of that same program. For best results, get a Windows guru friend to help you set up any DOS programs to run in Windows — or buy *Windows For Dummies*, this book's cousin.

Chapter 4

Making Your Software
Talk Nice to Your Modem

. . . .

In This Chapter

▶ Discovering what modem software actually *does*

▶ Changing modem software settings

▶ Creating a dialing directory

▶ Recording what you do on-line

▶ Fiddling with file-transfer protocols

▶ Top Ten Popular AT Commands

. . . .

*W*ithout modem software, modems merely sit still, being expensive. Only modem software can call a modem to life. In fact, modem software makes you sit up straight in your chair, too: It's the orneriest breed of software you'll ever butt heads against.

Being dubbed *communications software* doesn't help. The software takes this moniker to heart, taking every opportunity to *communicate* with you about settings, modes, and all sorts of gibberish.

Luckily, most people can survive by changing only a few settings here and there. This chapter traces the long lineage of babble dished out by general-purpose modem software. For service-specific software, like Prodigy or America Online, head to the Table of Contents and find the appropriate chapter.

Starting Your Modem Software

Unfortunately, there's no one fast and hard rule for loading *any* program, much less a telecommunications program. So, Table 4-1 provides a handy chart.

Table 4-1 Starting Windows and Mac Modem Programs

To Do This	Do This
Start a program in Windows	Double-click the program's icon in Program Manager
Start a program on the Macintosh	Double click the program's icon on the Desktop

If you're running DOS, type the program's name at the little C : \> thing — the *DOS prompt* — and press Enter. (First, be sure to move to the directory where you installed the program.) Table 4-2 shows what to type.

Table 4-2 Typical DOS Program Entrances and Exits

DOS Modem Software	To Start	To Exit
Crosstalk Communicator	**ccm**	Alt-Q
Hayes Smartcom Exec	**exec**	F2, X
ProComm Plus	**pcplus**	Alt-X
ProComm, shareware version	**procomm**	Alt-X
Qmodem Pro	**qmpro**	Alt-X, **Yes**
Qmodem Test Drive (freeware)	**qm**	Alt-X, **Yes**
Telemate	**tm**	Alt-X
Telix	**telix**	Alt-X

Quitting Modem Software

Here too, there's no one, universal command for exiting your modem software. Windows and Mac programs let you click (in Windows, double-click) with your mouse a *close window box* in the top left corner of the window, like the one in Figure 4-1.

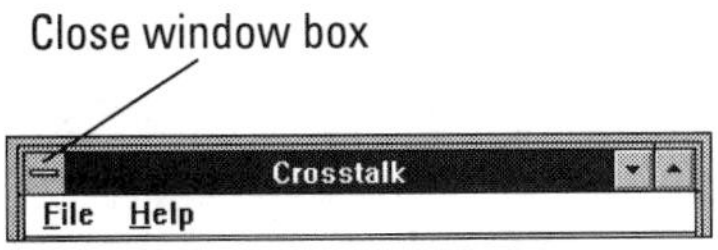

Figure 4-1:
The close
window box.

Close window box

Using Windows with no mouse? Eeek! Seriously, Windows is more fun with a mouse. In the meantime, quit your program by choosing the File menu (press Alt-F). Then press X for Exit.

DOS programs generally respond to a key combination, where you're called upon by the contortionists who designed the program to perform finger yoga by pressing two or more keys simultaneously. Refer to Table 4-2 for some common escape routes.

Terminal Mode Versus Settings Mode (Does This Software Have a Schwarzenegger Complex?)

Modem software is a two-headed beast. One head is called *terminal mode*, where any commands you type go straight through your Terminal window to your modem. The other is called *settings mode*, where you type instructions to the *software* — the modem doesn't hear them.

You need to know how to access terminal mode from your own software. This is where you can test your modem and perform other feats.

Terminal mode

Terminal mode means one thing: The stuff you type goes straight through to your modem. When your software is set up in terminal mode, your monitor simply shows the information currently moving to and from your modem.

If the modem is not talking to another computer's modem, you won't see anything exciting on the screen. In fact, the screen probably will be blank.

✔ You use terminal mode when your modem is *talking* to another computer: You've dialed up another computer, and it's sending stuff across your screen. Chances are, you'll see the other computer's *menu*, welcoming you aboard and making you choose between a bunch of confusing options.

✔ Figure 4-2 shows terminal mode in Qmodem Pro, a DOS modem software package. Here, I've dialed the local number for Time by using an *AT command*, a direct way of telling my modem to do something. (Skip down to the Technical Stuff icon to find out more about AT commands.)

✔ Some modem programs start you up in terminal mode automatically. This is especially true of DOS programs. Basically, anytime you're looking at OK or NO CARRIER, you're in terminal mode.

✔ Other programs start up in settings mode, described in the next section. To find terminal mode, try pressing Esc or a function key. If you find yourself in a dialing directory, pressing Esc usually drops you into terminal mode.

✔ Some GUI (graphical user interface) programs for the Mac and Windows make it tough to find terminal mode. You can try selecting New from the File menu. No luck? Try clicking the close window box in the upper left corner. You may need to scream for your guru, or even resort to looking in the software's manual.

✔ Some programs call terminal mode the *Online Menu* or something different. It's all the same concept: What you type goes straight to the modem itself, and "sees through" the modem program.

With modems, the word *terminal* pops up repeatedly. This goes back to the Olden Days when the method for accessing the early (large) computers was by typing into the early (smaller) *terminals* wired to them. The terminal was just a cheap keyboard wired to a monitor.

Different brands of monitors displayed information in different ways, however. So, when people logged on to the other computer, they had to tell it what sort of terminal type they were using. Only then would the information look right on the screen.

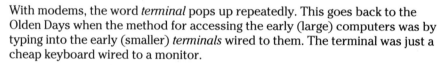

Figure 4-2: Terminal mode is where you can talk directly with your modem.

```
QmodemPro 1.50 Compiled 06/15/93
Copyright (C) 1992,93 Mustang Software, Inc.

You are now in TERMINAL mode

ATZ
OK
at
OK
atdt 853-1212
RINGING

NO CARRIER
```

Today, different computers expect different types of terminals to call in. So, modem software lets your computer *emulate* several different terminal types.

Changing your terminal settings is one of the most common adjustments you'll make to your modem software. Terminals still exist, by the way; they're mostly used with large computers called *mainframes*. You'll see terminals in use at airports, where the perky reservations agents confirm that, yes, the flight's at least two hours late.

Settings mode

Terminal mode is fine and dandy for sending your keystrokes to the modem. But what if you want to type a command to the *modem software*, not to the modem? What if you want to tell the modem software to start capturing that important party schedule that's flowing across the screen? You need to switch to *settings mode*.

Settings mode is a catchall term for all the settings that let you command your software to do various things. Settings mode lets you adjust your software's colors or add new entries to its *dialing directory* (see Figure 4-3).

Figure 4-3:
Dialing Time again, this time from a Dialing Directory, part of the software's settings mode.

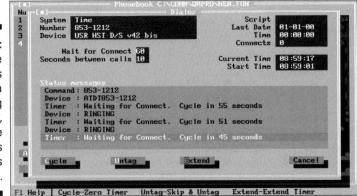

Settings mode lets you choose your modem type and decide which settings are best for a particular on-line service or BBS (electronic *bulletin board system*).

✔ Some programs refer to settings mode as the *Offline Menu* or something different. Both mean the same thing: talking to the modem software rather than sending your keystrokes through to the modem itself.

✔ You can be on-line (connected with another computer/modem), however, yet switch to settings mode to access your software's commands.

When the information you're typing is controlling the modem software —
moving menus around — you're in settings mode.

When the information you're typing is appearing on the screen — being sent
over your modem to the other computer — you're in terminal mode.

Fine-tuning the dialing commands

Most modem software contains a dozen other settings to fine-tune the way your
modem dials. Table 4-3 summarizes the most common settings you'll see.

Table 4-3   Yet Another Round of Pesky Dialing Settings

Setting Name	What It Does	Can I Ignore It?
Initialization command or Modem Init	It sends a series of commands to the modem every time you start the modem software or use the Initialize Modem command to reset it from your software.	Yes; your software probably provides a "generic" modem init command.
	Every line of commands sent to a modem must start with **AT**.	If you want the modem to perform specific commands each time it dials anywhere, however, this is where to put them.
	Always end the string of commands with the **^M** characters.	
dialing command or Dial Prefix	Specifies the command that tells your modem to dial a number. It sends the characters **ATDT** (for **AT**tention, **D**ial, **T**one) to make the modem start dialing; uses tones.	Yes; all software handles this automatically.
	If you have pulse phone service, use **DP** instead. (Most software assumes everybody has Touch-Tone service.)	
	The characters **DP** (for **D**ial, **P**ulse) make the modem use pulse dialing.	

Do not believe everything you hear or anything you say.

Setting Name	What It Does	Can I Ignore It?
dialing suffix	Commands the modem to start dialing. The default is always **^M**.	Yes; most modem software provides the default **^M** characters. But if your modem requires a different end-of-command character, it goes here.
hangup command	Tells your modem to disconnect: it's typically **ATH0**.	Yes; your modem software handles it.
auto answer command	Tells the modem to answer all calls automatically. If you want it on, type **ATS0=1**, and the modem will answer all calls after one ring.	Yes; your modem software handles it by setting this to "off" usually. (Unless you have a dedicated modem line, you don't want your modem answering all calls.)
wait for connection	Tells your modem not to hang up before *x* number of seconds.	No; set this to 30 seconds, at least.
pause between calls	Tells your modem how long to wait before dialing again.	No; set it for 2 seconds or less unless you enjoy waiting pointlessly.
redial	A value you specify limits the number of times your modem should try the number.	No; limit the number of redials to 15 or so. Consider calling another, less-busy BBS or on-line service.

Testing Your Modem and Software

Your modem is connected and your software is installed. Now it's time to make sure that the two are communicating. (Backtrack to Chapters 2 and 3 if you still need to install your modem or software.) To test your modem and software, follow these steps:

1. **Turn on your computer and monitor if they're not already on.**

2. **Turn on your modem. (Internal modems come to life automatically with Step 1.)**

If you have an external modem, some lights should come on. Modem brands differ, but look for **MR,** for **Modem Ready**, at the very least.

MR should come on when the modem is plugged into a power supply and turned on, regardless of whether it's connected to your computer or a phone line. It tells you the modem is working. It's difficult to generalize about modem lights, however. For example, a common indicator light like MR goes by **DSR**, for **Data Set Ready**, on ZyXEL brand modems.

Refer to Chapter 17 for more on external modem lights.

3. **Start your modem software.**

Starting the software should bring another light to life on external modems — typically, it's called **TR**, for **Terminal Ready**.

4. **Enter your software's terminal mode.**

Described earlier, terminal mode should look like a nearly empty screen; sort of like your word processor's screen before you've typed much. You may see the word OK on-screen.

5. **Type the following on a line by itself, and then press Enter (be sure to use all uppercase or all lowercase — not mixed —letters).**

```
ATV1
```

The "word" you typed, ATV1, should appear on the screen. If you don't see what you typed, head for "Troubleshooting the Modem Software" at the end of this chapter.

You're talking directly to your modem, sending it something called an *AT command*. (This is one of the most basic commands: **AT**, for ATtention, and **V1**, for "send me verbal result codes.")

If you got the modem's attention, it should answer. This should appear on the screen:

```
OK
```

Congratulations. Modem and software are on speaking terms.

If you don't see anything, head to "Troubleshooting the Modem Software."

Modem software cushions you against AT commands and worse stuff

Modems live and breathe by commands like ATV1. In fact, you *could* control the modem yourself, simply by typing code words like **AT** to the modem. But who wants to remember more than 100 different codes?

That's where modem software comes in: It lets you choose tasks from simple menus. When you choose a task, the software converts your choices into AT codes and then sends them to the menu.

- AT commands are a set of instructions you can use to grab your modem's ATtention — and then tell it do something. AT commands control everything your modem does, from the speed at which it connects to how many times it redials a number.

- Most people push their modem around by using their modem software instead — letting *it* worry about which AT command to use.

- You can type direct AT commands at your modem from your software's terminal mode. If you get the command right, and the modem is set to send you "verbal" result codes, your modem types <u>OK</u> back to you — in a very satisfying manner. If you type the command incorrectly, or type nonsense just for fun, it sends back a message saying ERROR. If you just see numbers, type ATV1 to get words instead.

- Your modem manual lists your modem's AT commands and what they do. Don't worry if you can't understand that part of the manual; hardly anyone admits to comprehending AT commands.

- AT commands come in two main varieties: Basic and Extended. Most Hayes-compatible modems share at least the Basic AT command set; it's one of the few standards in the modem world. Many modems also share the Extended command set. From there, however, it's an AT command free-for-all, with each brand touting a plethora of specialty commands that control its modem's unique features.

- Modems that aren't *Hayes-compatible* may not offer the full set of agreed-upon, standardized Basic AT commands developed by a company called — you guessed it — *Hayes*. This matters because your software is expecting to command a Hayes-compatible modem. Non-Hayes-compatible modems are bad news, especially for new users. Refer to Appendix A, "Buying a Modem," for more about Hayes compatibility.

- Many people *prefer* using AT commands to change the way their modem behaves. It's difficult to tell these people apart from normal folk just by looking.

- Head to this chapter's Top Ten Most Popular AT Commands for some AT commands you can try.

- Anytime you type an AT command, be sure to use all uppercase (ALLCAPS) or all lowercase letters — don't mix 'em or it won't work.

A wee bit o' knowledge

A *bit* is a *binary* dig*it*, the smallest unit of information your computer can store. Everyone thinks computers are so smart, but they can only count two digits: 0 or 1. Bits are the digits in this *base two*, or *binary*, counting system.

Computers send out electronic pulses, so the bits are really tiny switches: When a bit is a 0, it's off.

If the bit is a 1, it's on. That's all there is to it: tiny electrons switching on or off.

Your term papers, to-do lists, and Mac burping noises all break down into 1s and 0s: bits o' data.

Cavalcade of Software Settings

So many settings, so little time! Eventually, everyone ends up making at least one or two adjustments to their modem software. (Some people tweak their settings just for fun.)

This chapter lines up the most essential settings in order of when you need them. Keep in mind that all modem programs are different; your settings may come under a different name (or not exist at all). Ah, the joys of generic modem software references.

Write it down!

Two general rules help with modem software settings: Don't change anything without having a good reason, and write down anything you change, in case you get stuck and forget what you did. Table 4-4 suggests a good format for your notes:

Table 4-4   When You Change a Setting, Write It Down

Date:	Setting:	From:	To:	Reason:
7/11	Terminal Emulation	TTY	ANSI BBS	The BBS menus had weird characters.
8/10	Screen Color	Blue	Neon Orange	The program looked boring.

Settings to make the software work with your modem

These are the settings you're most likely to fiddle with before calling someplace.

Choosing or changing your modem "type"

Most communications software contains a list you can browse for your own modem brand, similar to the one in Figure 4-4. Choosing a modem from the list lets your software take the most advantage of your modem's capabilities — by configuring something called a *modem initialization string*.

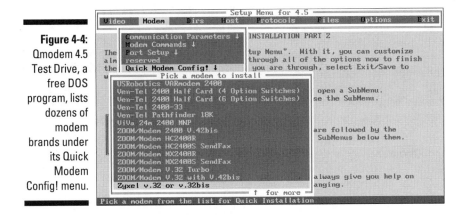

Figure 4-4:
Qmodem 4.5
Test Drive, a
free DOS
program, lists
dozens of
modem
brands under
its Quick
Modem
Config! menu.

✔ The more modems supported by your software, the better your chance of finding your modem in the list. This is something to keep in mind when shopping for modem software.

✔ Different brands of modem software don't always treat modems the same way. If you're using two different packages of software, don't be surprised if each one uses a different initialization string for the same modem.

✔ If you don't find your modem brand on the list, choose the "Generic Hayes Compatible" item that best matches your modem speed.

✔ Your software may not offer such a list of modems. Often, shareware modem programs don't have a list; instead, they ask you to fine-tune the modem settings yourself. Refer to the following technical section for help with this.

Ignore this stuff about modem initialization strings

Adjusting your software's modem setup strings can be a fine occasion for bribing a Modem Guru to come over. Barring that, plenty of sources for modem settings exist.

✔ Your modem's manufacturer probably has a BBS you can call to download *modem drivers* (special programs that fine-tune the relationship between your modem and your software).

✔ No BBS? Call the modem company's tech support number and ask for the best "modem initialization string" for your specific software brand. Ask your modem company whether it offers a *forum* for customer support on CompuServe or on one of the other commercial on-line services.

✔ Call your software company's BBS; modem software companies strive to write drivers for each new modem that comes out. Most of these companies also offer forums on CompuServe and other services.

✔ People on local BBSs are usually more than willing to look up the optimum settings for your particular situation. Make your e-mail messages complete and readable to any potential gurus by specifying your modem's brand, model, and speed; any error correction, data compression, or other fancies; your software's name and

version number; your computer model; any problems you're having; plus any other details that will help. (Refer to Appendix A, "Buying a Modem," for ways to tell whether your modem has error correction, data correction, and other features.)

✔ If someone offers a suggestion, it will look like a series of letters and numbers starting with the letters *AT.* Type the characters exactly as they appear (in the e-mail message, or wherever you found it). The command string is actually a number of AT commands strung together.

After you uncover some modem settings to try with your software, look for a menu or command in your software called "Modem Initialization" or something similar. You'll need to refer to your software's manual (or ask your guru) to find out exactly where to type the command string.

Figure 4-5 shows the screen you use if you are trying to use Hayes Smartcom II/Macintosh with a modem brand other than Hayes. In this program, you type the command string on line 1, "Setup 1."

The modem software that came "bundled" with your modem probably doesn't contain drivers for other modem brands. This shouldn't present too much of a problem for you; it's only an inconvenience if you try to use it with a different modem brand.

Connection type

Your software wants to know what type of connection you'll be making. Figure 4-6 shows Crosstalk for Windows' Connection choices.

Here are some terms you may see in your software's connections setting:

Serial, Modem, Local Modem: To dial up other computers with your modem, choose the setting that sounds closest to one of these.

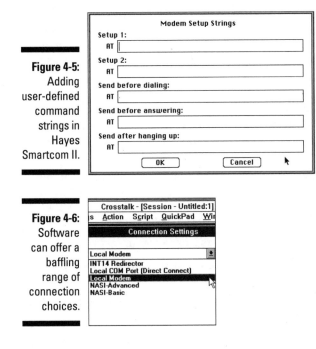

Figure 4-5:
Adding
user-defined
command
strings in
Hayes
Smartcom II.

Figure 4-6:
Software
can offer a
baffling
range of
connection
choices.

Direct: Choose Direct when you want to connect two adjacent computers with a cable, not with the phone line. After changing this setting, you need to plug a special, *null-modem cable* into the serial ports of each computer. Or, you can unplug your serial cable from the back of your modem and attach a *null-modem adapter* to it; then plug the other end of the cable into the other computer's serial port. Don't try stringing a plain serial cable between the two computers: the signals are crossed and it just won't work.

(Wouldn't you rather walk a floppy disk over to the other computer? Nerds like to call this "Sneakernet.")

NASI, INT14, NACS: These are choices for connecting to a modem on a network computer. Your network administrator (or some equally harried-looking person) should handle this connection type.

COM port

The COM Port setting is how you tell your modem software which unique serial port "door" your modem is using to get data into and out of the computer.

> ✔ Your modem can't work with your software until they agree on the modem's serial port assignment. That's why your software usually asks this crucial question as you install it.

✔ Mac software almost always assumes that your modem is connected to your Mac's modem port. If you chose the printer port instead, you need to tell your software. (Stick with the modem port if possible; that way, you won't forget to tell your other software programs what port is being used.)

✔ If you switch from an external modem to an internal model, or otherwise need to change a serial port assignment in your software, a few programs make you run the installation program again. Chances are, though, you won't have to. The settings you need may be hiding under menus called Connections, Modem Setup, Port Assignments, or even General. Figure 4-7 shows Crosstalk for Windows' port assortment.

✔ If your software already performed the "AT Test" at the beginning of this chapter, it already knows your modem's correct COM port. If you haven't already filled in the Quick Reference Card's Serial Port chart, do so now, to get it down on paper. Now rejoice and turn the page quickly.

✔ Most software lets you change serial ports from a menu.

✔ Many restaurants let you order cereal from a menu, and your mouth is your cereal port.

✔ You assigned your modem to a specific serial port when you first installed your modem. Refer to Chapter 2 for a refresher.

✔ If your software doesn't "talk" to your modem on the serial port you specify, try some of your software's other port settings.

✔ If you can't get your modem software to recognize your modem on any of your serial ports, head to Chapter 13's section "Now I Gotta Learn About COM Ports."

Figure 4-7:
Most
software
lets you
change
serial ports
from a
menu.

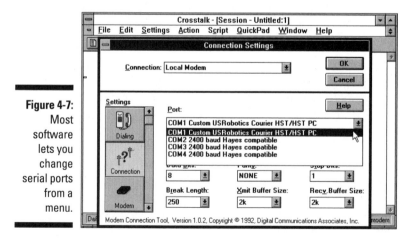

Baud, data rate, and that bps stuff

Baud, *bps*, and *data rate* are used interchangeably in the computer world. All these terms refer to how fast your modem can talk; specifically, how fast it sends and receives data bits. (Nerds may get on your case for using the term "baud," however, because it's technically wrong.)

✔ Here, you should enter **2400 bps**, or your modem's maximum speed, if under 2400 bps. 2400 bps is the *default*, or automatic, setting most software chooses.

✔ Most software asks you to adjust the data rate (plus some other settings — we're coming to them next) for each place you call. Data rate may be one of the blanks you fill in when adding a new entry into your dialing directory. Or your program may ask you to create a new setting for each place you call; data rate is one of the hoops you'll jump through.

✔ Have a 9600 bps or faster modem? If the on-line service or BBS you're calling supports faster speeds, you can "turn up" your data rate before you call the next time. It's best to make the first connection at a moderate speed.

✔ An advanced modem may surpass its "advertised" data rate, achieving higher *throughput* due to *data compression* and other technical wonders. Refer to the following technical sidebar for more information.

Is bps the same as BBS?

No ... but, good try! *Bps* stands for *bits per second*, the flow of data squeezing through the phone line. Bps is the standard way of measuring a modem's speed. A modem transmitting at *2400 bps* is sending or receiving about 240 characters per second, or almost 2400 words per minute.

Disabling MNP5 when transferring compressed files

If your modem offers MNP5 data compression, you should know where to find the AT command that turns it off. If you're sending or receiving a compressed file — one with a file name ending in .SIT, .SEA, .ZIP, .LHA, or .ARC — MNP5 will actually *slow down* your file transfer. For my modem, the command to disable MNP5 is

AT&K3

That is, **AT**, ampersand (**&**), **K3**. Of course, your modem's command may differ, so you may want to consult your manual. Refer to Appendix C for more on data-compression software.

Many BBSs offer special "high-speed" phone numbers equipped with fast modems. Callers with faster modems or special modem brands are urged to use the matching number. Keep awake during the sign-on announcements so that you can flag any special numbers that pertain to you. Refer to Chapter 11 for more information on BBSs.

TECHNICAL STUFF

High-performance, data-compression stuff to skip

Some advanced modems offer a feature called *data compression*. You can tell whether your modem has it because the box or manual will say "MNP5," "V.42*bis*," or both. If your modem has data compression, you'll get better performance if you set your software's *DTE rate* to four times the speed of your modem.

Data compression works sort of like packing for a vacation and stuffing your socks inside your shoes to save room. With your modem, the compressed data travels at normal speed over the phone line. But, because there's more data in a particular chunk, your modem needs to work *faster* to get all this stuff from modem to your computer.

That's vastly oversimplified, of course. The important thing is that you can get the most out of data compression by setting your software's data rate to be *four times* the speed of your modem for all your calls. Table 4-5 does the math for you.

Now you need to *lock down* your modem-to-computer rate (nerds call this the *modem/DTE* — for *data terminal equipment* — rate). That way, your modem doesn't go changing to try and please you. In terminal mode, type

    AT&B1

and press Enter. You're locked down until you turn off your modem or restore the modem's factory settings.

This higher bps rate affects only the rate of data transfer between your computer and modems, not the rate between sending and receiving modems. The high rate keeps the modem's buffer full so that it makes better use of data compression.

To save this (or any) setting in the modem's memory, type

    AT&W

and press Enter. (Refer to the "Top Ten Most Popular AT Commands" section at the end of this chapter.)

Table 4-5	Data Compression? Use These Data Rates
Modem Speed	*Set Data Rate to This*
2400 bps	9600 bps
9600 bps	38,400 bps
14,400 bps	38,400 bps

Should I admit to having a baud?

Baud is not bawdy, at least when it comes to something as dry as modems. Simply put, *baud* refers to how many times the modulated signal (your data) changes its state on its journey through the phone lines. The confusion starts because almost *everybody* uses *baud* when they're really talking about a modem's transmission *speed*, or *bps*.

The terms *baud* and *bps* were once used interchangeably to describe modem rates. It was correct to describe a particular modem as being 300 baud *and* 300 bps, for example. After modems surpassed the lower speed limits, however, the terminology blurred. Today, few modems sport the same figure for their baud and bps rates.

I'd love to, but I promised to help a friend fold road maps.

That factoid doesn't stop anyone from using the terms to mean the exact same thing: modem speed. Even modem ads and manuals may use the term "baud" when they really mean "bps."

Data parameters: parity, data bits, and stop bits; stop it!

After two modems have agreed on how fast they'll be talking to each other, they need to agree on the language. That comes under the heading of *data parameters*. Specifically, they need to agree on three things: *parity*, *data bits*, and *stop bits*. You don't have to understand what this stuff means; just make sure that your settings match those of whomever you're calling.

Modems usually adjust in speed automatically and on the fly to keep up a good conversation. But if their data parameters don't match, they simply spout gibberish at each other.

- The most typical settings for any modem connection you'll make are **8,N,1**: **8** data bits, **No** parity, and **1** stop bit. Most of the commercial on-line services (and all of the BBSs) you'll see expect to meet you at 8,N,1.

- CompuServe and a few other mainframe computers (like those at some universities) use **7,E,1**: **7** data bits, **Even** parity, and **1** stop bit.

- Anything else is extremely bizarre; I've never seen it.

- Most software stashes these parameter settings under the Connection menu. Some store the data parameters in the Dialing Directory, instead. Figure 4-8 shows the data parameters window that pops up when you click the Port Settings item in ProComm Plus for Windows' Dialing Directory.

- Figure 4-9 shows QModem Pro's "Quick Picks," oft-used data parameters that are all ready to go.

Feel free to gloss over this bit of weirdness

Recall the detour into *bits* earlier in this chapter? Well, computers like to save time by sending data around in groups of 8 bits, called *bytes*. (Single bits are too teeny to bother with.)

So your sales letter races around in your computer's innards and chugs up your printer's cable in these 8-bit bytes. Modems have to deal with the serial port, however, which can't handle the whole 8 bits of data chunks at once. Instead, your software breaks the bytes back down into bits, and sends them one at time, or *serially*.

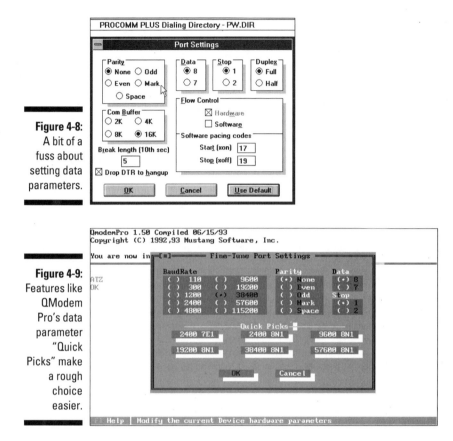

Figure 4-8:
A bit of a fuss about setting data parameters.

Figure 4-9:
Features like QModem Pro's data parameter "Quick Picks" make a rough choice easier.

How do the bytes appear at the other end of the modem transmission? Still broken down into single bits. But special "framing" bits are sent along with the data, spelling out where one byte ends and another begins; as well as checking everything for errors. When you set up your data parameters, you're setting up these "working" bits.

Parity is an older, built-in way to check for errors in a modem transmission. Most systems use No parity checking.

Modems agree to send their data in chunks that are either 7 or 8 bits "wide." Most systems agree on chunks of 8 *data bits*.

Start bits are always sent; these tell the system that a byte is coming.

Stop bits are sent to tell the system that a byte has been sent. Without stop bits, all the bits would run together, like text in early Latin. Almost without exception, you'll see 1 stop bit required by the system you're calling.

Flow control, handshaking, and other groovy practices

Data bits can pile up during a modem transmission; *flow control* eases the pressure, telling everyone to take five and do some deep breathing. *Handshaking* is another name for flow control. Two types of flow control lighten up the load: *software flow control* and *hardware flow control*.

Modem software generally stashes the flow control setting under Modem; some programs keep it with the Terminal settings. Your software asks you to type an **X** in a little box, or otherwise toggle the setting of the one you prefer, as in Figure 4-10.

Figure 4-10:
Setting flow
control in
Crosstalk for
Windows.

✔ Software flow control, also known as XON/XOFF, sends special software characters that say, "Hey, cool it, man!

✔ Software flow control takes extra time to send those take-it-easy characters with all the other data, so it's less efficient than hardware flow control, which operates outside the data flow altogether. Software flow control is only good up to 2400 bps.

✔ Hardware flow control, also called *hardware handshaking*, works in tandem with a *hardware handshaking serial cable*. It sends little "cool it" currents down two of the cable's wires (the RTS and the CTS wires, if you're into acronyms). Another name for hardware flow control is RTS/CTS; your software may use this term instead.

✔ Hardware flow control is recommended if you have an advanced, error-correcting, and data-compressing modem.

✔ Hardware handshaking lets you use the fastest method for transferring files back and forth: a file-transfer protocol called YMODEM/G. (But only with an error-correcting modem.) Even so, you're better off sticking with a safer protocol like ZMODEM — which is almost as fast. Refer to the section "File-Transfer Protocols and Other Formal Occasions."

✔ Never rely on hardware handshaking unless you're sure you have a genuine, okey-dokey hardware-handshaking cable. If you're not sure, opt for software handshaking instead. Remember, for speeds over 2400 bps, you *need* this cable: software handshaking won't work.

AAcckk!! II""mm iinn hhaallff dduupplleexx

Dialing Directories and Other Settings for Calling Someplace

Every modem program provides a place to type and store the phone numbers of BBSs and on-line services you want to call.

Two paths of modem software, Grasshopper

Two main methods for recording your numbers reflect the two divergent paths of modem software philosophy. There's the *dialing directory*; and there's the *setting*, or *service*.

The dialing directory: numbers for the masses

A dialing directory is like having your address book on your computer (if all your friends and acquaintances were computers). Programs with dialing directories are usually DOS programs, or they borrow heavily from the DOS communication software tradition that deems it handier to have all your phone numbers in one vast list.

The top picture in Figure 4-11 shows a typical dialing directory in ProComm, a DOS shareware program. On the bottom is a rare Macintosh dialing directory in a program called ZTerm, also shareware.

✔ The Alt-D key combination usually brings up a DOS program's dialing directory.

✔ To add a number to a directory, press the Add command (or your program's specific command to do this). Then type the BBS's name, phone number, speed, and all the rest in the spaces provided.

✔ To dial a number from a directory, highlight it and press Enter. Or you may need to punch a Dial command or OK instead.

✔ The best dialing directories let you put several entries in a dialing *queue* by marking them with a spacebar or a key combo. The modem dials each number in turn; if one's busy, it tries the next number in the queue. This sort of mass dialing is great when you're in the mood to call some BBSs and you don't really care which one you connect with first.

✔ Another sign of excellence is a dialing directory that lets you add several numbers under the same entry. This is great for the larger BBSs with several phone numbers: If one of the numbers is busy, the software tries another one, going down the list until it connects (or runs out of numbers).

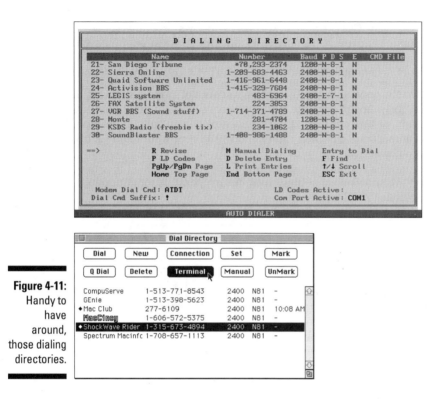

Figure 4-11:
Handy to
have
around,
those dialing
directories.

✔ Software with dialing directories typically offers hundreds of entries in each directory and an unlimited number of directories. You could group all your gaming BBSs in one directory, for example, and call it something distinctive, like *games*.

✔ You can switch between directories, because each one is a separate file in your communications software.

The GUIs: to each a service or setting

The other breed of modem software (Mac and Windows programs) creates a different file for each place you call. That's like having a separate address book for each of your friends and acquaintances. Your software may call this file a *document*, *service*, or *setting*, but it's all the same thing.

✔ To create a new service or whatever your program calls it, look under the File menu, and choose New. This should do the trick.

✔ To call an existing service, look under the File menu for Open. Then highlight the one you want and click with your mouse or press Enter. Your software may differ, of course.

> ✔ Generally, GUI programs (Mac and Windows programs) go for the settings/
> service approach. It's slightly inconvenient, because you have to answer
> tons of questions for each place you call. (The better programs offer some
> sort of default setting so that you can get around most of these queries.)
>
> ✔ On the good side, the GUI settings school of modern software offers you
> slightly more flexibility. You can easily set up a different modem, COM
> port, or flow control setting for each place you call, for example. (Of
> course, no one in their right mind would ever want to *do* such a thing, but
> you *could* ...)

Figure 4-12 shows what settings look like in Crosstalk for Windows, top, and
MicroPhone Pro/Mac, bottom.

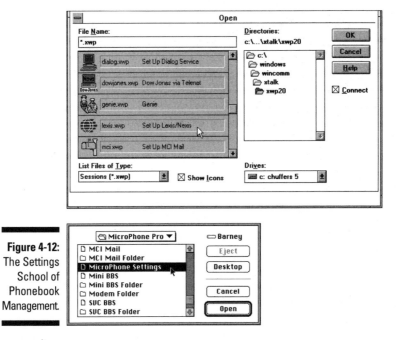

Figure 4-12:
The Settings
School of
Phonebook
Management.

Beating "call waiting"

Call waiting is convenient, no doubt about it. When you're talking to somebody,
that beep tells you somebody else wants to reach you.

Unfortunately, if your computer's talking to another computer, that beep will
throw it off-line.

To please its modem-wielding customers, the phone company lets you temporarily disable your call waiting service before making a phone call. To turn off call waiting for a phone call, add the characters in Table 4-5 *in front of* the phone number in your setting or dialing directory. This should work for you; if not, contact your phone company and ask how to disable call waiting.

Table 4-6	Disable Call Waiting Before You Dial	
Type of Phone Service	*Type This*	*Sample Entry*
Touch-Tone service (common)	*70,	*70,1-213-555-1212
*Pulse service (rare these days)	1170,	1170,333-4444

* *Note:* Dialing out with pulse phone service requires a separate AT command: ATDP. Your modem software probably supports pulse dialing.

Scripts, recorders, and other robotic devices

Most modem programs come with a "recording" feature. Whenever you start a modem procedure that you do often, you can record what you and the other computer do. The software watches as you dial and log on to a service; then it creates a *script*, based on the text your modem sends and receives. (You have to remember to turn off the recording command, however.)

- ✔ You can save the script with a filename. Then, each time you want to call that service, tell your software to run that particular script. The modem simply repeats everything that happened the first time, sparing you from typing all the log-on and password commands.

- ✔ Scripts are convenient for services that make you type long series of characters and carriage returns before they let you in.

- ✔ Modem nerds get very excited about scripting *engines*, the powerful script modules inside some modem programs. These people can get the modem to dial somewhere, collect mail, send and receive specific files ... make coffee each morning, feed the cat ... all unattended.

- ✔ Procomm Plus for Windows comes with a demo that shows a chess game they designed using Procomm's scripting language. It's pretty cool.

Settings to Control the Way Things Look On-Screen

Modem programs set aside heaps of settings to control the way the screen and commands look to you. These can range from changing your screen colors to designing brand new menu bars and buttons to play with in your software. Some of the more common settings follow.

- Many of the GUI programs let you mess with how fonts and other text look on the screen.

- *Line feeds*, or *line wrap*, determines whether your program adds carriage returns to text coming in from your modem. Stick with the default settings until you have a reason to change them. If everything appears on one line, turn on line feeds.

- *Local echo* determines whether you see characters on the screen. Turn this off, unless you can't see what you're typing. In that case, toggle this on.

- *Duplex — Full* or *Half:* Many programs call the echo setting Duplex. Turn this setting to Full Duplex, unless you can't see anything on the screen, then toggle this to Half Duplex. If you see everything repeated twice, switch to Full. Duplex governs whether data is sent and received in both directions simultaneously (Full) or one way at a time (Half).

Terminal types and when to change 'em

All modem programs include at least two different terminal-emulation settings. The problem is, they stick these settings in the most unpredictable places.

Terminal settings are important: They should match, or at least agree with, those of the host computer. Terminal settings cloak your computer so that it impersonates the terminal type expected by the computer you're calling.

Table 4-7 sets out some of the more commonly encountered terminal settings, and tells you when to use them.

Dedicated terminal emulation programs

You'd think that the number of terminal-emulation settings contained in the average modem program would be enough to meet any host computer head-on. But some folks need to connect with bizarre mainframe computers or other nonstandard hosts.

These people need *dedicated terminal emulation programs*, with settings that go beyond the run-of-the-mill assortment in the average program.

- ✔ VersaTerm is an extremely capable Mac terminal-emulation program.

- ✔ Crosstalk Mk.4, Hyperaccess/5, Procomm Plus, and Relay Gold are good considerations for specialized terminal emulation on your DOS computer.

- ✔ Check with your system administrator to find out which program he or she recommends.

- ✔ Most terminal emulators allow users to switch keyboard functions and adjust other elements via software commands, to better meet the host's expectations. Ask your system administrator for any handouts or manuals that will help you determine what changes (if any) are needed on your program to reassign function keys and fool the host computer in other ways.

Table 4-7	Terminal Settings You'll See
Terminal-Emulation Setting	*When You Should Use It*
TTY	When nothing else works; pretends your computer's a teletype machine
ANSI BBS	When calling most BBSs
VT-100, VT-102, and so on	When calling on-line services
IBM PC	For calling BBSs with doorways that require this emulation

Settings for Getting Stuff into and out of Your Computer

While you're on-line, you can review what you saw a few screens ago with a *scrollback buffer*, or record every minute of an on-line session into a file on your hard disk with a *session log*. You can use a *text editor* to compose a message and then send (upload) it as a file to the host system by using a *file-transfer protocol*.

Modem software contains these and other ways to get stuff into and out of your computer. Figure 4-13 shows some of the ProComm Plus for Windows settings for saving stuff you might otherwise miss on-line. Here, text on the screen or in the scrollback buffer can be saved to the Windows Clipboard, a capture file, a text file or other document, or sent directly to your printer.

Figure 4-13:
ProComm
Plus for
Windows'
settings for
saving stuff.

PROCOMM PLUS for Windows

File **Edit** Scripts Communication Window Help

Copy Text   Ctrl-Ins
Copy Link
Paste Text   Shift-Gray-Ins

Screen to
Scrollback Buffer to | Clipboard   Alt-V
File Clipboard   Alt-= | Capture File
                             | File...   Alt-G
Scrollback/Pause   Alt-P | Printer   Alt-L
Clear Screen   Alt-C
Reset Terminal   Alt-U

Log file or capture

Your program may be capable of creating a log file. Some software calls this the capture command. When you invoke the command that turns on a log, everything that takes place on the host system is "recorded" in the log file on your hard disk. You save the file with a name. Then, when you're off-line, you open it and review it anytime you want — when you have more time (and the rates are cheaper).

The scrollback, or capture buffer

Often called a *capture buffer*, this feature lets you pause text flowing from the host computer to your screen and actually reverse back through the text to take a quick peek back at a particular section. Very useful!

✔ If your program has a scrollback buffer (most do), you don't have to press any keys or invoke a command to start it. The last few screens of text are automatically available to you from the buffer.

✔ Figure 4-13 shows a menu in ProComm Plus for Windows offering to save any text on the screen or in the buffer to a file, the printer, and so on. Your software usually provides a way to set up the buffer to contain as much or as little text as you'd like (within limits: buffers may impact some of your computer's valuable memory).

✔ In Windows and Mac programs, you access the scrollback buffer with the scrollbar to the right of your terminal window. To see some text that whizzed by you, simply position your mouse pointer on the scrollbar's scroll box, hold down the mouse button, and move it up, as in Windows' Terminal (see Figure 4-14). To exit scrollback mode, slide the scroll box back down to the bottom right corner of the screen. (Most GUI programs also let you press a key combination to exit scrollback mode.)

```
┌──────────────────────────────────────────────────────────────────┐
│ ▬                           Terminal - PNET.TRM                ▼ ▲ │
├──────────────────────────────────────────────────────────────────┤
│  File  Edit  Settings  Phone  Transfers  Help                   ▲ │
│ From ray Thu Aug  5 09:08:09 1993                               █ │
│ Date: Thu, 5 Aug 93 09:08:08 PDT             Move scroll box up │
│ From: ray (Andy Rathbone)                    to go back a few screens
│ To: tbone                                                         │
│ Subject: Scone heads                                              │
│                                                                   │
│ Tina, although Franz Kafka and Jorge Luis Borges were raised in vastly
│ different cultures, they used similar styles to explore universal themes. At
│ least, that's what Ian said at the coffee shop yesterday while stuffing two
│ scones into his shirt pocket. Both authors tried to grasp an often
│ indecipherable universe, by writing realistically -- indeed, almost
│ matter-of-factly -- about fantastic events. In Kafka's "The Great Wall of
│ China" and Borges' "The Lottery in Babylon," for example, both men explored
│ abstract concepts through the use of fictional narrators and by emphasizing
│ differences in social class structure.                            │
│                                                                   │
│ Plus, the plots of both stories depict a society that lives according to
│ guidelines set down many generations before by long-forgotten rulers. And,
│ like Ian, the wives of both authors complained about the scone crumbs
│ constantly clogging the dryer's lint bag. What do you think?      │
│                                                                   │
│ >read> c                                      Move scroll box forward (down)
│ To: ray                                       to return to your place on-line
│ Subject: Re: Scone heads                                          │
│ Cc:                                                               │
│ Bcc:                                                              │
│ Enter text, end with '.' alone.                                  │
│                                                                   │
│ Ray, few have stopped to link modernistic and post-modern literature to scone
│ crumbs. We may have hit upon the reason for the crumbling of societal
│ structures (and scones) as we know them.                          │
│                                                                   │
│ Tina      I                                                       │
│                                               Scroll box        █ │
│                                                                 ▼ │
│ ◄▬▬▬▬▬▬▬▬▬▬▬▬▬▬▬▬▬▬▬▬▬▬▬▬▬▬▬▬▬▬▬▬▬▬▬▬▬▬▬▬▬▬▬▬▬▬▬▬▬▬▬▬▬▬▬▬▬▬▬▬▬► │
└──────────────────────────────────────────────────────────────────┘
```

Figure 4-14: The scrollback buffer makes it easy to review messages that have scrolled off your screen.

✔ DOS users must hunt for the command that displays the buffer. You'll have to press Esc or perform whatever command exits out of scrollback mode before you're back in present time with the other system.

✔ Even though you're happily scrolling back through your buffer, the host system is still sending you text (or waiting for you to type something). Don't keep the other system waiting too long. You can get thrown off some systems after a few minutes without any input. If you feel you've missed something crucial but can't find it in your buffer, you may have to log off and call back again. (A buffer can only hold so many past screens of text.)

✔ Turn on your software's capture or log command before you call a new BBS or on-line service. You can save time and money by keeping essentials like file lists, message topics, announcements, alternate phone numbers, pricing, rules and hints, and other stuff on your hard disk.

Screen dumps and other niceties

Your software may call them *screen dumps* or *snapshots*, but they boil down to the same thing: a command that takes a "picture" of the information on a current screen during an on-line session. Generally, you can save a screen dump with a file name; call it something descriptive so that you'll remember what's in there.

Printing while you're on-line

Your software can send screen dumps, capture buffers, or scrolling text directly to the printer, but this method can be slow and cumbersome. First, save information to a file. Then you can log off, review it at your leisure, and decide whether it's really worth printing. The printer is much slower than a modem.

ASCII and you'll receive-ee (you can sendee, too)

All modem programs contain a command to send and receive *ASCII*, or plain *text* files, between two computers. (Text transfers predate all other ways of sending and receiving files, so this is one fairly universal feature. *Finally!*)

- ✔ ASCII text transfer is an okay way to send or receive casual messages. Many BBSs and on-line services store articles and other text files you can receive by text transfer.

- ✔ Transferring text is *not* the same as uploading or downloading a file, technically. Text transfers involve less error checking than true file transfers, and they're not as fast.

- ✔ With text transfers, you see your message as it scrolls by, whereas in uploading, you get some sort of a thermometer telling you how things are progressing.

- ✔ Text transfers can't send and receive program files. These require official file uploads or downloads using a *file-transfer protocol* like XMODEM or ZMODEM. If you're not sure what type of file it is, avoid ASCII transfers. Refer to the very next section for file-transfer protocols.

File-Transfer Protocols and Other Formal Occasions

One of the best parts of having a modem is being able to send and receive files of all types between your computer and the other system.

As you may have suspected, your modem software plays a pivotal role in file transfers, from starting the transfer to providing a range of protocols used to do the job right.

✔ A file-transfer protocol is a strict, official set of rules agreed upon by the sending and receiving computers. They keep phone noises and other errors from creeping into the data that's being transferred, among other things.

✔ Like data parameters, file-transfer protocols on your end and the other system must match!

✔ Host computers differ in how they begin a file transfer. Generally, you need to nudge the host computer, and then specify the filename you want to send or receive. The host asks you to choose one of its file-transfer protocols to use, and then it waits for you to perform the commands on your modem software to choose a matching protocol and start the file transfer rolling.

✔ Commands to start the file transfer at your end vary, but DOS programs usually use the PgUp key to upload; the PgDn key to download. (What? Logical command names?) A window pops up, listing all the protocols from which to choose. Choose the same one you specified on the BBS or on-line service.

✔ Try the command to start uploading or downloading "for practice," just to see what protocols your software contains.

✔ Mac programs generally start a transfer from the File menu's Send or Receive menus (see Figure 4-15).

✔ Don't turn off the MacBinary setting in your modem software. It helps Mac file transfers go more smoothly.

✔ Windows programs stick the file-transfer commands in various places. Crosstalk uses the Action menu, and ProComm Plus for Windows has you click little folder pictures with up or down arrows on them. You may need to consult your manual to figure out your software's file-transfer "start" command.

✔ Special *batch* protocols let you select several files to send or receive. You specify the file names on the host computer, for example, and then select the protocol. The file names are automatically specified on your computer, ready to be transferred in succession.

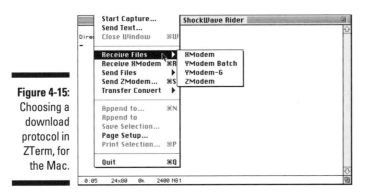

Figure 4-15: Choosing a download protocol in ZTerm, for the Mac.

✔ Don't forget the protocol you told the host computer you wanted to use; you'll need to pick the same one at your end. If you make a mistake, nothing truly heinous will happen ... just ... nothing will happen. Table 4-8 lists the most common file-transfer protocols and when to use 'em.

Still Other Settings

You'll probably spot other strange settings lurking in your modem software. Here's a field guide.

Exiting the software temporarily, or, what's Shell to DOS?

Most of the DOS modem programs have a Shell to DOS command. You can invoke the command while you're poking around in your software — or even while you're on-line. It temporarily dumps you at the DOS prompt, where you can look for a file, make a subdirectory, do other stuff, or just sulk.

✔ Come back when you're ready by typing **EXIT** (and pressing Enter) at the DOS prompt.

✔ Windows users can roam around the Program Manager or even look at other Windows programs from their modem software. Windows tolerates and even boasts about this _task switching_ capability. But it's probably not a good idea to drop to DOS from Windows while you're on-line. You may forget where you are and never ... come back.

✔ Mac users can also open other programs and do stuff on the Desktop or in the Finder while they're on-line or off-line in their modem program. Mac users are used to this _task switching_, just by opening and closing windows and folders. And, of course, it's not referred to as "shelling to DOS." Ick.

Table 4-8	A Typical Array of File-Transfer Protocols	
Protocol	**Distinguishing Features**	**Remarks**
XModem	Supported by almost every modem program and host computer	Tried and true; go for it when you can't find ZModem available
	Slower than many protocols due to its error-correction scheme	
XModem-CRC	Faster than plain-wrap XModem, and more reliable	Be sure that you're using the same XModem variety as the host system
YModem	Faster than the XModem clan	Slightly better than XModem
	Allows for multiple (*batch*) file transfers	
YModem-G	Very fast	Slick but slightly suspect for all but the most advanced conditions
	Contains no error correction, so it's used only between two error-correcting (V.42) modems	
	Hardware-handshaking cable and flow control advised	Generally, avoid this; ZModem is only a *teensy* bit slower and *much* safer
	Allows for multiple (*batch*) file transfers	
ZModem	Fast and reliable, found in most software and on most hosts	Crash Test Dummy of the protocol world
	Allows for multiple (*batch*) file transfers	Use it!
	Recovers from an interrupted file transfer, and *starts where it left off* (honest!)	
CompuServe B	CompuServe's fast protocol	Supported by about 50% of the modem programs, but only used on CompuServe
Kermit	Very slow and powerful	Where's Miss Piggy?
	Used with many mainframes and other larger computers	
	Named for Kermit the Frog.	

Settings for reading and writing messages

Most people who want to read or write messages do so on-line, using the host computer's built-in message "word processor." Even so, many modem programs contain editors and other programs that let you read and write messages.

Having such programs built into your modem program comes in handiest when you're logged *off* the other system. Reading and writing messages when you're off-line is one of the best ways to save money (and write better messages, because you have more time to compose yourself).

The editor

Most modem software contains a powerful subprogram called an *editor*.

- Editors are like miniature word processors, except they don't bother with all the fancy typefaces and tab codes.

- Editors are handy for when you need to save your words as plain ASCII text — for ASCII text transfers, for example. Refer to Chapter 5's e-mail section for more information about editors.

The off-line mail reader

Off-line mail readers are outside programs you obtain and set up to work with your modem software.

- Off-line readers log on, quickly retrieve all your mail and whatever else you've specified, and then log off.

- As the name implies, these programs let you read mail after you're logged *off* the BBS or commercial on-line service.

- Processing your mail on your own time is a sure-fire way to save on connect charge$.

- Off-line readers go a long way toward automatically responding to messages, as well. A split-screen mode is especially desirable, because it allows you to read and respond to the message at the same time, on one screen. Refer to Chapter 11 for more on off-line reader programs.

- Qmodem Pro, a DOS program, contains a built-in off-line reader.

Settings for Faxing

Many of the newer modem programs can send faxes. A few can even receive them. You need a modem with fax capability before you can use these features. Refer to Appendix A, "Buying a Modem," to learn about standards for fax modems.

You can only fax *files* from your computer — stuff that's already stored inside your computer. That includes text files from your editor; documents from your word processor; spreadsheets, charts, and reports from your spreadsheet program; a drawing from your "paint" program. You get the idea.

✔ If you want to send someone an existing, printed document — like a newspaper clipping, magazine article, or a letter from Uncle Marv — you need to hunt down an actual fax machine. (Actually, you *could* buy a gizmo called a *scanner*, scan the article, and then send it. But that requires scanning software and all sorts of complexities. Wouldn't you rather just mail it?)

The *fax module* within a modem program does not contain as many features as a dedicated, fax-only program. For example, many fax settings "built-ins" can only send (not receive) faxes.

If you need to send an occasional fax, however, the simple capabilities in your modem program will be sufficient. All programs differ, but they usually have at least a few of the following features in common.

Your own private fax service

If you need to send a very important fax and you're not sure how well your modem program will do, you can have one of the commercial on-line services send a fax to any fax machine in the nation.

You leave your fax as an e-mail message. Then follow the commands on the particular service to have it sent as a fax, instead of as e-mail. The service also can print and mail any message you desire. (Some services even hand-deliver messages.)

On-line services charge a fee to send faxes, usually a few dollars per page. Of course, that's in addition to any connect charges for the time you spend on-line with the service.

Cover sheet

Before you send a fax, most programs give you a *cover sheet* form to fill out. Here's where you type in the fax-ee's name and a subject line.

- ✔ The software generally asks you to fill in *your* name, phone number, and fax number the very first time you send a fax. It can "plug in" your information automatically for subsequent faxes.

- ✔ WinFax Pro comes with a separate cover sheet program that contains dozens of examples — ranging from silly to serious. Plus, you can design your own, even adding your custom logo.

Resolution, or mode

Yet another mode? Here, your software's asking you to choose how fine the fax should look to the recipient. Choose Fine or the equivalent in your program if it's an important fax. Why not choose Fine all the time? Because those faxes take longer to send, driving up the long distance bill.

Printing

After you've finished composing your letter in your word processor, saved the file, and filled out the fax cover sheet, faxing is usually a matter of specifying the documents to send and choosing the "print" feature in the fax module.

However, some fax programs can send only "text files." That means they can't handle any fancy headlines, margins, or tables your word processor has added, without sending you through complex "translation" hoops first.

Dedicated Fax Software

People who are serious about using their fax modems usually end up buying a dedicated fax program.

These programs contain extra features, like fancy cover sheets, powerful fax "phonebooks," and the capability to "broadcast" a fax to many machines at once.

Often, a dedicated fax program offers a more sophisticated *receive fax* function, where incoming faxes scroll in the background, making it possible for you to still get some work done instead of sitting there twiddling your thumbs.

Using Your Modem as a Giant Remote Control Device

If you're on the road and you need a file that's on your computer, you don't need to make a U-turn and head for home. With some foresight, you can call your computer with a modem and grab what you need, through your modem software's *host mode* or a *remote-control program*.

Host mode

Most modem software offers a *host mode* that lets you or someone else dial into your computer.

Host mode looks and acts like a mini-BBS. The caller types a user name and a password. After getting through the log-on sequence, the caller sees a welcoming screen that says something like, "Welcome to Procomm's Host Mode." Then the caller sees a menu offering choices like Upload a File, Download a File, Leave a Message, Chat, and the usual BBS stuff.

> ✔ Your modem needs to be turned on, and your modem program must be set to host mode before it can take a call. That's why accessing host mode while you're away requires some advance planning.

> ✔ Before you (or someone else) can call your program's host mode, you need to set it up. You decide things like how far the caller can "range" within your computer by setting host mode to "open" or "restricted." Then you tell host mode which directories are accessible to a caller. (If you choose to restrict host mode, no caller — not even if it's you — can access any directories outside the restricted area.)

> ✔ If a friend wants to try out your host mode, determine in advance whether you'll require him or her to type a special password in order to "get in." (If you're the one trying to get in and you've forgotten the special password you set up in host mode, you're stuck without access.)

> ✔ Host mode can be confusing — even more so than other features. You have to pay attention to settings like *auto-baud detect*, for example, which ensures that your modem will adjust to the speed of the calling modem. Successfully receiving a file can require lots of advance tweaking and usually a few voice calls between you and your friend. For these reasons, try to find a modem pal to help you with host mode.

Remote-control programs

An entire breed of programs exists to let people call their computers from afar and run programs. Remote-control programs like pcAnywhere, Carbon Copy, ReachOut, Remote2, Co/Session, and others work like enhanced host modes.

- ✔ From your computer, a connection through one of these programs makes it look as if you're sitting at the remote computer. Besides file transfers, you can run any program or do stuff just as if you were using that computer. For example, you can load a word processor, fine-tune a letter, and then send it to the printer connected to the remote computer.

- ✔ With remote-control programs, "half" the program is installed on the host computer, your office PC, for example; the other goes on your "local" computer.

- ✔ These programs usually recognize and work with other "remote on/off" programs that can actually contain a way to switch on the host computer automatically.

- ✔ If remote control sounds like it may meet a need for you or a coworker, check with your office system administrator to see whether there's already a remote-control program in place. You'll still need to install the "other half" of the program on your home computer or laptop, however.

- ✔ Before you go out and buy one of these programs, seek advice from BBS pals, user group members, or even your modem company's tech support staff.

AppleTalk Remote Access is a powerful remote-control program that comes with PowerBook portable Macs.

Troubleshooting the Modem Software

The following tips should solve any minor software problems you encounter. If you suspect your modem and software aren't communicating — or if you encounter deeper problems — help awaits you in Chapter 13, "Common Modem Mysteries."

I Typed AT to test my modem but I didn't see what I typed.

Your modem is not "echoing" back the commands you type. There are two ways to fix this: with DIP switches or through an AT command. Refer to the "tiny switches" sidebar later in this chapter for more on DIP switches.

DIP switch: Turn on the "Echo off-line commands" switch.

That's the best solution, because it fixes the problem permanently.

AT command: From your software's terminal mode, type **ATE1** and press Enter.

Unfortunately, you'll have to type **ATE1** each time you turn on your modem, making this less than convenient. (Or, tell your software to add **E1** to your modem's initialization AT commands.)

My software says it can't find a CTS signal.

Turn on your modem.

It says NO CARRIER? Now What?!

When you see the words NO CARRIER, it means you're not connected to the other computer anymore. You're off-line. This is a normal message you'll see when you tell the BBS or on-line service you want to hang up. Pressing the commands in your modem software to hang up gives you the NO CARRIER response as well.

CARRIER is simply the name nerds gave to the signal that the two modems sing back and forth to each other while exchanging your Shrimp Mousse recipes, sound files, and e-mail.

Top Ten Most Popular AT Commands

For your modem software enjoyment, welcome to the nerdiest section of Tens in this entire book: the Top Ten Most Popular AT Commands. You use these AT commands by typing them to your modem through your software's terminal window. Before you do, head back to the Technical Stuff section on AT commands earlier in this chapter for a refresher.

The Earth is like a tiny grain of sand, only heavier.

My external modem has some tiny switches. Why?

External modems made by U.S. Robotics and some other manufacturers sport tiny DIP switches. (*DIP* stands for *Dual In-Line Package*, but knowing this won't get you cuts in line at the cafeteria.)

Like everything else in the modem world, DIP switches and their functions vary from brand to brand. (Somehow, this doesn't exactly surprise you, I'll bet.) In fact, many modems don't come with any DIP switches at all.

- Look on the back for a recessed area with some tiny "light switches" inside it. Some modems wear them inside a "well" on the modem's underbelly, instead; flip the modem over to see the little DIP switches nestled down in there. Figure 4-16 shows my modem's switches. Hello, tiny switches.

- DIP switches are simply on/off switches, like the light switch in your bedroom. Modems with DIP switches let you choose between changing a setting by flipping a switch or by sending an AT command to your modem.

- For example, switch #4 on a U.S. Robotics modem controls whether you see commands you type to your modem. But I can easily type **ATE1** in terminal mode to pull the same "switch" for this echo command.

- If this sounds redundant, you're onto something. Not only do you have DIP switches and AT commands that do the same thing, but your modem software can probably "switch" some of the same settings as well. Beginning to see why modem software is such a bear?

- Never change a DIP switch without writing down when and why you made the change. There's usually an AT command that lets you return to your modem's "factory settings," but why fool around? In case you get stuck, your modem's manual should contain a diagram of default "factory" settings.

- If there's a setting you want to permanently change, the first thing to check is the DIP switch (if your modem has them, of course). Next, try changing the appropriate setting in your modem software. Finally, you can resort to looking up the proper AT command in your modem's manual and sending that in your software's terminal mode.

- Any AT commands you type override the DIP switch settings, until you turn off the modem. (The DIP switches determine the modem's power-on default settings.)

- The thoughtful, sensitive, Alan Alda-type modem manuals suggest DIP switch settings preferred by various popular software packages.

- External modems occasionally sport DIP switches. But internal modems *almost always* have either DIP switches or other switches called *jumpers*. These serve a different, much more crucial purpose than DIP switches on external modems: An internal modem uses DIP switches or jumpers to set the modem's COM port. (Head to Chapter 13 for more on changing an internal modem's COM ports.)

- Remember: Not every modem has these DIP switches.

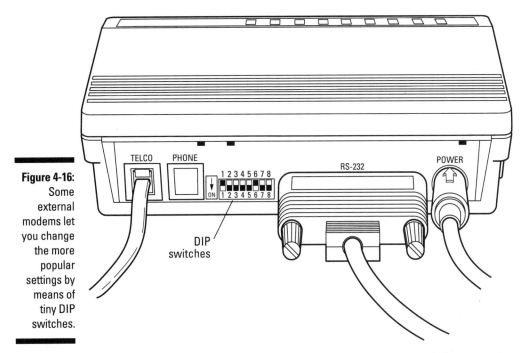

Figure 4-16:
Some
external
modems let
you change
the more
popular
settings by
means of
tiny DIP
switches.

You *can* type them on one line, separately. But most people string AT commands all together. When they do, they leave out all but the first AT letters, like so:

```
AT&B1&E1&M0&DT234-5678
```

AT commands aren't completely delegated to the pocket-protector crowd. Sometimes, typing a quick straight-to-the-modem command can be quicker than heading back to the software and wading through menus.

Remember: Anytime you talk to your modem by typing an AT command, you have to type either in ALLCAPS or in all lowercase letters. Don't mix them.

AT&W

Saves any AT commands to your modem's memory.

If you've changed your modem setup by using AT commands and you want to save those changes, use AT&W.

ATZ

Brings back your modem to the settings in memory.

Use this whenever you think your modem is behaving strangely. Also, you can turn your modem off and on to slap its hand.

AT&F

Loads the "factory settings" your modem came with out of the box.

ATDT555-5555

Dials the phone if you have Touch-Tone service (the phone number goes after the ATDT command). The hyphen is optional.

ATDP555-5555

Dials the phone if you have pulse (rotary) service.

ATH

Hangs up the phone (*drops carrier*, nerds would say).

A/

Repeats the preceding command. For example, you can make the modem redial the last number it tried just by typing **A/** on a line by itself.

ATE1

Tells the modem to show on-screen the commands you type while the modem is off-line (in command mode).

ATA

Leaves your modem ready to answer an incoming call (for a short period — as long as you don't do anything else). You use this command when your friend decides to try out his or her shiny new modem by calling your modem. (When you see RING RING RING on the screen, you simply type **ATA**, and then press Enter.)

ATMn

The little *n* next to **ATM** stands for a number; together, these control the modem's speaker.

A **1** leaves the speaker on until you connect (the typical setting).

A **0** turns off the speaker (handy for late night modem sessions).

A **2** leaves the speaker on for the duration of the connection (not a Good Thing).

A **3** is the best setting, if your modem offers it.

AT&B1

"Locks down" the modem/DTE rate you've selected in your modem software, to get the most out of a high-speed modem with data compression. (Refer to the Technical Stuff, "High-performance, data-compression stuff to skip" section earlier in this chapter.)

AT&B0

Sets the modem/DTE rate back to the default, variable rate. This works only with high-speed modems.

Sure, they're super nerdy, but AT commands can be sort of fun when you're in the mood and all. For a vast source of even more entertainment with AT commands, simply pick up your modem's manual. (Well, okay ... open it, too.)

Part III
Making the Modem *Do* Something

The 5th Wave   **By Rich Tennant**

"WELL, I NEVER THOUGHT I'D SEE THE DAY I COULD SAY
I DIALED IN A MODEM VIA A STAT MUX INTO A DEDI-
CATED PORT ON A COMMUNICATIONS PROCESSOR,...
BY ACCIDENT."

In this part...

Arousing round of Interstellar Annihilation or a quick stock market quote? A modem lets you do just about *anything* ... after you know where to go and what to type when you get there.

This part of the book contains the stuff everybody's all excited about: sending and receiving files and electronic mail, discussing special interests in forums — and the other staples. Best bets for business, games, travel, and other interests line up here, too.

You'll find out the difference between local BBSs and commercial on-line services, and how to get around on each of the major ones. This part concludes with a chapter on the Internet, the Mother of all Computer Networks.

Chapter 5
Who Ya Gonna Call?

*P*eople have different shopping comfort levels. Just about everyone feels intimidated walking into one of those ultrachic "shoppes" — their names are usually spelled funny, and they always *smell* funny. Even if you throw caution to the wind and do venture in, the sales clerk makes you feel as though there's a green thing wedged between your teeth.

Like stores, on-line services vary, both in their "merchandise" and in their atmosphere.

This chapter explains why people go on-line and also recommends a few *best bets* among BBSs, commercial on-line services, and the Internet. Just think of this chapter as your Yellow Pages — where your modem's fingers can do all the walking.

A clean desk is a sign of a cluttered desk drawer.

Understanding the Places On-line

You're probably not too familiar with terms like *on-line service* or *BBS*. But chances are good that you've been shopping before. Well, if the various places on-line were stores, they'd sort of compare to the following types of places to shop.

I hope that the following descriptions give you some idea of what these on-line places are like, but there's really no substitute for getting on-line and just doing it.

The commercial on-line service

Calling a commercial on-line service is like going down to the mall. Services like Prodigy, CompuServe, GEnie, and the rest offer tons o' goods and services under one big "roof."

✔ The mall is easy to get to. You can always find parking, although the on-line malls charge an hourly rate. You can shop for stuff, cruise people, or just hang out.

✔ Like malls, commercial on-line services are easy to reach and have branches everywhere. Their phone lines are hardly ever busy because they offer local access numbers throughout the United States. Some of the larger ones, like CompuServe, offer international connections.

✔ Both malls and commercial on-line services offer a few ultraspecialized places. You'll find banks, travel services, art galleries and museums, post offices, and other cool places on-line.

✔ At malls, you can head to a department store's electronics department and watch the evening news on TV. Or you can browse the endless bookstores and magazine racks. On-line services offer a wealth of news and publications. Many also provide a way to search for specific news tidbits.

✔ Just like arcades in malls, you'll find computer games on commercial on-line services. Some of the newer games let you compete against lots of other "shoppers" at the same time. While playing games, you can meet friends and catch up on the latest gossip.

The local bulletin board system (BBS)

Local BBSs have lots in common with your neighborhood thrift shop.

✔ For such a small place, a thrift shop is always busy, and there never seems to be any parking when you swing by to check it out. Like thrift shops, BBSs are generally smaller services. BBSs offer fewer phone lines because most of them are run by hobbyists who pay their own phone bills, so calling a BBS can be an exercise in patience and endurance for all but the most dedicated redialers.

✔ Thrift shopping and BBSing can be intriguing, yet extremely disorienting. Each "shop" sports a different layout, so it's sometimes hard to find stuff. That's because the people who set up and run BBSs (they're called *sysops*, short for "system operators") can choose from more than 20 different types of software. And even these evolve, often becoming customized beyond all recognition.

✔ In thrift shops and BBSs, you can't predict just what you'll find. Other customers (and callers) bring in so much of their own stuff to "sell." In addition, the guy or gal who runs such a place can often be quite a character.

✔ The regulars appear to hold certain privileges that only long-time membership brings. Many old-timers share a bond forged just by "shopping" at the same place, at least until they break away and start their own "shops."

✔ Bigger thrift shops like Goodwill and Salvation Army can be a bit more organized, and they certainly offer more stuff — often at higher prices. The same goes for the larger, pay-BBSs that vie with some commercial on-line services in the range of stuff they offer (and the rates they charge).

✔ Chapter 11 covers BBSs in more detail.

The Internet

It's been called "The Universe." Catchy, but if it were a place to shop, the Internet would be all the tempting roadside stands you pass as you crisscross the globe down an endless blacktop highway.

✔ If you already have a "car," that is, a free account, it costs nothing to zoom around from stand to stand. If not, you pay a hefty "rental fee," plus mileage (connect charges).

✔ Roadside stands carry only the freshest, most desirable goods, being right at the source, usually. You have to help yourself, though, and getting stuff home is rarely easy.

✔ You must remember to leave a well-marked trail if you want to return. Finding a stand on a lonely, country road can be difficult, and so is finding specific files or data on the Internet.

✔ Chapter 12 talks more about what to expect on the Internet.

Electronic Mail

Electronic mail is also known as e-mail. It's a quick way to communicate.

For example, imagine that you're sitting at your computer composing a memo. Now imagine sending that memo directly to its recipient's computer just by pressing Enter. That's e-mail.

E-mail allows you to send and receive messages almost instantaneously over a network or over ordinary phone lines. E-mail software can be found in office computer networks, BBSs, the Internet, and commercial on-line services.

What is electronic mail for?

More and more people are turning to e-mail instead of faxes for instant communication. A fax you send is actually just a *picture* of your letter or memo: Your recipient can't edit your words or include them in other documents without turning to special text-scanning software. You can, however, edit e-mail messages, forward them to other people, and even save them to a file on your hard disk.

E-mail is quicker than the postal service; even Federal Express can take 24 hours. E-mail takes only minutes or, at most, a couple of hours to arrive. Plus, e-mail leaves a record, just like paper memos. You can refer to your saved messages to verify what was said instead of shuffling through sticky notes trying to recall the outcome of a phone conversation.

Unlike a phone call, e-mail doesn't depend on the recipient being near the phone, or home. Like voice-mail messages left on answering machines, e-mail can be received any time at the recipient's convenience. Better yet, people can reply with their own e-mail back to you — anytime they want — avoiding the telephone tag of phone messages.

Who uses electronic mail?

Many government offices and educational sites use e-mail to correspond. Consumers and businesses are also starting to learn its value. For example, many companies offer technical support for their products via e-mail. Explain your glitch in e-mail to tech support, and you'll probably find a response the next time you call. Kids can even send e-mail to pen pals (keyboard pals?) across the globe, right from their computerized classrooms.

Who offers electronic mail?

Nearly every on-line service offers e-mail, from a local BBS to a national service like DELPHI to the international network of computer networks known as the Internet. A few specialty services like MCI Mail focus almost entirely on e-mail.

How does e-mail work?

Generally, when you join a service, you receive your own *mailbox*. Messages you receive from others await you here, and your mailbox can store copies of messages you send as well. When you call a service, the first message you see tells you whether you have mail.

Each service has its own *editor* that works as a miniature word processor. To send a message, start the editor and compose the message. Some editors make you compose the message and then decide whom to send it to; others let you fill in the recipient's name first and then fill out the message. Some services let you send the same message to multiple recipients or even *attach* a file to a message.

Editors: They're everywhere

Editors come built into BBS software and as part of commercial on-line services, but you can also get your hands on some very good shareware editors that serve as stand-alone, no-frills word processors.

Computer gurus use editors for programming, and for editing AUTOEXEC.BAT files and other unmentionables. But you can use an editor to write messages while you're not on-line. For example, you can compose a reply to your editor, save it as a file, and upload it as a mail message when you're on-line.

✔ Some on-line editors have a *quote* feature that lets you paste portions of a sender's original message into your reply. Message traffic can become heavy, and it's considerate to jog your correspondent's memory by peppering your replies with snippets of his or her message.

✔ Few people bother to take advantage of the built-in editors in most modem software. Maybe that's because many BBSs and on-line services contain their own, powerful mail editors.

✔ If your program doesn't offer an editor, don't despair. Some excellent editors are available for downloading from your local BBS or a commercial on-line service. Windows comes with a built-in editor called Notepad. A text editor called EDIT comes built into DOS Version 5 and later.

✔ Using editors and other special programs is the smartest way to manage mail and save on connect charges.

Microbiology Lab: staph only!

What does e-mail cost?

In general, e-mail comes as one of the benefits of membership on a BBS or on-line service. You usually won't be charged extra to send or receive e-mail. (Exceptions do exist, of course, so be sure to ask about e-mail charges before you subscribe.) Mail specialty services (like MCI Mail) do charge a small sum for each message you send plus a yearly fee or one-time sign-up charge.

Best bets

Look for an e-mail system that can reach the widest variety of people you need to reach. For example, the e-mail on a local BBS will probably only reach other members of that BBS, but that may fit your needs pretty well, at least initially. CompuServe's e-mail, on the other hand, reaches CompuServe's one million subscribers, as well as the Internet, MCI Mail, and beyond. If you're on the Internet, you can reach just about anyone who's on-line anywhere through its global grasp — often at little or no cost to you. If the only service you seek on-line is e-mail, however, you'll save the most money by opting for MCI Mail or one of the other e-mail specialty services.

Not too long ago, you and your pals had to join the same service if you wanted to exchange e-mail. If Kathy used MCI Mail and Marv subscribed to America Online, Marv had to sign up for MCI Mail as well in order to reach Kathy.

All that's changed now. Most on-line services offer e-mail connections to all the others, thanks in part to the growth of Internet mail *gateways*. Even the majority of local BBSs now offer *echoed*, or networked, mail through which your e-mail message can reach callers on BBSs that belong to one or more local, regional, national, and even international networks, like FidoNet, for example. Refer to Chapter 11 for more on BBSs.

Drawbacks

Like any instant means of communicating, e-mail presents some pitfalls. Here are a few tips from Miss On-line Manners.

Watch out for the common blunder of sending messages without thinking them through. And *never* assume that your e-mail is private. Many hapless victims have suffered terrible embarrassment — or lost their jobs — after their so-called private e-mail gripefest about their boss got loose on the office network.

Typed words lack the subtle nuances of our spoken communications. That's why people use punctuation marks to create little happy faces, called *emoticons* or *smileys,* to sprinkle inside messages. To view a smiley, tilt your head to the left:

```
Your basic smiley:        :-)
```

You'll also see lots of funny-looking code words inside e-mail. You can save time (and connect charges) by abbreviating common phrases with acronyms like ROFL for "rolling on the floor, laughing," or <g> for "grin." Smileys, acronyms, and other symbols help you convey irony, sarcasm, and other tonal undercurrents that are difficult to get across in stark print. (To see the Top Ten Smileys, head to the end of this chapter. A gaggle of acronyms awaits you in Chapter 15.)

Don't overuse smileys and acronyms: You don't want to look giggly. And NEVER TYPE IN ALL CAPS. Doesn't it look like I'm shouting? That's what people reading your messages will think, too. When you want to emphasize a word, use a pair of *asterisks* instead.

Uploading and Downloading Files

Uploading a file means sending a copy of a file from your computer to another computer. *Downloading* means bringing a copy of a file or program from another computer over to your *own* computer. Even capturing information as you read it on-screen can be considered downloading.

What is uploading and downloading for?

As one of the most popular aspects of modem use, downloading lets people sample a wide variety of files from around the world. People can download shareware programs (see Chapter 1), articles, and collections of messages left by other callers in on-line *forums*, or special-interest groups. Technical *patches*

Data highway

A message reading, "You all suck" recently greeted commuters along Connecticut's Interstate 95. It seems a high-school student hacked his way into the highway department's computer and changed the electronic sign boards that normally warn of fog and road conditions.

After taking hours to clear the message, highway officials and police were blitzed with a new message, this one attacking the governor. After he was caught, the youth told police that the highway computer didn't ask him for a password.

The meek shall inherit the Earth — but not its mineral rights.

or upgrades for existing programs can be found on hundreds of customer support BBSs offered by computer companies. (You can find a list of some of these companies at the end of Chapter 14.)

For purely practical uses, uploading a document, spreadsheet, or database file is the quickest way of sharing work with a coworker in another locale — provided there's a modem on that end, too.

Who uploads and downloads?

When people have a file they'd like to share, they upload it to another computer. Folks who want to sample a variety of new programs download files from other computers.

Downloading files gives you access to the very latest programs and program updates. Broadly speaking, the almost limitless capability to download files is what drives most people on-line in the first place.

Who offers uploading and downloading?

A standard feature found on nearly every computer you call, uploading and downloading are the heart of almost any BBS or on-line service.

Most on-line places stick the files in specific areas, usually divided into categories such as "Weird Mac Sounds" or "Goofy Windows Wallpaper."

How does uploading and downloading work?

After reviewing lists of files and picking out a few plums to copy onto your computer, you tell the host computer to send the file(s), and then start the download from your own modem software. Uploading is basically the reverse: you describe the type of file you'd like to share, give it a file name, and perform the commands to start sending the file from your modem software.

On-line services differ in the specific commands for uploading and downloading, but most offer helpful menus to make the process easier.

Don't get bent out of shape if it takes a few days before you see the shareware game you uploaded on the public file area of a BBS or an on-line service. The sysops are probably scanning your file for viruses before giving it a home on their system.

What does uploading and downloading cost?

Depending on where you go, you can pay anywhere from nothing to several dollars to download a file.

Local BBSs vary. Some let you download as much as you want. When the call is local and the BBS doesn't charge a membership fee, the files you download are free. Many BBSs use a ratio system: you're asked to upload a file for every few you download (usually four downloads for every upload). Some BBSs keep a file-size tally: you can download so many megabytes for every megabyte you upload. A simpler approach is to donate money to help with the BBS's upkeep and costs. Usually, the more you donate, the closer you approach *unlimited downloads*.

With the commercial on-line services and pay BBSs, any hourly charges are in effect for downloads. So, depending on your modem's speed, those neat-o files can cost you some bucks. (Consult the various on-line service chapters for tips on saving money on-line, as well as for details on rates.)

Files abound on the Internet, and unlimited downloads usually apply. Depending on how you get the Internet, however, your personal file storage space on your computer may be limited.

TIP

Even systems that charge an hourly rate usually suspend their charges while their users are uploading files. As soon as the file's been uploaded, the clock starts again.

Best bets

CompuServe probably has the largest number of files available for downloading, but GEnie and America Online aren't far behind. The Internet offers perhaps the broadest spectrum of files, but finding and downloading what you need may be challenging to all but the most hardened computer weenies.

Drawbacks

Many people, overly influenced by bad publicity, consider BBSs to be hotbeds of computer viruses. Because of all the bad rap, BBS sysops are usually among the most careful purveyors of files. In fact, I've heard of computers contracting viruses and other malicious programs from shrink-wrapped software as well as from disks handed to you by friends or coworkers. In general, it's smart to be careful with new files, whatever their source. Check newly downloaded files for viruses the minute the files land on your computer. Head to Appendix C for more on what to do with these pesky new files.

Don't download too many files at once and become overwhelmed. Try out a few programs at a time and get to know them. (Register any shareware files you find yourself using, too.) Then go back for more.

On-Line News

Most newspaper companies send out their newspapers once a day, but they're constantly collecting information throughout the day. Much of this flowing news stream is considered on-line news.

What is on-line news for?

Newspapers couldn't function without the on-line news, known as the *news wire*. On-line services know that news is a big draw, so most boast of one or several news areas.

You know it's time to pay the shareware author when...

The only thing easier than downloading and trying out a shareware program is forgetting that it's not free. Here are a few guidelines for knowing when the trying should end and the buying should begin:

You know it's time to register when

- ✔ The money that went toward printing out the manual would have covered registering. Twice.

- ✔ You've memorized the guilt message that pops up each time you start the software, imploring you to pay.

- ✔ You find yourself asking your computer guru to reprogram the shareware to get the guilt message to disappear.

- ✔ The program saved you much more money than registering the software would have cost.

- ✔ You find yourself sharing "this great little time-saver" with everyone you know.

- ✔ You've missed out on dozens of program updates because you didn't register and get on the mailing list.

- ✔ You press the correct keys to bypass the guilt message before it's even displayed.

- ✔ You've been "trying it out" for a year or two now.

Who uses on-line news?

People who want the latest details on a news story turn to on-line news. But more and more folks read the constantly flowing news wire for sheer pleasure. Business people scan news for specific topics, and news hounds chase hourly updates.

Who offers on-line news?

Most commercial on-line services and many of the large, commercial BBSs offer on-line news as part of the package. Some offer the Associated Press, or "AP news wire." Others offer United Press International (UPI), Reuters, or electronic news publications like *USA Today*. Prodigy has its own newsroom. You can also find specialty, targeted news — like Newsbytes News Network, dishing computer-industry dirt — on many services and BBSs.

For really specialized needs, a number of news services can serve up customized news to you. Daily they scan thousands of periodicals and papers for you, clip out items on topics you've preselected, and fax or e-mail them to you. MCI's Information Advantage can deliver a customized, electronic newspaper. And CompuServe's Executive News Service lets users type in key words; the service then scans all the incoming news wires for your topics and saves any matching stories in a folder for your review. An on-line service called News in Motion takes a global approach, translating and mixing some 50-60 of the week's top stories from international papers and news wires and preparing them for a weekly, ten-minute download.

Many leading publications and television news networks maintain a presence on the various on-line services. Members can corner editors and anchors to discuss news and media coverage in an interactive setting. For example, America Online offers regular CNN conferences where you can meet CNN anchors. The service features feedback areas for the *San Jose Mercury News*, *TIME Magazine*, and the *Chicago Tribune*, as well. Most of the on-line publications solicit readers' feedback and dissenting viewpoints. Plus, writers' and journalists' forums on almost every on-line service are frequented by professionals and *wannabes* alike.

What does on-line news cost?

Most news is tossed in at no charge as part of basic on-line membership packages. Depending on the level of coverage and customization you require, you may pay extra hourly charges or, for the hard core newshounds, up to thousands of dollars per year.

Best bets

Of the general on-line services, CompuServe offers the most news — Associated Press, United Press International, Reuters, and several international news systems. CompuServe also has the Executive News Service mentioned earlier.

In addition, CompuServe offers Reuters news photos: That color picture you saw in this morning's newspaper was probably downloaded and viewed by hundreds of people yesterday afternoon.

Serious news junkies who can't get enough should try Dow Jones/News Retrieval, MCI's Information Advantage, News in Motion, or one of the other specialty services.

Drawbacks

News can only be found on pay services like CompuServe, GEnie, America Online, Prodigy, and others. Some of the larger BBSs also carry news. You won't find news on any free local BBSs, so reading the news can be expensive, much more so than a 25-cent local paper.

On-line Information

Actually, an on-line service's entire content could fall into the category of on-line information. Narrowing it down a bit, on-line information usually falls into one of two categories: *information databases* or *forums* peopled by folks with

Is the USENET News, er, news?

Increasing numbers of BBSs offer access to a phenomenon formerly associated with the Internet: USENET News, where users can read messages called *articles* in thousands of topical *newsgroups*. Although newsgroups focus on newsworthy subjects in the arts, sciences, and other fields, many others focus on bizarre topics (like the newsgroup called "alt.pave.the.earth").

Reading the news in topics that interest you is a good way to keep up with trends and issues — as long as you have enough time to sift through extraneous material to find the real gems. Broadly speaking, USENET News belongs more fully to the on-line information category, described next, than to "hard on-line news."

common interests. On-line information can be an electronic library of sorts, containing back issues of newspapers and magazines, photos, government documents, forum messages, and just about any other information somebody's typed into a computer at one time or another.

Forums, USENET newsgroups, and other special-interest gathering areas are an important aspect of on-line information. Here, members exchange views and information in public messages that are similar to group e-mail. Many forums offer months or even years of past messages for downloading and browsing. Other forum activities include live conferences and specialty file areas. Refer to the various on-line chapters for more on forums, their equivalent names, and tips on how to find the ones in your area of interest.

What is on-line information for?

One of the biggest benefits provided by on-line services is the capability to store information on-line for users to access in a rapid, organized fashion.

Being on-line instead of in print allows information to stay much more current. For example, the publishers of Grolier's Academic Encyclopedia, a basic service on Prodigy and several other on-line services, update the text every six months. Try that with a printed book.

Who uses on-line information?

Anybody who's used a library will appreciate the convenience of on-line information. Most information can be accessed by typing in key words. Entering **CAT AND FLEAS**, for example, dredges up anything written about those little black things that pester your favorite feline.

Businesses can search for trademarked names; journalists for past coverage of a news subject. Fiction writers can research a locale or era for realistic details. As more pictures, sounds, and even film clips make their way onto computers, they'll make their way onto on-line services where people can find and download them.

Who offers on-line information?

Many of the general on-line services offer a few databases full of information. Specialty information services offer a fuller range at a higher cost. Refer to Chapter 10 for more on these resources.

From the local BBS's "Save the Earth" forum to the USENET newsgroups found in increasing numbers of BBSs on-line, forums and their message archives are important components of almost any on-line service's information resources.

Certain computers on the Internet serve as special information archives containing files, articles, library catalogs, and other resources on any topic you can imagine. An increasing number of searching tools like Archie, Gopher, and WAIS are making Internet searches faster, easier, and more rewarding.

What does on-line information cost?

Some commercial on-line services, like CompuServe, charge additional hourly rates for access to their forums. Almost all the information databases levy extra charges. Expect to pay astronomical fees on the specialty, information-only services like Nexis or IQuest. Charges vary depending on the specific database you're after, so be sure you know the rates before you start your search. Also, don't forget to ask about lower, after hours rates.

Most information on the Internet is available as part of your regular access fee (this could even be free to you if you have an account through work or school). BBSs usually offer access to forums and any message archives as part of regular membership.

Best bets

The easiest way to find the information you're looking for is to search through the largest pot of information. Knowledge Index, offered after-hours on CompuServe, serves as a budget database gateway to hundreds of databases where you'll find everything from the Bible to Government documents to back issues of *Barbie* magazine.

A tougher but even more rewarding (and possibly cheaper) information resource awaits you on the Internet. But where do you go and how? Talk to Internet gurus you meet in various Internet forums. Your Internet provider may also prove to be a helpful resource. If you have kids in college or even high school, try asking them for recommendations.

For the most reasonable rates, start your search in on-line forums — even on the free, local BBSs, especially if they offer the huge numbers of users found in the nationwide networked, *echoed* forums. Head for forums that attract people who may have already retrieved similar information. Ask them for pointers on where to search.

Tips for searching on-line databases

Searching on-line databases is largely a matter of being familiar with the particular searching *engine* offered by the database. Here are some tips for saving money on-line:

✔ Download any *how-to* files on searching techniques. If instructions appear as help screens or menus, capture and save them to a file by using your modem software's capture or log commands. Familiarize yourself with the basic commands before you head back on-line to start your search.

✔ Prepare for a database session by joining any forums that focus on on-line databases. Discussing your search with database veterans you meet can yield great tips and leads.

✔ Leave out prepositions, articles, and other needless verbiage from your key words. They increase precious search time, racking up higher rates for you.

✔ Generally, you can narrow the number of matches by sticking an AND in your key words, like NASA AND MARS PROBE. Database experts recommend starting with the narrowest terms and branching out from there because databases rarely charge for searches that don't yield any matches, or *hits*.

✔ You can almost always expand on items to search by using OR in your key words, like TOFU OR SOYMILK.

✔ Adding a NOT usually excludes a term, like DINOSAURS NOT BARNEY.

✔ Remember that your database's commands may vary, but the ones previously mentioned are fairly universal.

✔ Publication databases like CompuServe's Magazine Database Plus let you search by article title, publication date, author name, and topic just by entering key words. Prepare yourself by jotting down as many key words in as much detail before you go on-line.

✔ Opt for specific, technical jargon words instead of general terms for the field you're searching in. Some databases offer lists of field-specific terminology as part of their help systems.

✔ Take advantage of any live, human tech support. Typing SOS brings in IQuest's librarians for some real-time help at no extra charge, for example. They may help you build a search or clue you into specialized terms and abbreviations used by a particular database.

Ask what reference sources are available to you as part of basic membership on an on-line service. In addition to dictionaries and encyclopedias, CompuServe, for example, offers Peterson's College Database, HealthNet, FundWatch On-line, and Consumer Reports as part of its basic pricing plan. The other big services offer similar reference works.

Drawbacks

Searching on-line databases can be ultra costly. Be sure that you're up on rates and fees before you go.

On-line Travel

With their modems, people can book airline flights, examine competing airline fares, or dig for details about their destinations before leaving.

What is on-line travel for?

Chatting about hometowns, recommending restaurants, or even booking a rental car or flight is an essential, popular part of being on-line for many members.

Who uses on-line travel?

Anybody who travels — or armchair-travels — can benefit from travel-related forums and services on-line.

Who offers on-line travel?

Most commercial on-line services offer at least some travel-related stuff. GEnie, DELPHI, Prodigy, America Online, and CompuServe offer EAASY SABRE, a computerized flight information and reservation system. Most offer news on discounted cruises and other insider information. Travel forums offer a wide network of folks who've been there before and are rarin' to talk about it.

CompuServe offers TrainNet, a train forum; a Florida forum; and a Scuba forum, for starters. America Online has BikeNet, among others. Access to large numbers of members makes Prodigy's Travel Club a fascinating place to discuss where to go and what to do. Even some of the larger pay-BBSs let you rent cars or make reservations on-line.

If your on-line needs focus mainly on travel, look into specialty services like Official Airline Guide (OAG), discussed in Chapter 10.

What does on-line travel cost?

EAASY SABRE is usually included as part of basic membership plans on-line. The Official Airline Guide levies an hourly surcharge if you access it through an on-line service, or it charges a monthly fee, plus hourly rates, if you join the stand-alone OAG service.

Tickets and reservation vouchers are mailed out to members, usually free of charge, although you may pay extra for weekend or overnight delivery.

Best bets

If you're seeking travel services as part of a general on-line service, try CompuServe, Prodigy, GEnie, or America Online. EAASY SABRE is easiest on Prodigy because it uses Prodigy's menus and pictures, but it's also slower for those same reasons.

America Online offers some highly focused *metro* services. Chicago Online, tied into the *Chicago Tribune*, offers TicketMaster ordering on-line (get the jump on that Bears/Giants game during that upcoming business junket), new house listings, Chicago arts schedules, and conversations with natives of the Windy City.

If travel is *all* you want, try OAG first.

Drawbacks

Nothing beats a good travel agent for currency, breadth, and depth of knowledge. But for access to "hometowners," "been-theres," and 24-hour convenience, on-line travel services have no equal. Travel agents don't charge for anything but the most customized itineraries, so why not use both?

Chatting and Conferencing: Making Friends On-line

Chatting is the act of typing "hello" and other pleasantries back and forth with a fellow on-line user in actual time. Chatting is akin to CB radio, using typed words instead of spoken ones. People drop in and out of chat sessions at random, sometimes using nicknames or *handles* instead of their real names.

As opposed to chatting, *conferences* usually involve more people; they're held in on-line *conference rooms* at prescheduled times. Conferences generally focus on a particular topic or guest speaker, while chat can be defined simply as random conversation.

What are chatting and conferences for?

Chatting provides a casual "free-for-all" atmosphere where people can let off steam and meet others. Some on-line services offer a sort of hometown bar atmosphere; others feature adult-rated chat with separate rooms devoted to every conceivable sexual persuasion. Private chat rooms are also available for people who just want to have a conversation in private.

Conferences are a mainstay of forums, bulletin boards, and other on-line special-interest groups. They provide a way for members to come together at a scheduled time to discuss specific issues or meet "speakers."

Who uses chatting and conferencing?

Anyone of a sociable nature will enjoy chatting. Conferences are a great way to meet others in your field as well as get to know the other members of a forum.

Who offers chatting and conferencing?

Chatting and conferencing can be found on almost every place on-line, from BBSs to on-line services to the Internet. The ImagiNation Network is a chat specialist. The capability to chat with opponents while playing games on ImagiNation has now become a feature on other on-line services as well.

What does chatting or conferencing cost?

Any hourly rates or basic subscriber charges apply while chatting. In fact, some on-line services levy extra charges for chatting and conferencing, so check before you go.

Best bets

For conferencing, you'd naturally gravitate to the on-line service offering forums relating to your interests. Chatting is a big part of every on-line service, so try them all and see who you meet. Most regions of the U.S. offer extremely chat-focused, pay-BBSs where singles and others meet.

Drawbacks

Commands used for chatting and conferencing can be a little more confusing than normal on-line navigation, so capture or download a list of the more common commands and practice a bit before you go on-line.

Because chatting isn't free and because people tend to lose track of time while doing it, addictive personality types may find this activity financially ruinous.

Finance and Business Matters

Banks, financial exchanges, and business news are just a few of the financial services found on-line.

What are on-line financial services for?

You can do your banking on-line and do away with writing out paper checks for good. Or you can follow a company's financial data through tips from forum members, as well as on-line business news and databases. Tracking stock quotes, trading shares, reading business newsletters, and paying bills are some of the popular finance-related activities available on-line.

Who uses on-line financial services?

Business executives, financial professionals, home business entrepreneurs, and everyday people use financial services on-line.

Who offers financial services on-line?

All the commercial on-line services offer some type of financial service on-line.

Specialty services like Dow Jones News/Retrieval, DIALOG Information Services, NewsNet, and several others offer a huge array of business and financial services.

What do financial services cost?

Charges for financial services range from low, like $14.95 per month for Prodigy's Strategic Investor, to high, as in $200 per hour for Dow Jones News/Retrieval's last-minute stock quotes and extensive financial databases.

On-line forums and local bulletin boards let you exchange tips and sob stories with fellow investors, sometimes for no extra charge beyond the cost of the phone call.

Best bets

Prodigy, GEnie, or CompuServe may fill your needs for financial and business news and services. If not, check out one of the specialty services available.

Drawbacks

Learning the commands for obtaining quotes and searching through business databases can be arduous. Check the "Tips for searching on-line databases" section for some pointers. Also, ask about reduced rates for off-hours access and other specials.

Shopping On-Line

You can buy flowers, gourmet candies, or even a new *Dummies* book by browsing the on-line services' *electronic malls* and filling out an on-line order form. Most people pay with their credit card number and have their orders sent parcel post to their homes and offices.

What is shopping on-line for?

Department stores, specialty stores, and catalog merchants can bring their wares before the on-line public in an easy and convenient setting, and they're often open 24 hours a day.

```
Cogito ergo spud: I think, therefore I yam.
```

The lure of saving money drives many shoppers on-line. Prodigy's prices on contact lens prescription refills, for example, are lower than most stores in my city.

Who shops on-line?

People who enjoy mail-order shopping find on-line shopping just as convenient and maybe a little more fun. Small-business owners can replenish office supplies at any hour. Family members can remember that special birthday or anniversary with a thoughtful gift — even at the last minute.

Who offers shopping on-line?

Look for on-line shopping on all the primary commercial on-line services.

What does shopping on-line cost?

Most places don't charge extra for members to browse their stores. In fact, on many services, shopping time is free of any connect-time charges.

Best bets

CompuServe and Prodigy rank high in the electronic mall department, but even the Internet offers a bookstore on-line. Ask your on-line pals about any current specials going around.

Drawbacks

You can't really see the product, or taste it, or try it on.

Also, if family members share the same account, as on Prodigy, you must be careful to enforce some rules when it comes to on-line shopping.

Gaming and Playing On-Line

On-line games are just like any other computer games, but with an important twist: You can usually play against somebody *else* instead of just a computerized opponent. You can play cribbage with Diane in Tennessee or shoot down Ted in his fighter plane in France.

What are on-line games for?

Just like any other games, on-line games are played for entertainment. Because most on-line services have games in which many people can play at one time, many people use these games for companionship: They can hold conversations while playing cards, for example.

Who uses on-line games?

Although males aged 28–40 used to dominate the on-line game arena, that trend is changing as fast as the computer-buying public is changing. Today, it's hard to tell who's holding the cards on the other end of the modem.

Who offers games?

Just about every on-line service — even BBSs — offer games, but the quality varies widely. Some offer simple, text-based games like Hangman. Other games offer graphics and sound that compete with off-the-shelf products for quality.

What do on-line games cost?

Although pricing varies, most on-line services charge the same amount for on-line games as they do for anything else. A few jack up the price a notch, but some even lower the price by offering bulk-rate specials.

Best bets

Although most on-line services offer games, The ImagiNation Network is by far the best bet with dozens of games designed for all ages. Adults can play Blackjack or drop down drinks in Larry's bar side-by-side with other players.

Smaller fry can choose from Paintballs, Boogers, and a host of other games. And just about anyone can pick up the joystick, strap themselves down into a Red-Baron model, and battle it out with other WWI aces in the virtual sky. Best of all, ImagiNation lets players chat while they play.

GEnie offers a great flight simulator, and late-comer CompuServe is trying hard to beat competing services with its Entertainment Center games. America Online offers Rabbit Jack's Casino, plus Quantum Space, and other "Dungeons and Dragons" types of games played through e-mail.

BBSs and the Internet offer other gaming possibilities. Ask around and see what fun stuff your on-line pals are up to.

Drawbacks

Most games are addictive, unfortunately. Be sure to keep a stop watch going while you're playing on the commercial services, or you'll rack up one heck of an on-line charge.

A sysop on a local BBS told me that he devoted 1-2 hours per day to his favorite on-line game, Interstellar Annihilation. And a friend who got involved with ImagiNation after I started writing this book has had to cut back severely on the many hours she logs playing Cribbage there.

Top Ten Smileys

Don't have the time or inclination to type it out? Express that feeling of joy, sarcasm, or disappointment with a smiley (or some other off-the-wall group of punctuation marks), instead.

A smiley, or emotional icon (emoticon), is used as "feeling shorthand" in e-mail and on-line chatting. Smileys have three basic components (and you have to tilt your head to the left in order to see any of them):

```
first, : ← eyes; then, - ← nose; then, ) ← mouth. Put 'em
all together and you get your basic, common garden-variety:
:-)     Smiling Smiley
```

Typical smileys

; -) Leering, winking, or snickering smiley

: - (Down-hearted, depressed, or blue smiley

8^) Sunglassed, suntanned, and smiling smiley

Less-typical smileys

: - 0 Amazed or shocked smiley

: - { Mustachioed smiley

: - d Laughing smiley, or smiley wearing a very big grin

: - p Smiley sticking tongue out

: - # Smiley wearing braces

: - x Smiley whose lips are sealed

Really weird smileys

{(:-)   Smiley wearing toupee

*<:-)#   Smiley is Santa Claus

C=:-)   Smiley is a chef

+-<|:-) Smiley is the Pope

*:0)   Smiley is a bozo

X-(Ex-smiley

Chapter 6

Prodigy and The ImagiNation Network

● ●

In This Chapter

▶ Logging on and logging off

▶ Hollering for help

▶ Finding stuff

▶ Sending electronic mail

▶ Downloading files

▶ Gaming and chatting on The ImagiNation Network

▶ Prodigy Tens

● ●

*P*rodigy. If someone walked up and commanded me to describe it in no more than four letters, I'd have to say, "PC TV."

Prodigy is like turning your personal computer into a TV set. (All that's missing is Connie Chung.) Maybe its resemblance to America's most beloved household appliance is what makes most new modem users try Prodigy first.

What makes Prodigy more TV-like than any other on-line service? It may have something to do with the robust colors and animation dancing across your monitor. Or "programs" like NOVA, National Geographic, and Victory Garden. It could be the ads that appear on almost every Prodigy screen. Or its strict control of content for "family" suitability.

Whatever the reasons, to date, Prodigy has appealed to millions of users. If you're eager to see whether Prodigy appeals to you, this chapter gets you hooked up and jumping.

Logging On

Connecting to Prodigy is pretty easy. (They planned it that way so you won't miss any of the ads.)

If you haven't installed Prodigy yet, hop to the end of this chapter for the dirt. Otherwise, DOS users can type **PRODIGY** at the `C:\>` prompt, like this:

```
C:\> PRODIGY
```

That's the word **PRODIGY**, followed by pressing Enter.

Starting Prodigy on the Mac is really a snap: There's nothing to type. Just open your Prodigy folder and choose Launch the Service.

Beware, however: Prodigy doesn't have much for the Mac. Most of it is oriented toward PC users. Prodigy's main source of file downloading, a service called ZiffNet, doesn't even work on the Mac. Besides, all the files are for that *other* type of computer. Sniff. Mac users can forget about The ImagiNation Network, Prodigy's new "place to go" for games and chat.

By all means, though, use your trial membership. And do stay if you like Prodigy. But most Mac users end up gravitating to America Online, a service that started out on the Mac. See you in Chapter 8!

🖊 When Prodigy first leaps to the screen, it asks you to type your member ID and password. You wrote these down in a top-secret place when you installed Prodigy, so shuffle through your pile of sticky notes until you find them. Then type them in.

🖊 Want a way to avoid typing in all that stuff each time you want to log on? Head to "Prodigy Tens" and read the AutoLogon section.

🖊 Prodigy uses special software sold in a Prodigy "kit." It comes in either DOS or Macintosh versions. You can't call Prodigy by using your plain, general modem software. Haven't signed up for Prodigy yet? Go to "How Much Is This Costing Me, Anyways?"

🖊 Prodigy is so family oriented that it lets you share your membership with up to five household members at no extra cost. The ID of the "Primary Member" — the one who foots the bill — ends in the letter A. Family members' IDs are identical, save for the last letter, which, you guessed it, end in B, C, D, and so on. To be on the safe side, however, make sure that Daughter Edwina and Great Aunt Elk all choose separate passwords.

✔ The "A" member — the Primary Member — can lock out unsuspecting family members from doing nasty things like putting exotic airline tickets on the MasterCard, as well as keep them from heading for the extra-extra charge services like playing the Baseball Manager games. From any screen in Prodigy, choose JUMP CHANGE ACCESS to learn more. What's JUMP? Head to the next section.

✔ Please, encourage your spouse or kids to log on with their own accounts. This message is brought to you by the Committee in Charge of Eradicating Confusing Messages, like the ones that say "From Brad" along the top, but are signed "Mary" along the bottom.

✔ Your first call to Prodigy takes place at a "medium" modem rate of 2400 bits per second — abbreviated *bps*, and sometimes called a *baud rate* — which is a measure of how fast the modem operates. (If modem speed were measured like blender speed, that would be "frappé.") At this slow rate, all the pretty pictures crawl across your screen.

✔ Bought a faster modem? Then log on to Prodigy and choose JUMP 9600 BPS SERVICE. Prodigy tells you about a special 9600 bps phone number in your area; then it sets up your Prodigy software to recognize your new, blazing speed the next time you log on. (This JUMP stuff is covered in the very next section.) At this point, Prodigy is not charging extra for 9600 bps service.

✔ Not sure what speed you're logging on at? Anytime you're on-line, you can check your speed by selecting the **T**ools item in the Command Bar and choosing **P**hone & Speed.

✔ Don't know what "Tools" or "Command Bar" means? See the following section.

Prodigy at a Glance

Like a desktop-publishing newbie, Prodigy is constantly fiddling with its graphics and screens. When you sign on to Prodigy, your screen will probably look a little different. A few main parts don't change, however, and Figure 6-1 dissects Prodigy's Main Highlights screen circa June, 1993.

What's on Prodigy's Command Bar?

The most powerful tools for getting around Prodigy cluster inside the *Command Bar* along the bottom of the screen (see Figure 6-2). You can use your Tab key or mouse to move to and select any area of a Prodigy screen, including the Command Bar.

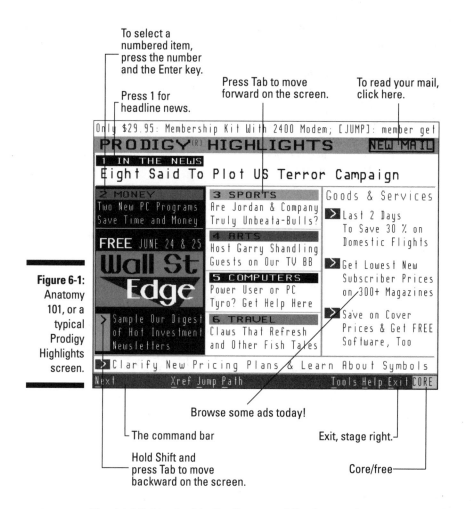

To select a
numbered item,
press the number
and the Enter key.

Press 1 for
headline news.

Press Tab to move
forward on the screen.

To read your mail,
click here.

Figure 6-1:
Anatomy
101, or a
typical
Prodigy
Highlights
screen.

Browse some ads today!

The command bar

Exit, stage right.

Hold Shift and
press Tab to move
backward on the screen.

Core/free

You highlight a tool in the Command Bar by moving over to it with your mouse
and clicking or by pressing Tab until the tool's little "square" starts blinking.
Then either click your mouse if the item is already highlighted or press Enter.
(Tired of the mouse and Tab key? Then press the first letter of the tool you're
after. That highlights it, too.)

JUMP!

One of Van Halen's most memorable tunes. Also the most essential tool on the
Command Bar — by far. You can tell, because I haven't been able to get this far
in the chapter without mentioning JUMP five times already.

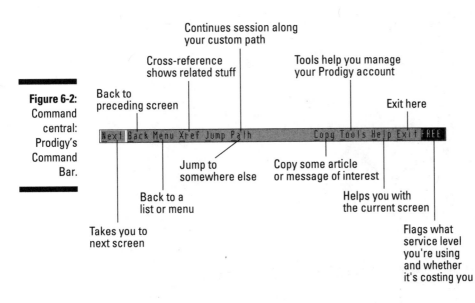

Continues session along
your custom path

Cross-reference
shows related stuff

Tools help you manage
your Prodigy account

Back to
preceding screen

Exit here

Figure 6-2:
Command
central:
Prodigy's
Command
Bar.

Jump to
somewhere else

Copy some article
or message of interest

Back to a
list or menu

Helps you with
the current screen

Takes you to
next screen

Flags what
service level
you're using
and whether
it's costing you

✔ Every section on Prodigy has its own special JUMPword. (Prodigy, like the rest of the computer industry, is not immune to WordsRunningTogetheritis.) A JUMPword is like a "pushbutton." By typing the JUMPword, Prodigy jumps to that area. Quicker and more sanitary than going through a maze of menus.

✔ How do you JUMP? Choose **Jump** from the Command Bar, and watch as a special Jump window pops up. Type your JUMPword in the little blank space, press Enter, and Prodigy zooms there.

✔ To "select" something in Prodigy, double-click it. Mouse-less? Press Tab until what you want starts blinking, and then press Enter. Or, try the number of the item, or its first letter.

✔ Looking for someplace but don't know its JUMPword? Select **Jump** from the Command Bar; up jumps a **Jump** menu like the one in Figure 6-3. See how the second item on the list says **Index of JUMPwords**? This handy item lists everything on Prodigy from A-Z (or "A's News" to "Zurich Weather," that is).

✔ Secret JUMPword trick: Just start typing the first few letters of any word or area you have in mind. Chances are that some similar words exist, and Index of JUMPwords hops up, ready to take you to them.

✔ The other top navigational secret is using Prodigy's Path tool to customize and automate your on-line sessions. To find out more about **Path**, hike to "Prodigy Tens" at the end of this chapter.

A penny saved is ridiculous.

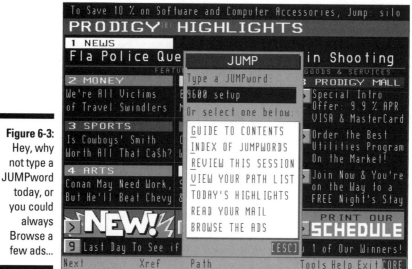

Figure 6-3:
Hey, why
not type a
JUMPword
today, or
you could
always
Browse a
few ads...

... and Logging Off

Pizza guy's here and it's time to log off? Luckily, almost every Prodigy screen offers a quick way out and off the service: It's the **Exit** tool on the Command Bar at the bottom-right corner of your monitor. If you don't see Exit, pressing Esc or selecting **M**enu should make it appear.

- ✔ Move your mouse to the **Exit** item on the Command Bar and click it. Non-mousers can Tab there or press E.

- ✔ The **Exit** command triggers a dialog box like the one in Figure 6-4. You know you're having a Prodigy moment when it asks you to choose between truly exiting, letting another member sign on, or viewing yet another ad.

So What if I Wanna See an Ad?

It's fun to poke a little fun at Prodigy for having ads. But they sometimes contain good information — especially if you don't watch much TV. Besides, they're entertaining: Some of Prodigy's best animation shows up in the ads. And Prodigy charges nothing for time you spend perusing its advertisers' wares.

- ✔ If an ad whizzes on-screen and catches your interest, select the **Look** box (see Figure 6-5). You're in ad country! Follow the commands on-screen to follow the pitch.

When puns are outlawed, only outlaws will have puns.

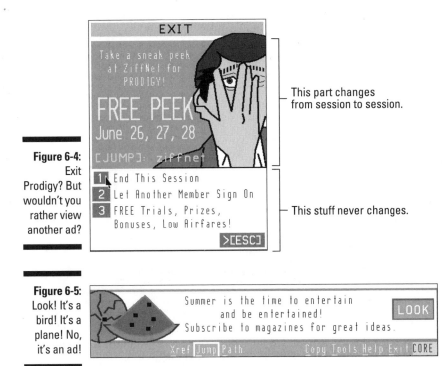

Figure 6-4:
Exit
Prodigy? But
wouldn't you
rather view
another ad?

This part changes
from session to session.

This stuff never changes.

Figure 6-5:
Look! It's a
bird! It's a
plane! No,
it's an ad!

✔ When you're looking at an ad, a new tool called **Z**ip appears in the Command Bar, as shown in Figure 6-6. Selecting **Z**ip zips you back to whatever regular Prodigy screen you were enjoying before you decided to watch the commercial.

✔ If you missed an ad that looked sort of interesting, select **J**ump from the Command Bar and then choose **B**rowse the Ads. An Ad Review screen pops up and lets you choose between stepping back through ads you've seen already or previewing fresh and zesty new ads.

✔ As you edge perilously closer to actually purchasing something on-line, another tool appears: **A**ction. It says, "Chaaarrge!" A comforting **Esc** command usually hovers nearby, offering an escape hatch away from any real financial commitments.

Prodigy wouldn't mind my comparing it to TV. It's actually planning even deeper TV tie-ins. As you read this, "interactive entertainment" is creeping from buzzword to household word as Prodigy works with cable companies to develop two-way TV for cable subscribers.

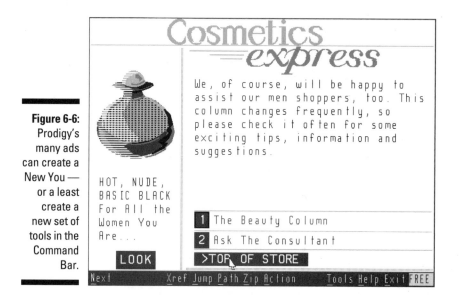

Figure 6-6:
Prodigy's many ads can create a New You — or a least create a new set of tools in the Command Bar.

After that's in place, you armchair sluggers glued to a double-header on ESPN can pull up a batter's stats by clicking buttons on a cable box. Educational TV buffs tuned into Discovery can pull up Prodigy's encyclopedia to learn more about pigmy chimps or get the latest on rain forest decimation.

Prodigy seems to be most excited about the commercial potentials of interactive TV. Contest hounds can click out answers to an ad's quiz and maybe win a lifetime supply of Coca-Cola. (Hardened nerds wouldn't click for anything less than a lifetime supply of Jolt Cola — "Twice the sugar, all the caffeine!")

"We're looking forward to the day when the viewer goes to the fridge during the program and hurries back in time for the commercials to start," says Prodigy's Steve Hein. Cool. Just try to keep the orange Dorito gunk off the cable box buttons.

Hollering for Help

Help is available from any Prodigy screen, in several ways. If general Prodigy stuff has you boggled, plenty of resources await.

- ✔ A **Help** tool lives in the Command Bar at the bottom of the screen. Help pops up info that is related to the screen or service area you're in.

- ✔ Press H or F1 and then Enter to make help come a runnin'. Mac users can also press ⌘-1 to activate the Help tool.

- ✔ Selecting the JUMP tool, typing **PRODIGY** into the box, and pressing Enter triggers a gusher of help sources.

✔ Choose JUMP ASK PRODIGY, and you can send a private message directly to the Prodigy gods. A reply to your question should await you within one business day.

✔ Choose JUMP PRODIGY CENTRAL to see all the Prodigy-related stuff collected in one big area.

✔ Choose JUMP PRODIGY EXCHANGE for a place to gripe and to read Prodigy's responses to other gripes.

✔ If you're really stuck, pick up the real phone and call Membership Services (with your voice, not your modem). They're currently at 1-800-284-5933, 7 days a week, between 7 a.m. and midnight, Eastern Time.

✔ All these Prodigy-related areas are free of timed charges, message charges, and other hidden costs.

✔ You could even read the manual!

Finding Stuff

Jump, hiding down there on the Command Bar at the bottom of the screen, is the best way to move around within Prodigy. But how do you know whether Prodigy even carries what you're looking for?

You can marshal the Command Bar's **X**-Ref tool. Its name sounds straight out of comic-book land, but this cross-reference tool brings up a list of places related to your current location.

Another tool, the **G**uide to Contents, helps you find stuff by dividing Prodigy into several distinct areas of interest, and then giving you their JUMPwords.

Table 6-1 shows you some other routes around Prodigy.

Table 6-1	Shortcuts for Finding Stuff on Prodigy		
Command	*Keys for DOS*	*Keys for Macs*	*Does This*
Index	I, F7	I, ⌘–7	Displays Index of JUMPwords
Jump	J, F6	J, ⌘–6	Opens Jump window
Menu	M, F9	M, ⌘–9	Takes you up one level in the menu structure
X-Ref	X, F8	X, ⌘–8	Lists related topics
Review	R, F10	R, ⌘–0	Lists last 26 places you've seen

Above all else: sky.

Private Messages

If you have private electronic mail, you'll know the minute you sign on: There's a New Mail box flashing at you from the top-right corner of the screen.

Select the box, and you'll zoom over to Prodigy's *e-mail* area. See the listing? Move to a box and press Enter (or double-click with your mouse) to read it: Aarggh — it's a junk-mail ad from an insurance agency!

Don't have any mail waiting for you? Well, start sending some. Head for Prodigy's mail area by typing **COMMUNICATION** or **MAIL** in the JUMPword box. Choosing Communication instead of Mail (and pressing Enter) lets you do more with your mail — things like setting up a mailing list and "broadcasting" one message to everyone on that list. Fun! Select Write a Message if you're in the Communicaton Center, or select Write if you're in the mailbox.

To send a private message to another Prodigy member, you have to know that member's Prodigy ID. Type the recipient's Prodigy ID in the **To:** box (see Figure 6-7).

When you're finished, select the **S**end box in the bottom left corner of the screen.

Depending on what type of membership you have, you'll either be charged 25 cents per message after the first 30 messages per month (Value Plan), or you'll be charged that much from the very first message (Alternate Plans). You'll find more "Plan" details spelled out toward the end of this chapter.

Figure 6-7: Catching up on correspondence.

```
┌──────────────────────────────────────────────────┐
│  ▭▭▭    ▭▬▬        WRITE A MESSAGE                 │
│  TO:                  SUBJECT:      1 OF 1         │
│  KRFF80A              Snack                        │
│ ┌──────────────────────────────────────────────┐  │
│ │ Andy,                                         │  │
│ │                                               │  │
│ │ Ready for a snack?                            │  │
│ │                                               │  │
│ │ Meet me downstairs at the fridge and          │  │
│ │ I'll whip up a quick batch of guacamole       │  │
│ │ brownies.                                     │  │
│ │                                               │  │
│ │ See you soon!                                 │  │
│ │                                               │  │
│ │ Tina                                          │  │
│ └──────────────────────────────────────────────┘  │
│ >SEND  >CLEAR  >PRINT  >TO: LIST  >OPTIONS         │
│ Next  Menu Xref Jump Path    Tools Help Exit ****  │
└──────────────────────────────────────────────────┘
```

The **O**ptions command in the bottom-right corner of the mail screen lets you count how many messages your household has sent in a given month. Get family members in the habit of checking **O**ptions *before* they send private messages — those "Hi, how are you?" "Fine, how are *you*?" exchanges are fun, but not when you're paying *extra!*

Posting Public Messages on Bulletin Boards

Moving to charging its customers hourly rates from a flat monthly rate has forced Prodigy to beef up its messaging capabilities in its bulletin boards (BBSs). You can now preview messages, as well as read and answer messages while you're off-line and not being charged — something your local, hobbyist-run BBSs have taken for granted for years.

These are some of the recent additions and improvements to the bulletin boards, Prodigy's special-interest areas where most public messages fly back and forth. Because bulletin boards are "Plus" services — costing extra — Prodigy designed these message "tools" to speed up members' sessions a bit. Member reaction has been curiously mixed.

- ✔ **Note Preview.** This command lets you see the first 51 characters (letters) in a message and decide whether you really want to bother reading the whole thing — and lets you skip all the "hi" and "thank you" messages on the bulletin boards. If you see an asterisk next to a note, it means that the note is short: 51 letters or fewer.

- ✔ **Import.** You can write a note in your favorite word processor or text editor and *then* call Prodigy. After you're logged on, you can use the **I**mport tool to import, or insert, your message where you want it. This way, you don't have to worry about any clocks ticking away in the "Plus" bulletin board areas while you compose your Great American Bulletin Board Message.

- ✔ **Export.** For the first time, there's a way to download a message or group of messages that interest you. Then you can log off, kick back, and read them at your leisure after you exit Prodigy. You can choose to save them forever on your hard disk or delete them after you've read them.

    Use Prodigy's **E**xport tool to save all the messages that interest you to a file on disk. Refer to other people's messages as you compose your witty retorts off-line. This gets around Prodigy's awkward habit of making you respond to a note without seeing what it says as you're writing. If you've never used an on-line service, you may not know what this is about.

- ✔ **Exclude List.** Avoiding the occasional on-line ogre just became easier with Prodigy's new Exclude List. Create a list of up to 15 Member IDs to avoid, and Prodigy filters out their notes and replies. Don't you wish Real Life came with the Exclude List option?

Downloading Files

Prodigy has never been known as a mecca for files. Still, it's possible to find and even download shareware on Prodigy — for an extra fee, of course.

- ✔ ZiffNet for Prodigy is a Custom Choice — meaning "extra cost" — service that offers shareware for downloading as well as computer news, reviews, and shopping tips. (ZiffNet, brought to you by the same people who publish *PC Magazine* and *MacUser*, runs this area and a more extensive one on the CompuServe Information Service.)

- ✔ The file download section harbors thousands of public domain and shareware games, "productivity" programs, and educational wares. You'll find demo programs of current commercial software, too. Many of these come ZiffNet-rated, ranking from 4-star "Best Values" to 1-star "Don't Bothers."

- ✔ The ZiffNet people screen all files for viruses and then *compress* them. (Hit Appendix C for a refresher on file compression.) So, to use a ZiffNet program you need to download ZiffNet's Download Manager (ZDM), which has built-in file decompression. Without ZDM, you'll need to download one of the compression/decompression programs discussed in Appendix C.

- ✔ One of the most common questions with Prodigy is "Where do my down-loaded files go?" ZDM solves that by listing them in one easy place, as shown in Figure 6-8.

- ✔ ZDM even offers to make a new home for your new program. It deletes the old, compressed file; lets you view the ubiquitous READ.ME files; and even asks whether you're ready to rev up the program.

- ✔ ZDM practically rubs out the need for Appendix C, "I've Downloaded a File: Now What?" in this very book. (Just kidding — unless you want to be limited to downloading from Prodigy all your life.)

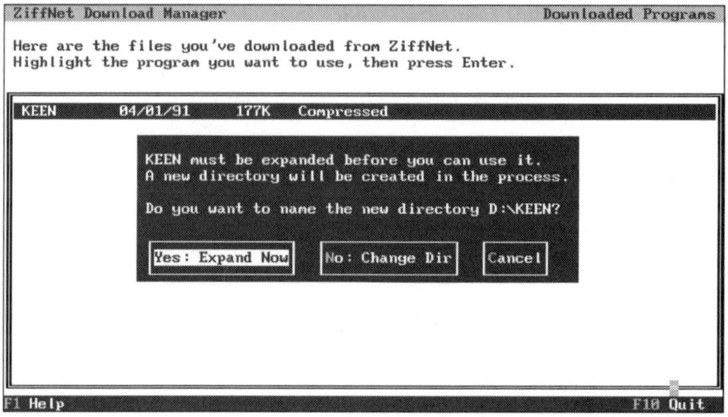

Figure 6-8:
ZiffNet's
Download
Manager
offers to do
everything
but pay the
shareware
fee.

```
ZiffNet Download Manager                          Downloaded Programs
Here are the files you've downloaded from ZiffNet.
Highlight the program you want to use, then press Enter.

 KEEN       04/01/91    177K    Compressed

         ┌─────────────────────────────────────────────┐
         │ KEEN must be expanded before you can use it. │
         │ A new directory will be created in the process. │
         │                                             │
         │ Do you want to name the new directory D:\KEEN? │
         │                                             │
         │ ┌─────────────┐ ┌──────────────┐ ┌────────┐ │
         │ │Yes: Expand Now│ │No: Change Dir│ │Cancel │ │
         │ └─────────────┘ └──────────────┘ └────────┘ │
         └─────────────────────────────────────────────┘

F1 Help                                              F10 Quit
```

✔ You'll have to join ZiffNet before you can download a file or read its articles and columns. (Unless you come upon a trial offer, as I did.) ZiffNet costs $14.95 extra per month for a base usage of three hours. Additional hours cost $6 an hour — roughly 10 cents a minute. Be sure to check for current pricing.

✔ A new twist: ZiffNet now restricts downloading to "A" members. To learn more, choose JUMP ZIFFNET.

✔ The Download screen tells you your file's size, as illustrated in the upper-right corner of Figure 6-9. Make sure that you have enough room on your hard disk to hold any files you want to download. You also see an approximate downloading time, given in minutes.

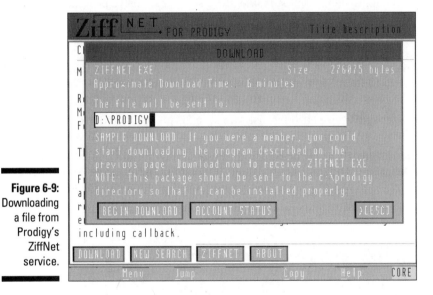

Figure 6-9: Downloading a file from Prodigy's ZiffNet service.

✔ If you do sign up for ZiffNet but you don't download the Download Manager program right away, watch carefully what the program is telling you. When the Download screen tells you the program's file name and which directory on your computer it's downloading to, write the information down! You'll need it after you've exited Prodigy and you're trying to find your new file.

Figure 6-10 shows the downloaded file parading conspicuously in my PRODIGY directory.

The other way to download software is when you order it on-line from one of Prodigy's advertisers.

✔ With the download in Figure 6-11, Prodigy has charged me a "shipping and handling" fee to download Pete Royston's Pro-Util, a program that makes Prodigy easier to use. That's on top of the program's $25 price.

Figure 6-10:
A down-
loaded file,
ZIFFNET.EXE,
in my
PRODIGY
directory.

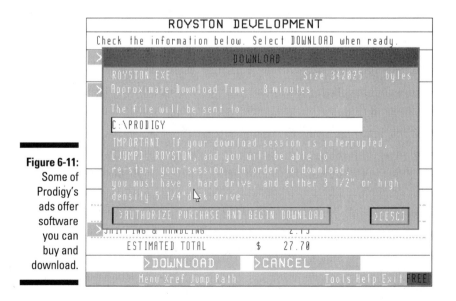

```
TLFD0000              78 06-28-93   6:47a
ZIFFNET  EXE     276705 06-28-93  12:27p
MODEMSTR SAV        250 06-28-93   6:52a
COMP                 19 06-28-93   6:57a
        26 file(s)      1711788 bytes
                       94738432 bytes free

D:\PRODIGY>
```

✔ A vendor called Download Superstore sells inexpensive programs you can
download in a similar fashion to the preceding example.

✔ Although it's not technically a "download," you can use Prodigy's **C**opy
tool on the Command Bar to grab almost any screen or article that looks
interesting. Save it either to your printer or to a file on disk (look for the
file to be hiding in your Prodigy directory). Prodigy asks you for your
preference the minute you choose to **C**opy something.

✔ Using **C**opy makes Prodigy ask you for a file name if you decide to **C**opy
the screen to a file on your disk. Choose a distinctive name you'll remember.

✔ In the bulletin board areas, Prodigy's **E**xport command can save lots of
messages to a file for reading later off-line (where the rates are cheaper).
Simply tell **E**xport the file name and follow the commands on-screen.

Figure 6-11:
Some of
Prodigy's
ads offer
software
you can
buy and
download.

```
            ROYSTON DEVELOPMENT
    Check the information below. Select DOWNLOAD when ready.
  >                    DOWNLOAD
    ROYSTON EXE                  Size 342825     bytes
  > Approximate Download Time:  8 minutes

    The file will be sent to

    C:\PRODIGY

    IMPORTANT: If your download session is interrupted,
    [JUMP] ROYSTON, and you will be able to
    re-start your session. In order to download,
    you must have a hard drive, and either 3 1/2" or high
    density 5 1/4" disk drive.

    >AUTHORIZE PURCHASE AND BEGIN DOWNLOAD        >[ESC]

  > SHIPPING & HANDLING              2.15

    ESTIMATED TOTAL          $    27.70
      >DOWNLOAD            >CANCEL
        Menu Xref Jump Path          Tools Help Exit FREE
```

Uploading Files

No way ... no uploading, that is, no sending programs or files to Prodigy from your computer. It's just not that kind of a place. However, the new Import feature does make it easier to send messages you wrote and saved to disk earlier, off-line, into a message you're composing on-line.

Chatting

No chance of chatting — typing messages back and forth simultaneously to other users — on Prodigy. Chatting with others will become an option after a special Custom Choice area called ImagiNation becomes available. To find out more about Prodigy's partnership with ImagiNation, see the section, "ImagiNation," later in this chapter.

How Much Is This Costing Me, Anyways?

There's no need to pay the full retail price for a Prodigy Membership Kit. In fact, you can probably *make* money purchasing Prodigy, judging by some of their latest offers!

 ✔ Last week, I saw a 2400 bps internal modem, phone cord, and Prodigy kit for $45.99 — with a coupon for a $50 U.S. Savings Bond glued to the box.

 ✔ Today I logged on and saw Prodigy offering to let you install your Prodigy software on a friend's computer, with Prodigy delivering an ID and temporary password on-line and mailing the rest of the membership kit for a $2 fee. Including a trial month, completely refundable!

 ✔ The coupon in back of this book contains a great deal for getting a Membership Kit sent to your home.

Getting started with Prodigy certainly won't drain your finances. (*That* starts when you begin calling the service and staying there for hours and hours and hours.)

Like the stock market, Prodigy is constantly changing. In fact, the Highlights screen — the first screen you see when you log on — changes every day. Even Prodigy's rates are subject to change.

 ✔ At first, Prodigy drew new members like flies with its low, flat monthly rate for unlimited use. Unfortunately, the accounting department wised up and realized it was losing money. Some users only called in occasionally. Other people (with very patient families, I'd say) spent up to 200 hours per month visiting Prodigy's resource-hogging bulletin board areas.

We have met the enemy, and he's all yours!

✔ Seeking to make the rates fair for everyone and to finally turn a profit, Prodigy has divided its services into "No timed-charges," "Core," and "Plus." And they've added a roster of extra-charge services called "Custom Choices."

✔ Confused? Welcome to the crowd. Table 6-2 summarizes Prodigy's current services and gives a partial listing of their contents. Table 6-3 lists prices.

Table 6-2   Prodigy's Confusing Rainbow of Services

No Timed-Charges	Core Services	Plus Services	Custom Choices
viewing ads and shopping	news and weather	bulletin boards	Strategic Investor
e-mail	games	DowJones Company News	Wall St. Edge
information about Prodigy	entertainment	EAASY SABRE (airline reservations)	ZiffNet
Prodigy Exchange bulletin board			
online-banks	education	QUOTE CHECK/TRACK (stock quotes)	Baseball Manager
BillPay USA	encyclopedia, Sesame Street,		Fantasy Football
PC Financial Network	Baby-sitters Club		Rebel Space
	everything else		Golf Tour

Notice that services ranked as "No-Timed Charges" may still levy a charge — it just means Prodigy's hourly "clock" isn't ticking while you're visiting that area.

In the lower-right corner of the screen, the words CORE or FREE or . . . will tell you how fast you're chewing up your paycheck.

For example, the actual minutes you spend reading or writing your personal messages — your mail — on-line don't count toward your hours on-line. But going over your monthly message limit still incurs a per-message charge. (The time spent composing messages is still free of hourly charges.)

Considering that you could get all your friends to sign up practically for free and then make all your long distance "calls" via local-call Prodigy messages, that's not such a bad deal.

If I save the whales, where do I keep them?

Table 6-3	Prodigy Pricing as of July 1993			
Pricing Plan	*Monthly Charge*	*Core & Plus Hours*	*Private E-Mail*	*Comments*
Value Plan	$14.95 per household	Unlimited core hours; 2 hours Plus Services	30 messages free; 25 cents each after that	*
Alternate Plan	$7.95 per household	2 hours Core and Plus combined	25 cents each from first	**
Alternate Plan 2	$19.95 per household	8 hours Core and Plus combined	25 cents each from first	***

Comments:

* Additional Plus hours:

$4.80/hr: 0-3 hrs
$4.20/hr: 3-6 hrs
$3.60/hr: 6+ hrs

** Additional Core and Plus hours:

see the "Additional Plus hours" rates

*** Additional Core and Plus hours:

see the "Additional Plus hours" rates

ImagiNation

One of the smartest things Prodigy has done lately is to add ImagiNation as a new Custom Choice. ImagiNation bills itself as the World of ImagiNation. It's a separate on-line service you can subscribe to, which is also making itself available to Prodigy members. ImagiNation specializes in interactive games, chatting, and gambling (see Figure 6-12).

✔ Prodigy's version won't offer ImagiNation's Adults-Only section (LarryLand) and will develop some Prodigy-only content.

✔ Special Game Point software you buy provides sort of a gateway from Prodigy into ImagiNation. The Game Point software requires a DOS-compatible PC, at least 7M of hard disk space, 640K of memory, and — highly recommended — a mouse.

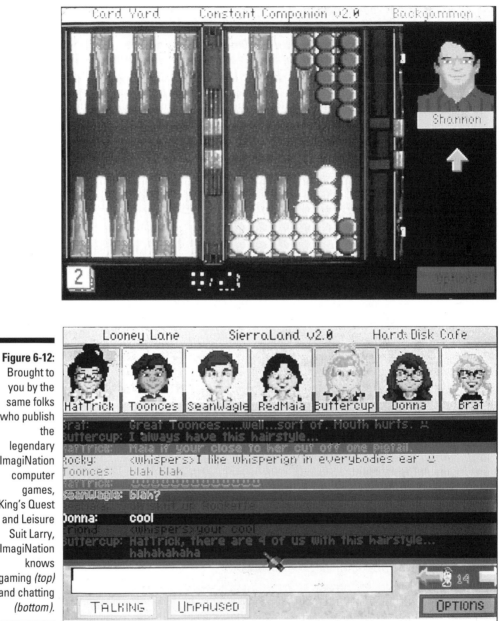

Figure 6-12:
Brought to
you by the
same folks
who publish
the
legendary
ImagiNation
computer
games,
King's Quest
and Leisure
Suit Larry,
ImagiNation
knows
gaming *(top)*
and chatting
(bottom).

✔ If ImagiNation is still not available by the time you read this, you can jump to Chapter 10, "Top Ten Other Guys to Call," to learn more about ImagiNation. Flip to the coupons in the back of this book to get a trial membership to the real thing.

Installing Prodigy on Your Computer

Prodigy is probably already installed on your computer, because many computers today come with Prodigy "loaded" on their hard drives. If not, don't be alarmed. Installing Prodigy is almost as easy as using it.

You need the following stuff before you can install Prodigy on your Mac or PC:

✔ A modem, installed and turned on

✔ A Prodigy Membership Kit

✔ Approximately 1M of hard disk space

✔ At least 640K of random-access memory, or *RAM*

Optional, but nice to have:

✔ Mouse (essential on the Mac)

✔ Hard disk. (The following instructions assume that you have a hard disk; if not, the Membership Kit's Handbook includes a detailed floppy installation guide.)

1. **Rip open the Membership Kit and notice the following:**

    • You get three little booklets: The Handbook, a guide to installing and troubleshooting Prodigy; The Phone Book, where you look up the number you'll use to call Prodigy; and the Member Guide, your Prodigy roadmap.

    • You get some Installation Disks; they're numbered Disk 1, Disk 2, and so on.

    • You get an ID and temporary password; they're on the "Get Started Now...It's Easy" card.

    • You get some pamphlets and other junk mail that you can happily wad into a ball and toss to your cat.

    If anything's missing, call Prodigy's service number at 1-800-284-5933 and holler.

2. **Insert Disk 1 into your computer's floppy drive and type these two commands:**

```
C:\> A:
```

Press Enter.

```
A:\> INSTALL
```

Press Enter.

You put your disk in drive B? Then make your first command **B:**, instead. Either way, press Enter after each command.

Mac users don't have to go through all this DOS prompt malarkey. Just insert the floppy and click the little folder to start installing Prodigy! Most of the tips that follow pertain to you Mac folks, too!

If confusion arises, the Prodigy Handbook is designed to help; it's clearly written and easy to understand. Honest! Remember, they've got to deliver bodies to their advertisers.

A Welcome screen pops up, asking you to switch on your printer, if there's one connected to your computer. If you don't have a printer, Prodigy runs fine without one. When you finally attach a printer, though, tell Prodigy about your new purchase. (How? By double-clicking the word SETUP instead of typing your ID number and password when the Prodigy connect screen comes on.)

3. **Pressing "any" key continues the installation screen's helpful messages (just press Enter instead of searching for the "any" key).**

Prodigy is smart and knows some stuff about your computer's secret innards already. But items that show up with a shaded box are asking for you to type in the answers.

One of the things you may need to answer is what COM port your modem is plugged into — especially if Prodigy can't figure it out for itself. If you're unsure about this COM port nonsense, you're not alone ... head over to Chapter 2 for a briefing.

In Figure 6.13, Prodigy figured out everything but my phone type.

Don't know whether your phone type is pulse or tone? Pick up your phone handset and dial a number. If you hear varying musical tones after each number, it's Touch-Tone. If you hear little clicks, it's pulse.

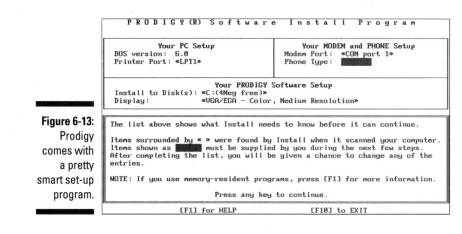

```
P R O D I G Y (R)  S o f t w a r e   I n s t a l l   P r o g r a m

          Your PC Setup                    Your MODEM and PHONE Setup
DOS version: 6.0                    Modem Port: «COM port 1»
Printer Port: «LPT1»               Phone Type:   ████████

                      Your PRODIGY Software Setup
Install to Disk(s): «C:(4Meg free)»
Display:             «VGA/EGA - Color, Medium Resolution»
```

Figure 6-13:
Prodigy
comes with
a pretty
smart set-up
program.

The list above shows what Install needs to know before it can continue.

Items surrounded by « » were found by Install when it scanned your computer.
Items shown as ████ must be supplied by you during the next few steps.
After completing the list, you will be given a chance to change any of the
entries.

NOTE: If you use memory-resident programs, press [F1] for more information.

Press any key to continue.

[F1] for HELP   [F10] to EXIT

Fine-tuning the Prodigy installation

If you don't like some of the guesses Prodigy makes, don't worry. When it's finished, the program gives you a chance to change any of the answers.

In my example, that's good news for me! A glance back at Figure 6-13 shows that Prodigy found only 4M of hard disk space on my drive C. That means I'll have to tell Prodigy to install itself on drive D, my other hard disk, when Prodigy gives me the

chance. That's what's happening in this figure. (Sure, there's room for Prodigy's 1M on drive C, but it's a Bad Idea to fill up a hard drive so "tight" — they're more apt to fail when they're jam-packed.)

Another default setting that's good to question is the display, which you can set to high, medium, or low sharpness, or *resolution*, depending on your type of monitor and video card.

```
P R O D I G Y (R)  S o f t w a r e   I n s t a l l   P r o g r a m

          Your PC Setup                    Your MODEM and PHONE Setup
DOS version: 6.0                    Modem Port: «COM port 1»
Printer Port: «LPT1»               Phone Type:  TONE

                      Your PRODIGY Software Setup
Install to Disk(s): C:(4Meg free)
Display:             «VGA/EGA - Color, Medium Resolution»
```

Since you have more than one hard disk, please select the one you want the
PRODIGY software installed on. It is to your advantage to select a disk
with at least 1.5 megabytes of free space. (Do not select a RAM disk.)

Select the hard disk you would like to use: `C: 5,132,288 bytes`
                                             `D: 96,485,376 bytes`
Note: Hard disks with less than
      811,000 bytes can not be chosen.

[F1] for HELP   [F10] to EXIT

When you're ready to continue the installation, use your arrow keys to highlight Continue, All Correct (it changes color). Then press Enter.

You'll hear your hard drive whir happily for a few minutes as it digests all this new information.

4. **When the whirring stops, a screen pops up, telling you to press any key to exit Install; press a key.**

5. **To start Prodigy right away, at the DOS prompt, type:**

```
C:\>PRODIGY
```

Press Enter afterwards.

6. **A Set-up screen asks you to type your member ID and password.**

   • You'll find these nuggets written on the "Get Started Now — It's Easy" card that came in the membership kit. In true Agent 007 style, your temporary password is concealed beneath the flap on the "Get Started ..." card.

   • Go ahead and type 'em in; you'll get the chance to change the dorky temporary password later, after you're on-line.

7. **Next, the sign-up screen asks for something called a "primary phone number."**

   You'll find this inside the Phone Book that came in your Membership Kit. If it helps, my Phone Book is a stunning shade of aqua.

   Leaf through the Phone Book until you come upon your state and city. Type the number at the blinking cursor, ignoring the area code unless you normally dial that to call your city (usually not).

   Press Enter to get the blinking cursor down to the Network Symbol box, and change this to **Q** or **Y**, depending on what the Phone Book lists next to your city's phone number.

   Follow the same steps to fill in your city's Alternate Number, listed in the back pages of the phone book.

   • Don't worry; you only do this "set-up" stuff once.

8. **The next screen asks you to select your modem's fastest speed. (Appendix A contains more information on determining modem speeds.)**

   Pressing Enter from this very screen starts dialing Prodigy. How exciting!

   After you're on-line, you may see a box in the corner that says Working. Prodigy is taking time to powder its nose and update its software on your computer. As with nose-powdering, this can sometimes take a while.

---

9. Prodigy's Enrollment screen comes up.

Prodigy asks you to fill in your name and address, read the Member Agreement, and make up your own, secret, cross-your-heart-and-hope-to-die password.

You impatient Type-A types can save reading the Membership Agreement for later. They're available on-line, and reading them doesn't levy a charge or time penalty. That's true for any Prodigy-related stuff on-line — even the Prodigy Exchange bulletin board, where people gripe about Prodigy's rates and stuff.

Guard your password carefully. Write it down in a secret place, or make it something you can remember.

Change your password often; use the Tools tool in the Command Bar to see how. And never use the same password on any other on-line service or BBS, in case someone somehow finds out what it is.

You'll *never* be asked for your password by a Prodigy employee. If someone on-line asks you to reveal it, ignore them. If they say they're from Prodigy, report them. Your ID, however, is not secret.

Congratulations! You know you're officially a member when Prodigy rolls out the Welcome New Member screen.

You can read more about Prodigy (yeah, sure!) or charge right into the thick of things like everyone else. Using your arrow keys or mouse, highlight the box next to Highlights. That'll dump you right at Prodigy's "front page."

Prodigy Tens: Ways to Make Prodigy Even Easier

Ironically, some parts of Prodigy that were intended to make it easier to use can make it more difficult. That's especially true if you're used to the automation, convenience, and power of a full-fledged communications program.

Here are some ways to make Prodigy a *real* no-brainer.

If worst comes to worst, you *can* turn most things off.

Start Prodigy automatically by using AutoLogon

Typing your Member ID and password each time you want to access Prodigy can be a pain. Consider making use of Prodigy's AutoLogon feature. It lets you type **PRODIGY** and a nickname at the DOS prompt, such as **PRODIGY TINA** (and then you press Enter, of course) — no more searching for passwords. From any screen in Prodigy, choose JUMP AUTOLOGON to learn more.

AutoLogon caveat: Use this feature only on a computer that you alone or trusted family members have access to.

Customize Prodigy with the Path tool

If you crave order and predictability, you'll love Prodigy's **P**ath. You set up a personal path, and it takes you on an automatic round of your top 20 favorite places on Prodigy, in order, each time you log on. If you always log on and then check the scores, check your stock prices, see the weather, and view your messages, you can automate those steps into a "path."

Stay on the path just by choosing the **P**ath tool in the Command Bar. It's okay to change your way of navigating, by using **J**ump or other tools. Just return to **P**ath when you're ready. Whatever you decide to do on-line, Prodigy "remembers" where you left off in your path.

View your Path

Prodigy gives you a generic Path as part of signing on. You can see what's on it and change it from any screen by pressing **V**, for **V**iew your Path List, and then heading down to **C**hange Path at the bottom of the Your Path List window.

Figure 6-14 shows Prodigy's idea of a suitable path for a 30-something female — based on the info I gave it when I signed up. Like a nervous guest at a cocktail party, Prodigy starts off with a safe topic: the weather.

Note: On weather, you get a selection of weather options — city, ski, international, and so on. If you want your city's weather (if it's one on the list), you can type *CITY* **WEATHER**, as in **LOS ANGELES WEATHER**, and press Enter.

Figure 6-14:
Homelife?
Arts Club?
Must be
Prodigy's
generic "30-
something
female"
path.

Change your Path

Fortunately, changing your path is easy. Whenever you discover a place you'd like to visit more often, select View and then Change Path; then press Enter (the Add command blinks by default). Press Enter again, as Previous Location is now blinking at you — and you want to add the place you just came from (your previous location).

Figure 6-15 shows my *real* (kinda nerdy, huh?) path.

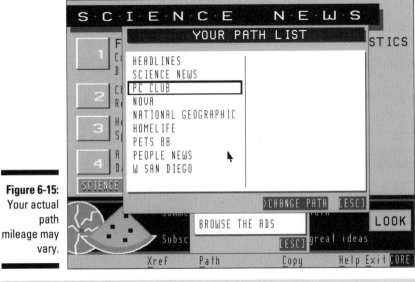

Figure 6-15:
Your actual
path
mileage may
vary.

Uh, yeah ... I *meant* to do that!

Use Pro-Util

Some on-line services make it easy to download all the messages in a particular forum and read — even reply to — them after you're off-line. Some don't. Until recently, Prodigy has been one of the "some don'ts."

Back in 1988, a guy named Pete Royston decided there had to be an easier way. He wrote a program called Pro-Util, and each version has made it easier for Prodigy's users to read messages and to print files to disk. Pro-Util contains a *type-ahead buffer*, which stores characters you've typed quickly — letting you move around faster on Prodigy, after you get up to speed.

Choose JUMP ROYSTON to learn more about Pro-Util.

Practical Things to Do on Prodigy

Sometimes Prodigy seems too huge to get a handle on. Here's some down-to-earth services to check out (and be sure to press Enter after typing any JUMPwords):

JUMP KIPLINGER'S to get the experts' opinions on real estate investments, insurance, and current financial trends.

JUMP PCFN BROKER to start playing the stock market. Buy low; sell high, by the way.

JUMP WEATHER to see what to pack for that upcoming business junket.

JUMP EAASY SABRE to book your own airline reservations and purchase tickets.

JUMP COMPUTER BB to ask other users about your possessed, evil computer.

JUMP COMMUNICATION to write notes to family members in other cities and save on long-distance phone charges.

JUMP CLOSEUP and keep up to date on current events and political scandals.

JUMP HOMELIFE and ask other Siamese cat owners what to do about fending off your pet's sudden mood swings.

Fun Things to Do on Prodigy

Where's all the fun stuff on Prodigy? Start with these:

JUMP WEATHER to look up the weather in your own city.

JUMP PARKER ON WINE to count how many ways this guy can describe the smell of rotten grapes.

JUMP GAMEWARE to collect tips on the top ten best-selling computer games.

JUMP TRAVEL CLUB and see what people are saying about your hometown.

JUMP COMPUTER BB to see what shareware programs are up for trade among Computer Club bulletin board members.

JUMP JANE DARE'S ADVICE and be thankful you don't have this many troubles.

JUMP NOVA, grab the kids, and do one of their scientific experiments. (If the kids don't grab you first, that is...)

JUMP SOAPS and catch up on Edgar and Jessica's steamy doings. Or Jane and Jessica's steamy doings!

The 5th Wave   By Rich Tennant

Chapter 7

CompuServe: On-Line Gigantor

*I*t seems like the CompuServe Information Service is everywhere. Buy a modem, and I'll bet there's a CompuServe membership leaflet lurking inside the box. You can hardly rip the shrink-wrap off a software box without stumbling across a stack of trial-offer pamphlets: One of them is bound to be for CompuServe. Buy a printer; find a CompuServe flier. Even computers come with trial offers.

No recent computer purchases? The start-up kits sold at the computer store contain everything you need to get going. Or follow the directions on the Special Offer for Dummies coupon in back of *this* book.

Incidentally, CompuServe's kits include the CompuServe Information Manager software, called CIM for short. This software, available for Mac, DOS, and Windows users, provides a graphical user interface (GUI, pronounced *gooey*) that does away with the "texty" displays, numbered lists, and other navigational horrors you'll see in many of the examples throughout this chapter. The CIMs are so good that CompuServe doesn't even want to *sell* you a kit without a CIM bundled inside.

Turn to the coupon in back of this book to see how to get on CompuServe and find out more about CIM. And think of the "olden" CompuServe examples here as interesting historical relics — artifacts of a time when GUI was something that you wiped off a three-year-old's face.

Either way, another world awaits you on CompuServe. And this chapter shows you how to get there from here.

Logging On

Logging on is just a way to make *signing on*, or *calling* a service, sound exciting. The following steps take you to CompuServe country:

1. Make sure that your modem is plugged in and turned on.

Are the modem's little lights on? If you have an internal modem, you'll just have to assume it's on — after you perform Step 2.

2. Turn on your computer.

3. Start your modem software.

Unlike Prodigy, CompuServe doesn't require any special CompuServe software. You can reach it from any communications software — or any brand of computer, for that matter.

Some special "CompuServe navigation" software, however, can help you blaze through CompuServe like a pro. If you bought a special start-up kit or if you want to keep your Mac or Windows "look and feel" while you're on CompuServe, you'll find more details at the end of this chapter, in the "CompuServe Tens" section.

4. Adjust your modem software's "communications" settings.

This part's a little tricky. Most modem software is set up to use something called **8,N,1**, which is described in Chapter 4. A troublemaker, CompuServe prefers to use the **7,E,1** settings. So, change your software's settings to **7** data bits, **E**ven parity, and **1** stop bit.

Also, make sure that your software's ECHO setting is off.

Confused? Check out Chapter 4 or refer to your software's manual for the dirt.

5. Set your software's *modem speed* to 2400 bps, or 2400 baud.

After you're comfortable on-line, you can beef up your speed to 9600 bps (and pay extra for the extra speed), if your modem goes that fast. But for now, start slowly.

6. Add CompuServe's name and phone number to your dialing directory.

Chapter 4 talks about your software's dialing directory — your "master list" of other computers you call. Type CompuServe's name into your software's dialing directory, followed by CompuServe's phone number. (Have "call waiting?" Check out Chapter 4 — you'll want to turn it off before calling another computer.)

That CompuServe phone number is usually written on the freebie leaflet, or displayed prominently inside the start-up kit somewhere. Or see the following tip.

What number gets me through to CompuServe?

CompuServe makes it easy to find a local access number. Tell your modem to dial 1-800-346-3247 at 2400 bps. (Those of you who enjoy those "word" phone numbers can treasure 1-800-FINDCIS, for _FIND_ CompuServe _I_nformation _S_ervice.)

When the modem connects, press Enter until you see the words HOST NAME on-screen (this bit of text is called the HOST NAME _prompt_, in on-line jargon). Then type **PHONES** and press Enter again. Follow the menu items on-screen, and CompuServe will tell you the phone number you need to call and then dump you back to the HOST NAME prompt. At that point, you can type **BYE** and give your software's "hang up" command.

Tip within a tip

Keep the number 1-800-FINDCIS in your laptop dialing directory so that you can find local phone numbers while sniffing the sheets in strange hotel rooms.

You can call CompuServe from any computer, as long as the computer has a modem and some communications software. You're the account holder, no matter where you're calling from. You can't foist part of your CompuServe tab onto Aunt Gertie during your next family reunion. The bill will show up on _your_ credit card, no matter whose phone you use.

You can reach CompuServe through several competing phone networks. Try to call by using CompuServe's _own_ access numbers, however; the other networks slap on some extra charges. If you do connect via the SprintNet, TYMNET, LATA, or DATAPAC networks, your log-on procedure may vary.

7. Tell your modem software to dial CompuServe.

Go to your modem software's dialing directory, or its "CompuServe Setting" — however your particular software is organized. Highlight the new CompuServe entry you just made and press Enter (or whatever command your software prefers). Your modem should begin to dial CompuServe. (Refer to Chapter 4 for help with modem software and dialing directories.)

8. Connect to CompuServe.

You're there! Now you just need to get past the hulking host computer guarding the gates to CompuServe. Your software will probably say something like Connect 2400 and then just sit there. Press Enter a few times to wake the guard dog out of its mainframe slumber. (Do mainframe computers count RAMs instead of sheep to get drowsy?)

A seminar on time travel will be held two weeks ago.

9. **When CompuServe says** Host Name:**, type** *CIS* **and press Enter.**

You can bypass the HOST NAME song-and-dance by pressing Ctrl-C (your Ctrl key and the letter C, simultaneously) instead of Enter when you first see that Connect message on your screen. Mac users press ⌘-C if Ctrl-C doesn't work. This jumps you right to the Step 10 User ID: prompt.

The Ctrl-C/⌘-C trick doesn't work with all types of computers or with every brand of modem software, though. And using the other, HOST NAME method puts you at the HOST NAME prompt when you log off, so someone else can log on with their user name and ID without making another phone call.

Type **OFF** or **BYE** to truly exit from the HOST NAME prompt.

10. **When asked, type your User ID number and press Enter.**

Your User ID is a bunch of numbers with a comma in the middle. It looks something like the one in Figure 7-1.

Figure 7-1:
You are
asked for
your secret
password.

```
03SDA

Host Name:  CIS

User ID: 76004,3267
Password: █
```

Although this book shuns memorization, it's good to remember your CompuServe User ID. Tell it to others and they can send you electronic mail (goodie!). Some really nerdy types even print their CompuServe IDs on their business cards.

11. **Type your password and press Enter.**

A final request pops on-screen: Password. Your password's sorta like something you had to whisper as a kid before the others would let you into the secret fort. It's *so-o-o* secret that nothing appears on the screen as you type it, as seen in Figure 7-1. Don't worry if you type it wrong; the host computer gives you another chance — actually, it gives you three chances before hanging up on you.

It's okay to pass out your User ID, but *never, never* tell anyone your *password*. No one on CompuServe will ever ask you for your password for legitimate reasons, so ignore any requests for this top-secret item.

To change your password, type **GO PASSWORD** and press Enter anytime you see an exclamation point (!) on-screen. Your new password should be a combination of two unrelated words separated by a symbol, like **glug&planet**, for example. (Don't use that one; and don't use your maiden name, birthdate, driver's license number, or anything else someone already knows about you.)

Finally, don't use the same password on Prodigy or on any other BBS or on-line service you call. And change all your passwords frequently.

Voilà! Your head's swimming with talk of IDs and passwords, but you've *finally* entered the CompuServe Zone. Like a wary traveler inching down an unfamiliar highway, you may want to pull out a road map. Luckily, one's waiting in the very next section.

CompuServe at a Glance

CompuServe's Welcome screen burps up lots of information. After the current time and date, CompuServe reminds you of your current membership plan. Mine's the "Executive Option," as shown in Figure 7-2. (Membership plans receive more coverage later in this chapter, under "How Much Is This Costing Me, Anyways?")

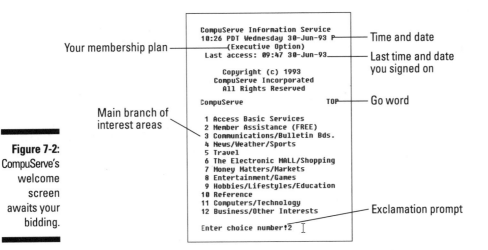

Figure 7-2:
CompuServe's welcome screen awaits your bidding.

Your membership plan — (Executive Option)

```
CompuServe Information Service
10:26 PDT Wednesday 30-Jun-93 P
      (Executive Option)
Last access: 09:47 30-Jun-93

    Copyright (c) 1993
  CompuServe Incorporated
    All Rights Reserved

CompuServe                    TOP

  1 Access Basic Services
  2 Member Assistance (FREE)
  3 Communications/Bulletin Bds.
  4 News/Weather/Sports
  5 Travel
  6 The Electronic MALL/Shopping
  7 Money Matters/Markets
  8 Entertainment/Games
  9 Hobbies/Lifestyles/Education
 10 Reference
 11 Computers/Technology
 12 Business/Other Interests

Enter choice number!2
```

Time and date

Last time and date you signed on

Go word

Main branch of interest areas

Exclamation prompt

Figure 7-2 deciphers some of the welcome screen's other enigmas, including shortcuts for moving to one of the service's interest areas. Typing the numbers to the left of the options menu takes you to that area.

How Do I Log Off?

CompuServe charges you by the *minute* for most of the "good" stuff, so it's important to know how to get out fast!

- ✔ Anytime you see an exclamation point, called the *exclamation prompt* in CompuServese, type **OFF** or **BYE** and press Enter, as follows:

  ```
  Enter choice number! off
  ```

- ✔ When you see a : symbol (they dubbed this the *colon prompt*, naturally) instead of the boisterous ! prompt, you can't type **OFF** or **BYE** to escape CompuServe. Instead, type Ctrl-C, and Mr. Excitement Prompt reappears. *Now* you know what to type: **OFF** or **BYE** (type in uppercase, lowercase, or mixed-case letters; either will do).

- ✔ Typing **EXIT** takes you "up" in menu levels. But only by typing **OFF** or **BYE** can you disconnect.

- ✔ Now you're off and back in reality — where it's duller but much less expensive.

Don't just tell your modem program to "hang up" when you want to leave CompuServe. CompuServe will keep charging you for 15 minutes or so until it realizes that you've left. Always turn off CompuServe's clock by logging off the right way — using the **OFF** or **BYE** command.

Hollering for Help

Thousands of people lurk on CompuServe 24 hours a day, but they don't dish out as much helpful information as CompuServe's computer. Here's how to tap CompuServe's computer on the shoulder when you're stumped by the current options.

- ✔ Like the BYE command, HELP is available at any CompuServe ! prompt. When you're faced with a particularly sticky menu, HELP guides you through the brambles. Just type **HELP** and press Enter, as in in this example:

  ```
  CompuServe Mail  Transfer Type

  1 ASCII

  2 ColorMail

  3 Binary

  Enter choice number! HELP
  ```

CompuServe's computer shuffles through its file cards and tosses you a few clues about your current predicament.

One way to quickly summon a ! prompt is to press Ctrl-C. Depending on their modem software, Mac users may have to whip out the trusty ⌘-C combo, instead. Those two commands catch CompuServe's attention whenever it's wandering down the wrong path.

✔ CompuServe is stuffed with helpful information, whether it's about CompuServe's myriad services, ever-changing rates, or other policies that puzzle you. Type **GO HELP** at any ! prompt to see a wide array of HELPful places to visit. CompuServe doesn't even charge you while you're bouncing around within its **GO HELP** areas. (Don't know about the GO command yet? Seek enlightenment in the section "How Do I Use the Menus?")

✔ "Gilligan's Island" begins with the zany cast boarding the S.S. Minnow for a "three-hour tour ... a threee ow-err tourr." CompuServe offers a tour, as well. It lasts closer to 20 minutes than three hours — if you don't stop to buy any souvenirs in CompuServe's "Shopping Mall." And, unless the power goes out in Columbus, Ohio, you won't be left stranded on a luscious tropical isle, either. ("They built a radio out of a coconut but they couldn't fix the darn boat!")

Type **GO TOUR** and press Enter at any ! prompt, Little Buddy.

✔ Turn on your modem software's capture command each time you access Help. Soon you'll have a handy Help file for the CompuServe areas you frequent most often.

✔ Lost? Then type **TOP** at any ! prompt. Eventually, CompuServe brings you back to its welcome main menu (the one in Figure 7-2). From there, typing **2** and pressing Enter whisks you to Member Assistance. There, you'll find a chunky menu listing everything you've ever wanted to know about CompuServe membership.

✔ You can leave electronic mail (*e-mail*, for showrt) for one of CompuServe's customer representatives by typing **GO FEEDBACK** and pressing Enter. Follow the prompts on the screen, and you'll be jotting down your questions or problems in minutes. You should get a personal response within two business days. (I'm not sure how long an impersonal response takes....)

✔ Type **GO QUESTIONS** and press Enter to get to the Customer Services menu. From there, you can select the Contact Customer Service item and find a voice telephone number, if you prefer yakking to typing. (Actually, if you prefer talking on the phone to typing electronic messages, maybe you should reconsider the modem "thing" altogether!)

✔ Hobbyists who band together to yak about their special interests head for *forums*. CompuServe has a forum for everything under the sun. (Even the sun has one — the Astronomy Forum.) When you're confused, band together with other confused people. They live on CompuServe's Help Forum. Type **GO HELPFORUM** and press Enter to get there. Refer to Chapter 5 to learn more about forums.

Don't like being charged while you're stumbling around CompuServe, trying to figure out what weird word does what? Head for the Practice Forum. Free of any charges, the Practice Forum lets you hone your forum-maneuvering skills until you're not embarrassed anymore. Type **GO PRACTICE** and press Enter to stop the clock and start practicing.

CompuServe's membership support areas are free, as is the Help Forum and the Practice Forum. Most other forums count as *Extended Services* and rack up a higher billing rate.

For more about forums, see the very next section. For cash-conscious details about rates, service categories, and membership plans, jump to the "How Much Is This Costing Me, Anyways?" section, near the end of this chapter.

How Do I Use the Menus?

CompuServe passes out more menus than the neighborhood Chinese take-out (although they're not as greasy). But endless strings of menus can slow you down. Luckily, CompuServe offers a way to speed things up. Table 7-1 shows the best ways to take "cuts" in line at the CompuServe banquet.

The GO Command

Table 7-1 lists some great menu shortcuts, but the ultimate shortcut is the GO command. From any ! prompt, type **GO** and a special *GO word*. CompuServe takes you there as quickly as it can.

Just about everyplace on CompuServe has its own special GO word. Don't know any GO words yet? You may already have some:

✔ The CompuServe Introductory Membership leaflet lists bunches of places to visit and their corresponding GO words.

✔ CompuServe's Directory brochure lists zillions of features and their GO words. For example, the Directory lists PASSWORD as the GO word for "Change Your Password." So, to change your password, you type **GO PASSWORD**.

An atheist is someone who has no invisible means of support.

Table 7-1   Top Ten CompuServe Command Shortcuts

Typing This at the ! Prompt	Does This
B	BACK one page.
BYE	Signs OFF of CompuServe.
EXIT	EXIT services provided by other systems, like Travelshopper; elsewhere, you can keep typing **EXIT** until you've exited any submenus you've entered and it logs you off CompuServe.
F	FORWARD a page.
GO EBERT	GO to a certain place. In this case, GO EBERT brings up Roger Ebert's Movie Reviews.
H or ?	Brings up HELP.
M	Returns to previous MENU.
N	Displays NEXT menu item.
OFF	Signs OFF of CompuServe.
P	Displays PRECEDING menu item.
R	REDISPLAYS a page; handy for seeing something that went by too fast to read it the first time.
S	SCROLLS. This makes information flow by in a continuous stream, without stopping at each new page. It's handy when capturing bunches of text to a file on your computer.
T	Returns to TOP menu of section you're in; type **T** repeatedly to get to CompuServe's Welcome Menu.

✔ CompuServe members automatically receive a free subscription to *CompuServe Magazine*, a colorful monthly geared toward helping members mine CompuServe's rich lode of entertainment and information. GO words are sprinkled liberally throughout.

✔ CompuServe publishes a booklet, *I Didn't Know I Could Do That on CompuServe*, and a pamphlet, *The CompuServe Directory*. Although some of this information occasionally goes out of date, most of CompuServe's larger information sources stay pretty constant. Both books are treasure-troves of GO words. *I Didn't Know* comes in CompuServe membership kits. You receive *The CompuServe Directory* in the mail after you join the service.

> ✔ Need to find a GO word? Use the FIND command, discussed in the very next section.
>
> ✔ Head to "CompuServe Tens" at the end of this chapter for some more GO words.

Each time you find an interesting area on CompuServe, note the GO word at the screen's top right. If it looks more like numbers than a word, it's a page number. (Each CompuServe "page" gets its own number — note that, too.) Type either one of them after the word **GO**, and CompuServe zooms right back to that area.

Several special navigational programs provide Mac, DOS, and Windows users with an alternative to CompuServe's barren menus and "terminal text-mode condition." One feature I like is the capability to set up a "Favorite Places" list. Also, *The CompuServe Directory* comes built into CompuServe's Information Manager programs, making it easier to find stuff on-line. Head to the "CompuServe Tens" at the end of this chapter to see how these programs can make CompuServe easy (well, *almost* easy).

Finding Stuff

There's no way to list all of CompuServe's thousands of services in this single chapter. But then again, there's no need to. CompuServe carries its own up-to-date index, ready for action whenever you're ready to explore.

This section covers CompuServe's FIND command and other ways to extract your particular needle from CompuServe's hefty haystack.

Also, check out Chapter 5; it lists the types of information and services you can unearth on-line — as well as which on-line services offer the best diggings.

The FIND command

After the GO command, the FIND command is the most popular way to search through CompuServe. While at any ! prompt, type **FIND**, followed by a word summarizing what you're looking for. Press Enter, and watch as a list of GO words pop up, each showing possible places to visit. For example, Figure 7-3 shows what happens when I decide to search for information on this summer's swimwear trends:

Figure 7-3:
FIND may
ask you for
a more
general
search
word.

```
Enter choice number! FIND BIKINI
No matches found. Enter topic (e.g. stocks) : swimsuit
CompuServe
 1 Graphics Corner Forum +            [ CORNER ]
 2 Graphics Plus Forum +             [ GRAPHPLUS ]
Last page, enter choice!bye
```

Some FIND words are more successful than others. In this example,
CompuServe couldn't find anything related to "BIKINI" and asked me to type a
more general term. The word "SWIMSUIT" proved more fruitful than "BIKINI."
CompuServe listed the GO words for two *graphics* forums, where I can *download*
(have CompuServe send my computer) some lovely photographs of swimsuits.
(Although I think the idea is to view the lovely tanned personages *wearing* the
swimsuits.)

I'll pass for now.

See the Plus symbols next to each forum name in Figure 7-3? The plus sign
means CompuServe is listing an Extended Service. Pay-as-you-go connect-time
charges kick in the minute you enter an Extended Service.

Forums are covered in "What *Are* Those *Forum* Things, Anyway?"
CompuServe's rates get their own section at this chapter's end.

Find a topic index

To see a list of every CompuServe topic relating to your search, type **GO INDEX**
at any ! prompt. The index lets you choose between searching for a specific
topic or seeing all of CompuServe's topics scroll by.

Turn on your modem software's "capture" tool and save CompuServe's entire
index on your computer's hard disk. (To find out more about telling your
modem software to capture text scrolling by on-screen, turn to Chapter 4.)

Then, when you're wondering whether CompuServe carries information about
something, check your captured index. That's cheaper than actually dialing up
CompuServe and plowing through the index yet another time.

What Are Those Forum Things, Anyway?

After a few sessions with the FIND command, you'll start to notice things called *forums* turning up in your searches. That's because forums contain people, files, and conversation — the lifeblood of any on-line service.

CompuServe is basically organized into databases, services, and *forums*. Forums are areas devoted to a specific interest — sort of like the high-school Biology Club or a city's Model Railroad Club. CompuServe's forums cover hundreds of subjects — ranging in scope from the Showbiz Forum to the Science and Math Education Forum to the White House Forum.

By "joining" a forum and following its menus, you can meet people with similar interests. Figure 7-4 shows the top menu of a typical forum.

Figure 7-4: Messages, file libraries, and conferencing: This must be a forum!

```
Modem Games Forum Menu
 1 INSTRUCTIONS
 2 MESSAGES
 3 LIBRARIES (Files)
 4 CONFERENCING (0 participating)
 5 ANNOUNCEMENTS from sysop
 6 MEMBER directory
 7 OPTIONS for this forum
Enter choice !
```

Forums come in two main varieties:

✔ **Computing Support forums:** CompuServe's forums make it the top on-line service for computing support. Most companies that make and sell computer stuff — both hardware and software — host their own forums on CompuServe. After joining that company's forum, you can find answers to your questions, find software updates and enhancements to download, or see how other members like a new product.

✔ **Special Interest forums:** Special Interest forums split into two more categories. The Professional Forums are based around careers — like the Journalism Forum and Law Forum. The General Interest Forums draw members who share similar hobbies and interest— like the Modem Games Forum or the Investment Forum.

✔ All forums count as Extended Services — except the CompuServe-specific ones like CompuServe Help, the Practice Forum, and the CompuServe Software Support forums. CompuServe charges an hourly rate for its Extended Services (read the section "CompuServe's Rates" to get the full scoop).

- ✔ When you first visit a forum, CompuServe asks you to "join." Don't be suspicious; it's nothing like the offers Ed McMahon dishes out. Joining a forum doesn't cost anything. And unless you join, you won't be able to download any files. This "joining" business just lets CompuServe keep track of how many people are joining the different forums.

- ✔ Forums are the easiest way to meet CompuServe members who share your interests. Each forum has a _Message Board_, where its members yak about their interests. If a message strikes a chord, respond by writing — called _posting_ — a message of your own.

- ✔ It's hard to get up the nerve to post that first message. So, just "lurk" for a while — read the messages but don't post any until you feel more confident. Either way, taking time to read the forum's messages is the best way to feel at home there.

- ✔ Hundreds of forum messages often stretch back for weeks or months. So, _capture_ them on your computer and read them after you've signed off of CompuServe. That's one way smart modemers save money.

  Chapter 4 covers your software's capture command in more detail.

- ✔ If you're intrigued by the idea of forums, head for the Practice Forum by typing **GO PRACTICE** and pressing Enter at the ! prompt. It's free, and your newly earned expertise can save you time and cash (on a forum, time = cash) after you get going in a real forum.

Forum messages

Reading the "give and take" between forum members on a topic can be as fun as eavesdropping! But posting messages of your own (and getting lots of other messages back) can be even more fun, after you overcome those initial "But will they like me?" doubts.

Before stepping onto the message boards, here's a quick review of messaging etiquette. (For the entire Miss E-Mail Manners on-line guide, turn to Chapter 5.)

- ✔ Lots of people _do_ eavesdrop, reading lots of messages without ever responding to one. And that's ... okay. They're called _lurkers_. It's okay to lurk. Better to lurk than to flame.

- ✔ Flame (verb): To insult, dress down, or otherwise intentionally annoy other callers through your on-line messages. Don't ever flame another forum member. If something in life has got you down, postpone your post until tomorrow. Don't take it out on forum bystanders.

- ✔ Using only capital letters in a message is considered SHOUTING. If you *must* emphasize a *word*, do it like *this* instead.

Think of all the marvelous human tones and groans, giggles and guffaws that help us express our disgust, delight, and other emotions in spoken conversation. Sadly, mere letters and numbers on-screen don't always project our true feelings. If you're aiming for irony, sarcasm, or even gentle teasing in a message, emphasize the spirit of your message by using *emoticons* — icons that express emotions.

Figure 7-5 shows people having fun with some "smiley" emoticons in a message. There's even a *winking* smiley ending the last line. Don't get it? Tilt your head so your left cheek is on your left shoulder and look again. (Turn to Chapter 5's Top Ten Smileys for the complete guide to smileys and stuff.)

Figure 7-5:
Turn your head to one side to read the funny punctuation marks.

```
Enter choice(s) or ALL !1
(#1,TigerShark) You might try paying the electric bills  :)
spent 10 minutes making faces with Ken Gayne a while ago  {:-)
(#1,TigerShark) Sure it wasn't ST ;-)
```

Reading forum messages

Forums house a variety of messages within sections. Figure 7-6 shows all the sections that have new messages since I last visited the Modem Games Forum. The number at the right of each line tells you how many different subjects, or conversation *threads*, the section has. The number after the slash tells you the total number of messages each section has.

Figure 7-6:
Section #2, Flight Sims & Games, has 99 subjects and 255 total messages in those subjects.

```
Modem Games Forum Sections Menu
Section names (#subjs/# msgs)
 1 General Information  (12/23)
 2 Flight Sims & Games  (99/255)
 6 Strategy/Conquest  (67/250)
 7 Sports/Racing  (1/2)
12 World Community  (2/7)
14 EMPIRE Tournament  (42/94)
```

After choosing a subject to follow, keep pressing Enter and follow the prompts on-line to follow the messages.

Note how very few forum conversations are between two people. Rather, people tend to jump in and out of conversations. Don't consider it bad form to reply to a message that's addressed to somebody else. Everybody just jumps on in, pitching in when they have something to say. That's what keeps everything so interesting.

If someone has something that's really private to say, they'll send it as private e-mail, discussed further on in this chapter.

Sending forum messages

You've *lurked* awhile in a forum and now you feel ready to post your own brand-new message. Sending a new message gives you a chance to introduce yourself to the other forum members.

The following steps show how to post your first message, out of the blue. To reply to someone *else's* message, see "Replying to a Message," in the next section.

1. **Select menu option 2, Messages, in the forum's top menu (see Figure 7-4).**

   (Either type **2** and press Enter to select Messages, or type **MES** and press Enter.) Another menu appears.

2. **Choose Option 4 to compose a message.**

   Type **4** and press Enter, and CompuServe tosses you its message-creating "scratch pad."

3. **Type your message at the 1: prompt on the screen.**

   Keep an eye on your right-hand margin as you type each line. CompuServe's editor does not have "word wrap," so you have to press your Enter key right before the line reaches the end of your screen. Just like an old-fashioned typewriter!

   If you don't press Enter, your message looks like it's running off the page to anyone trying to read it.

4. **Try not to ramble! Type /EXIT when your message is finished, and then press Enter.**

   Make sure that you type **/EXIT** on a line all by itself, or CompuServe foolishly assumes that you're trying to include that word in your message.

5. **Choose Option 1 to post your message on the forum's message board.**

   Now's your chance to edit your message if you've goofed. To do so, you choose Option 3 (TYPE) to view your message. Press Enter and then choose Option 2 (EDIT) to replace a line.

6. Address your message.

When CompuServe asks to whom you want to send the message, type **ALL** and press Enter, unless you have a specific forum member in mind. If you want to send the message to a certain user, you type that member's User ID. Want your message to go to the Forum's leader? Type the word **SYSOP** (short for "system operator").

7. Type a subject.

When CompuServe asks you for your subject, type something short (under 24 characters) and relevant to your message. Typing **STUFF** just doesn't cut it. Press Enter.

8. Type a section number.

Because the forum wants to put your message in the most appropriate section, it gives you the current list of sections (automatically). Scan the topic list and choose the number of the subject closest to your message. Type the number and press Enter.

Figure 7-7 shows you the section choices that came up in the Modem Games Forum.

```
 1 General Information
 2 Flight Sims & Games
 3 Submarine/Naval
 4 Tank Sims & Games
 5 Space Sims & Games
 6 Strategy/Conquest
 7 Sports/Racing
 9 Board/Card Games
10 Arcade Games
11 Other Modem Games
12 World Community
14 EMPIRE Tournament
15 BBS/Network Games
```

Figure 7-7:
A selection of "sections" where I can place my message.

9. Make sure that you're finished.

CompuServe now asks whether your section number is correct. Is it OK? If so, press Y and then Enter to send your posting. Figure 7-8 shows a finished message. (Words in bold type are added to explain what the lines mean.

Figure 7-8:
Anatomy of
a forum
message.

```
Message number/section:  #: 82785 S1/General Information
Posting date/time:       02-Jul-93  16:37:30
Subject line:            Sb: Best 1st modem game?
From line:               Fm: Tina Rathbone 76004,3267
To line:                 To: ALL
Body of message:         What's the first modem game I should
                         get when I buy my new super-zoom modem?
                         Thanks in advance,
                         Tina
```

Replying to a message

Want to respond to a particularly interesting message? After reading the message, type **CHOICES** and press Enter at the ! prompt to bring up a menu like the one in Figure 7-9.

Figure 7-9:
Replying to
an existing
message.

```
Modem Games Forum Read Action Menu

   1 REPLY with same subject
   2 COMPOSE with new subject
   3 REREAD this message
   4 NEXT reply
   5 NEXT SUBJECT
Enter choice !
```

The most common choice, Option 1, lets you add your reply under the message's same subject. Remember, subjects are known on-line as *threads*, because they're like the "thread" of a conversation. If you want, you can choose Option 2 to start a new thread of your own. If you're responding to a message, though, it's best to keep your message in the same thread so that others can follow.

In any case, compose and post your message as in the preceding example.

Head to the Practice Forum to try out all the many messaging options available on CompuServe.

What do they mean by "conferences"?

Forums hold *conferences* from time to time. The name is stuffy, but it's just another word for a scheduled meeting of forum members. Conferences stimulate members through live discussions of hot topics and occasional live visits by famous (or infamous) people — all pertinent to the forum's special interest.

Check out Figure 7-10 to see a typical conference announcement.

```
PLAYING ONLINE:  Join us in the MTM Gaming Lobby to connect
your modem capable games with other users for head to head
play. Type GO MTMLOBBY to access the service and GO MTMGAMES
for more information.

TUESDAY NIGHT CONFERENCE:  Join us on Tuesday at 9:00 PM
Eastern time in Conference Room 1 for fun, discussion,
trivia, and anything else you can think of. Meet new and old
friends!

FRIDAY NIGHT PARTY:  Everyone is invited to join us for our
weekly "Thank Goodness It's Friday" Conference in Conference
Room 1. The fun begins at 9:00 PM Eastern Time. Come and
join us for live chatter and lots of surprises!
```

Figure 7-10:
The Modem
Games
Forum
announces
two con-
ferences.

✔ Typing **GO CONVENTION** and pressing Enter lists all of CompuServe's upcoming conferences. Type the same GO word to participate or to reserve a place in a conference.

✔ In a conference, everyone sits around and types at each other, watching as the messages all appear on the screen.

✔ Conferences require that you type special commands; you can find these explained in one of GO CONVENTION's menus. It's a good idea to practice your conference commands before going "live" and maybe irritating other members with goofs and questions while they're trying to "hear" the forum's guest speaker.

Files, or what do libraries have to do with anything?

Forums are great for swapping messages or yakking in "real time" during conferences. But the most beloved thing in CompuServe's forums is its library of files. And the forums are the best place on CompuServe to find and download files.

✔ Files in the computing-related forums tend to be fixes to programs, updates, shareware, and stuff. Files in special-interest forums contain more articles, essays, and more "texty" stuff relating to the members' interest.

✔ Most forums pack, or *archive*, their most interesting message threads into files you can download. That's an easy way to catch up with past messages — at your leisure — after you're off CompuServe. Many forums offer lists of all their files packed into one file for downloading, too.

✔ To quickly grab a forum's messages, type **MES** at any ! prompt; turn on your modem software's capture feature and then type **SCA QUI** (for "SCAN, QUICK") and press Enter. Don't forget to turn off "capture" when the messages stop scrolling down your screen.

Figure 7-11 shows a menu for Library 2, Flight Simulators & Games, in the Modem Games Forum.

Figure 7-11:
Most forums
have file
library
menus like
this one.

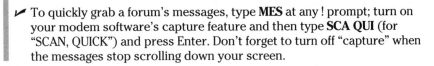

```
Modem Games Forum Library 2
Flight Sims & Games
 1 BROWSE Files
 2 DIRECTORY of Files
 3 UPLOAD a File (FREE)
 4 DOWNLOAD a file to your Computer
 5 LIBRARIES
Enter choice !
```

To save money on-line, the first file you should grab is a library's file list. After you sign off (and no charges are accumulating), you can look it over and decide on other files of interest.

Turn on your modem software's capture buffer. Choose CompuServe's Library Menu Option 2, for DIRECTORY (shown in Figure 7-11). Be sure to turn off your software's capture command when it's finished scrolling.

Figure 7-12 shows Microsoft Windows' built-in Terminal modem software capturing a forum's Directory.

```
┌─────────────────────────────────────────────────────────┐
│ ─              Terminal - CIS.TRM                    ▼ ▲ │
│  File  Edit  Settings  Phone  Transfers  Help           │
│                            ┌──────────────────────┐      │
│ 153045:  Polish WP Charact │ Send Text File...    │ic/E. European │
│                            │ Receive Text File... │      │
│ 153465:  TrueType Polish F │ View Text File...    │ic/E. European │
│                            │                      │      │
│ 153502:  MS Russian Copepa │ Send Binary File...  │ic/E. European │
│                            │ Receive Binary File..│      │
│ 152587:  Serbia-Yugoslavia │ Pause                │lish │
│                            │ Resume               │      │
│ 152810:  English Teacher I │ Stop                 │lish │
│                            └──────────────────────┘      │
│ 152885:  CALL                          S 6 / English     │
│ 153441:  english language school?      S 6 / English     │
│ 152559:  Modern Japan                  S 7 / East Asian  │
│ 152694:  Oriental Languages            S 7 / East Asian  │
│              10 replies                                  │
│                                                          │
│ ┌─────┬──────┬─────────────┬──────────────────────────┐ │
│ │Stop │Pause │Bytes: 10240 │Receiving: LANGUAGE.TXT   │ │
│ └─────┴──────┴─────────────┴──────────────────────────┘ │
│ ◆                                                      ◆ │
└─────────────────────────────────────────────────────────┘
```

Figure 7-12: Telling Windows' Terminal to "capture" the Library File Directory.

Commands for Uploading/Downloading Files

Grabbing cool files from CompuServe for your computer (*downloading*, as nerds say) starts with a menu like the one in Figure 7-11. Here's what all the menu options mean:

- 🗸 **1 BROWSE.** The best way to search for files, this option can dredge up files related to a specific subject, or "key word."

- 🗸 **2 DIRECTORY.** This command merely spits the names of the files across your screen, along with the date they first appeared on that forum.

- 🗸 **3 UPLOAD.** To send a file from your computer to CompuServe, choose this command.

- 🗸 **4 DOWNLOAD.** To grab a file off CompuServe and send it to your computer, choose this command.

- 🗸 **5 LIBRARIES.** This option lets you switch to other sections in the library to search for files.

How do you know which files are best? Well, dredge up a list of files by using the BROWSE command and then look for the word COUNT in the description. There, you'll see the number of people who've downloaded that particular file. The more popular the file, the more people will have downloaded it.

Downloading a File from CompuServe

Downloading a file from CompuServe is a two-way negotiation, with you playing the Henry Kissinger go-between. First you tell CompuServe the name of the file you've selected for downloading; then you need to tell your computer about the new file's name.

CompuServe asks you to pick a "language" both computers can agree on to conduct the transfer of your chosen file from CompuServe to your computer. This language, or *file-transfer protocol*, is something you have to tell your computer about, too. (For details on how to tell your modem software to start receiving the download, transfer yourself to Chapter 4.)

Files on CompuServe are squished — *compressed* — by special programs. That makes them smaller, which means it doesn't take as long (and cost as much) to download them. Unfortunately, it also means you'll need a program to uncompress them after they're on your computer; otherwise, they won't work. Flip to Appendix C, "I've Downloaded a File: Now What?" for more information on what to do with those "ZIPped" or otherwise-compressed files.

Follow these steps to Downloading Bliss when using the XModem protocol:

1. **If you've already browsed and you found a file you want, choose Option 4, DOWNLOAD, from the desired Forum's Library Menu.**

2. **When CompuServe asks you for the file's name, type that in, exactly as it appeared in the Browse or Directory listing.**

    You can see my file's name: FREN.JOK. (Some French jokes.)

    ```
    File name: FREN.JOK
    ```

    You can type from BROWSE after any listing that captures your fancy. Proceed to Step 3; you don't have to type the file's name just then.

3. **Choose a transfer protocol from the Library Protocol Menu CompuServe tosses up on the screen.**

    Type a **0** (zero) at any time during the file transfer to cancel if you change your mind.

    For better performance, if your modem software supports a protocol called CompuServe B, or better yet, CompuServe QB (or Quick B), choose that on CompuServe's Library Protocol Menu and on your modem software's download menu. Otherwise, choose dependable ZModem. If your software doesn't include *that*, get new software (but stick with XModem in the meantime).

TIP

After you download a file, CompuServe starts any additional downloads you perform during that session by using the original file-transfer protocol you chose. To change the download protocol, type the following after the file you want is listed within BROWSE:

DOWNLOAD PROTO:*XXX*

The *X*s stand for the various protocols you can use:

PROTO:XMO	XModem
PROTO:CAP	ASCII text capture
PROTO:B	CompuServe's B protocol
PROTO:QB	Quick B protocol
PROTO:KER	Kermit
PROTO:YMODEM	YModem

4. **Type a file name for your computer, if you're asked to do so (some protocols ask; some don't).**

   Some modem software fills in the file path information for you. Others make you do all the work, telling where the file will end up on your computer.

5. **Start the "Receive" at your end by typing or selecting your modem software's "download" command.**

   You'll see some numbers "counting" on-screen (sort of like a gas pump's gauge). It's telling you how much of the file's data has streamed down from CompuServe into your computer. When the downloading is finished, the numbers stop growing. The final number, or *byte count*, as CompuServe likes to call it, should match the file's size.

Congratulations! You're the proud owner of a new file you can enjoy on your computer. Let's check out FREN.JOK (see Figure 7-13). Hope those connect charges didn't pile up too high for this puppy.

Figure 7-13:
As you
can see,
FREN.JOK is
a winner.

```
French Joke:    --Dites-moi, comment peut-on mettre quatre
                  elephants dans une 2CV?
                --Deux devant et deux derriere.
Translation:    --Tell me, how do you get four elephants in a
                  VWBug?
                --Two in the front seat, and two in the back.
```

A hangover is the wrath of grapes.

The File Finders

Finding a particular file can take a lot of time and money. For example, how can you tell which forum carries which files? CompuServe's File Finders solve that problem. The File Finders let you describe the file you're after; then CompuServe checks all the libraries of dozens of forums. Afterwards, it spits up a list of files matching your "key words," as well as a list of what forums you should search. Table 7-2 lists CompuServe's most popular File Finders, and how to get to them.

Table 7-2 CompuServe's File Finders Whiz You to File-o-Rama	
To Find Files for These:	*Type This:*
IBM-Compatible Computers	GO IBMFF
Macintosh Computers	GO MACFF
Graphics (all types of computers)	GO GRAPHFF

What "key words" can I use in my file search?

The File Finders start their sleuthing at the merest mention of a key word that relates to the file; a file name or extension (those three letters after the dot in a file name); the date a file was submitted to CompuServe; or even the file submitter's User ID number. A key word can even be a word that describes what the file does or what program it works with. After the search is over, CompuServe gives you a list of files that match your search criteria — if any do.

When CompuServe members upload files from their computers to CompuServe's forums, they're asked to supply a few key words describing the file to potential downloaders.

Can I start my download then?

Unfortunately, you can't just point to a file name in the list and ask to download that one. You need to jot down the file's name, the forum's name, and what *library* of that forum the file is in. Have you forgotten what a library is? Check out the section in this chapter called "Files, or what do *libraries* have to do with anything?"

CompuServe's special navigating software is called the CompuServe Information Manager, or CIM. CIM comes in DOS, Mac, and Windows flavors. These programs are great helpers in getting around what can be a vast, confusing service.

Maybe the niftiest thing about CompuServe's special navigational CIM programs is that they let you download a file right from the File Finder area. No more need for sticky notes with scribbled jottings of forums and library numbers. (To learn more about the CIMs, skip to "Making CompuServe Easier.")

Sending and Receiving E-Mail

CompuServe lets its members send 60 three-page electronic messages (*e-mail*) per month as part of its Basic Services. To learn more about sending mail, type **GO MAIL** at any ! prompt and follow the simple menus on-screen.

- ✔ Practice sending mail in the oft-mentioned Practice Forum, first.

- ✔ If you have mail, you see a message telling you so when you first log on. Type **GO MAIL** to read it and reply if you like.

- ✔ You can keep an address book on CompuServe of User ID numbers and member names you send mail to frequently.

- ✔ You can send program files as well as text messages through CompuServe's mail service. When prompted, choose the *binary* file option to send a program file, like a spreadsheet or a Word document, for example.

- ✔ You can tell CompuServe to send a mail message to someone "Receipt Requested." That way, CompuServe notifies you when the addressee gets your urgent message. Like typing in all uppercase letters, return-receipt is sometimes considered rude by on-line cognoscenti.

- ✔ CompuServe members can send mail to people on the Internet, a scientific/educational network; MCI Mail subscribers; AT&T Mail members; and most fax machines. (Some of these services carry minimal extra charges.) To find out more about these CompuServe Mail intricacies, type **GO MAILRATES**.

Here's the place to check for instructions on how to send e-mail to other on-line services.

Chatting

CompuServe offers this weird service where otherwise perfectly mature adults type bizarre messages addressed to no one and everyone, at random, for hours — in a constantly scrolling operation more resembling CB radio than computer communications.

This what-is-the-world-coming-to service is called *chatting*, but CompuServe runs with the CB radio analogy and calls it CB Simulator.

Newcomers to CB are advised to call CB Band A (**GO CB-1**) and go into Channel 2. The commands are the same ones used for the live *conferences* discussed earlier.

How Much Is This Costing Me, Anyways?

"Getting" CompuServe isn't like "getting" Prodigy: There's no special software to buy first.

To get started, you need a temporary password and User ID. Then just set up your general modem software to suit CompuServe's requirements (detailed in "Logging On" at the start of this chapter).

✔ New users find the password and User ID to call CompuServe inside the Introductory Membership leaflets distributed with many computer hardware and software products. Typically, these memberships include a free month of "Basic Service" (worth $8.95), plus a $15 credit toward "Extended" and "Premium" Services, as well. (Extended Services are worth anywhere from $6-$16 per hour, depending on the modem speed at which you call; Premium Services' costs vary.)

✔ Or, you can buy a CompuServe Membership Kit wherever software is sold. The kits, in Mac, DOS, or Windows versions, include the CompuServe Information Manager (CIM) software, manuals, and a credit for the first month of Basic Service. Kits run about $45.

✔ There's a coupon in back of this book with a toll-free number you can call if you want to join CompuServe right away.

If you join CompuServe through a Trial Membership leaflet or coupon, the first thing you should do is download one of CompuServe's own navigational (CIM) programs in a version to suit your computer. You are charged $25 to download the program, but you receive that same amount in credit for Extended Services on-line.

You aren't charged anything for the time you spend in the CIM support forums — whether you're reading messages, asking for help, or downloading the program. The CIM programs make CompuServe easier to use.

CompuServe's rates

CompuServe recently launched its flat-rate, Standard Pricing Plan for unlimited usage of its Basic Services. But it still offers a choice between two *types* of plans: the Standard one and another, "Alternate Plan." Plus, there's the Executive Service Option and some features that are dubbed "Premium Services." And even "Standard Plan" members still pay an hourly rate for many services.

Aarrgghh! Which are which? Table 7-3 dredges the muddy depths of CompuServe's rate structure. Table 7-4 lists CompuServe's hourly connect rates.

The hourly connect rates mentioned in Table 7-4 vary, too, depending on your membership plan:

Table 7-3	CompuServe Has a Plan for You!			
Membership Plan:	*Monthly Charge:*	*Unlimited Services per Month:*	*Basic Services*	*Extended Services:*
Standard Pricing Plan	$8.95	Unlimited Basic Services	No limit	Hourly connect rates
Alternate Plan	$2.50	Unlimited Membership Support Services	Hourly connect rates	Hourly connect rates
Executive Option	$10	Unlimited Basic, Services, plus extra databases and discounts and stuff (good for financial types)	No limit	Hourly connect rates

Table 7-4	Hourly Connect Rates on CompuServe			
Plan:	*300 bps*	*1200 bps*	*2400 bps*	*9600/14,400 bps*
Standard	$ 6.00/hour	$ 8.00/hour	$ 8.00/hour	$16.00/hour
Alternate	$ 6.30/hour	$12.80/hour	$12.80/hour	$22.80/hour

Speed of Modem Connection appears above the bps columns in Table 7-4.

So what does all this table stuff boil down to?

It means that if you have a Basic membership and you call CompuServe with a 2400 bps modem, and you decide to visit one of CompuServe's forums and download some software, you'll pay your regular $8.95 per month charge, *plus* $8 per hour the minute you set foot in the forum.

Now you know why I've been urging you to try the *Practice* Forum (freebie) so much! Other free forums are the CompuServe Software (CIM) Support Forums and the CompuServe Help Forum.

Forums are hot, it's true, but Basic Services offers a lot for $8.95 per month. Take a look at Table 7-5.

Table 7-5	**Currently Showing in Basic Services Near You**		
News, Sports, & Weather	*Reference Library*	*Shopping*	*Games & Entertainment*
Associated Press Online Hourly News	American Heritage Dictionary	The Electronic Mall	Black Dragon
This Day in History	Consumer Reports	Shopper's Advantage	CastleQuest
Accu-Weather Maps/Reports	Consumer Reports		Classic Adventure
National Weather Service	Complete Drug Reference		Enhanced Adventure
UK News Clips	Grolier's Encyclopedia		Hangman
UK Sports Clips	Handicapped Users' Database		Hollywood Hotline
UK Weather	HealthNet		Roger Ebert's Movie Reviews
	Peterson's College Database		Science Trivia Quiz
	Rehabilitation Database		ShowBizQuiz Soap Opera Summaries
			The Grolier Whiz Quiz

(continued)

We now return to our regularly scheduled flame-throwing.

Table 7-5 *(continued)*		
Money Talks	*Communication Exchange*	*Travel and Leisure*
Basic Current Stock Quotes	Ask Customer Service	Department of State Advisories
FundWatch Online by Money Magazine	Classified Ads (to read)	Travelshopper and EAASY SABRE
Issue/Symbol Reference	CompuServe Mail	Visa Advisors
Mortgage Calculator	Directory of Members	Zagat Restaurant Survey
	DOSCIM Support Forum	
	Help Forum	
	MACCIM Support Forum	
	Navigator Support Forum	
	Practice Forum	
	WinCIM Support Forum	

CompuServe Tens

"Help! I'm lost in CompuServe and I'm paying a zillion dollars a minute!!"

Sometimes CompuServe feels more like some sort of cyberspace Twilight Zone than an on-line service. But stick it out; it's *way* cool. Here are some Tens and Tips to help you feel a bit more grounded.

Making CompuServe easier to use

Yes, Virginia, there is a Santa Claus — and he delivered WinCIM right around Christmas of '92. Since that day, my CompuServe merriment factor has surged way past the chimney tops.

TAPCIS, Navigator, and other programs that make CompuServe easier exist; and they have an equally loyal following among their users.

A harp is a nude piano.

Download or order CompuServe Information Manager

Versions: DOS, Macintosh, and Windows.

GO Word: GO CIM

File Names to Download: Follow the prompts in the CIM Forum for the version you want.

Primary Virtues: Type a Find word into a window, and the program dials CompuServe and takes you right to the spot. Lets you download files without worrying about file-transfer protocols and other repulsive relics. Enables you to download a forum's entire message board for reading off-line; read and write mail off-line; prepare and send files to people on your mailing list with the press of a button; start a download right from File Finder.

Comments: Need I say more? Okay, I will. The CIM programs are free. You pay a fee to download them, but it's turned around as credit on your account. Figures 7-14 through 7-16 show screens from the CIM programs.

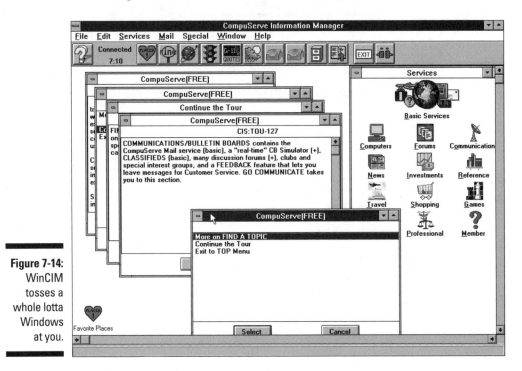

Figure 7-14:
WinCIM
tosses a
whole lotta
Windows
at you.

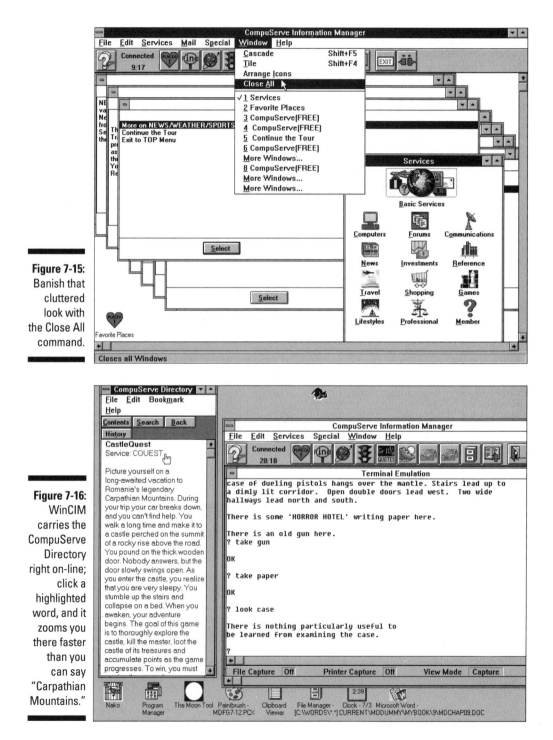

Figure 7-15: Banish that cluttered look with the Close All command.

Figure 7-16: WinCIM carries the CompuServe Directory right on-line; click a highlighted word, and it zooms you there faster than you can say "Carpathian Mountains."

Download or purchase Navigator

Versions: Macintosh

GO Word: GO NAVIGATOR

File Names to Download: Choose options 4 and 5: Download...

Primary Virtues: Set up your CompuServe sessions before you call; Navigator goes on-line and, well, *navigates*. Tell Navigator to make the rounds of Forums, Mail, the stock database, and most of the other stuff on-line. Then log off and read and respond to information off-line, where the connect charges are less devastating.

Comments: Great for people who spend a lot of time in Forums (and want to spend less, so to speak). Users should make sure that they get plenty of use out of the program, because downloading it from the CompuServe store (GO ORDER) slaps a hefty $70 (*on sale*) charge on the ol' plastic money card.

Download TAPCIS

Versions: DOS

GO Word: GO TAPCIS

File Names to Download: TAP.EXE and TAPDOC.EXE, Library 1

Primary Virtues: Save money by using TAPCIS to automate your CompuServe sessions. This one's been around a long time and lots of users are there to help with questions and problems. TAPCIS contains more power to automate your on-line sessions than CIM, but it's more difficult to set up.

Comments: This one is *shareware*, so you have to pay for it by registering with the author. Need a refresher on shareware? Turn to Chapter 1.

Download OZCIS

Versions: DOS

GO word: GO OZCISDL

File Names to Download: Choose options 3 and 4: Download...

Primary Virtues: This is free to download; ordering by the mail costs varying amounts, depending on whether you're a single user or a business.

Comments: OZCIS offers sophisticated navigating tools but can be tricky to set up.

Synonym: a word you use when you can't spell the other.

Ten practical things to do on CompuServe

Access Basic Quotes as part of Basic Services and track the Stock Market (GO BASICQUOTES).

Research the prices and features of new cars and coffeemakers in the Consumer Reports section (GO CONSUMER).

Calculate your personal finances (GO FINTOL).

Send President Clinton or Vice President Al Gore e-mail suggestions for running the country (GO WHITEHOUSE).

Find out how the people who live aboard yachts take their dogs for a morning walk on the Sailing Forum (GO SAILING).

Buy a used car and save some money by browsing on CompuServe's Classifieds (GO CLASSIFIEDS).

Set up your own news "clipping" service to capture all the news mentioning subjects you're keen on (GO ENS).

Read the hourly news updates from the Associated Press (GO APV).

Research college costs and party potential (GO PETERSON).

Ten cool things to do on CompuServe

Discuss/debate the latest movies with Roger Ebert (GO EBERT).

Search The Electronic Gamer for hints or complete "walkthroughs" to almost any computer game released in the last three years (GO TEG).

Head for the Bacchus Wine Forum to participate in a live, on-line wine- or beer-tasting session with people around the world (GO WINEFORUM).

Chat with those fluent in Spanish, French, German, and other languages in the Foreign Language Education Forum (GO FLEFO).

Chat in actual time, anytime, with actual humans (GO CB-1, Channel 2, for newcomers).

Drop by the Art Gallery Forum to view Smithsonian art collections (GO ARTGALLERY).

Test your trivia knowledge in Grolier's Quiz Whiz (GO WHIZ).

Earn your own CompuServe rating on the Chess Forum (GO CHESSFORUM).

Play CastleQuest and relive the olden days of computer games, BG — Before Graphics! (GO CQUEST).

Chapter 8

America Online

America Online likes to say it emphasizes "people" over "computers." And, for an on-line service, that's pretty true.

America Online has borrowed heavily from the Macintosh computer's "friendly" look: replete with windows, menu bars, and little pictures. As with a Macintosh, America Online works better when you use a mouse. Eeek! ... A *computer*-type mouse ... even in the DOS and Windows versions.

Most important, through the special software you install on your computer in order to call the service, America Online grants you "blinders." It substitutes little buttons and windows and pictures for any technoid aspects of modeming that might intrude on your good times on-line. And it does this so well that, to borrow a phrase from Macintosh marketing droids, it just may be the on-line service "for the rest of us."

GUI Stuff to Chew On

If Prodigy's special software makes your screen look more like a TV, America Online's special software transforms your monitor into a true *GUI*, or *graphical user interface*. Pronounced *gooey* to rhyme with *phooey*, GUI is one of the more "cutesy" computer buzzwords. With a GUI, you point at little pictures and "menus" and click your mouse instead of typing commands.

Can I yell "Movie!" in a crowded firehouse?

Gooey may sound goofy, but it's certainly better than dealing with harrowing modem experiences like "transfer protocols" and "modem init strings."

America Online's GUI protects you from having to know anything about your modem or modeming. In fact, it makes America Online more relaxing than other services. (Unless you're subconsciously adding up your per-minute on-line costs, however ... tick ...tick ... tick ... .)

✔ Macintosh computers and Windows are other examples of GUIs. Actually, America Online started out specifically for the Macintosh. It has kept the Mac's look — windows, icons, and buttons — through each new version.

✔ Buy a mouse (if you don't already have one — you don't need two) before trying to use a GUI. Even America Online's DOS version uses little on-screen buttons and menus.

You're in Moof! Country

One of the signs of America Online's Mac roots is the omnipresent "Moof!" outcry. "Moof!" invades chat rooms and peppers messages. "Moof!" sounds punctuate chat sessions.

What's a Moof!? It's the sound of the Dogcow, a mysterious creature who lives deep inside the Mac's File menu, under Page Setup, Options. See him? The bovine bowser?

He's a working Dogcow, shown in Figure 8-1, helping Mac users everywhere orient their printouts. Is he dog or cow? Do you feed him kibble or hay? No one knows for sure.

Figure 8-1:
Caution:
Dogcow at
work.

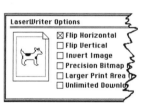

Whatever his family tree, legend has it that Apple Computer's corporate corridors ring with the Dogcow's cries 'round about midnight. Mooooof!

This chapter uses unavoidable GUI-mousey terms like *click*, *double-click*, and *select*. That's because it's best to get around America Online's graphical screen with a mouse.

To *click* an *icon* (those little pictures that represent "real" stuff), you move the mouse stealthily (mouse on mousepad, not in the air!) until the *mouse pointer*, or *cursor*, touches the "target" icon. Then press and release your mouse's left button. Gotcha! (Mac mousers will only see *one* button, lucky devils.)

Sometimes, one click is not enough; you'll need to hunker down and give that button two clicks in a row: a *double-click*. If your mouse has more than one button, go for the one on the far left.

To *select* text on the screen, *highlight* it: Move the mouse until the cursor sits next to what you want, then press and hold down the left mouse button. Then, while holding down the button, drag the mouse until the item is covered. Finished? Release the mouse button, and the item is "selected." From there, you can delete or copy the item.

America Online generally makes you click or double-click icons, menu items, or other pesky controls on the screen before it'll do what you want it to do.

Play with the software's menus and windows for a few minutes before dialing the service for the first time. It may not seem so simple right this minute, but this GUI stuff really does make your computer easier to use ... .

Logging On

 ✔ If you already have the America Online software on your computer, head to the following steps.

 ✔ The first time you log on, you have to select a screen name. America Online "suggests" terrible names like **TINA68014**, but you can choose anything you want as long as it hasn't been used. Keep trying until you hit one.

 ✔ Don't have it installed yet? After you get somebody else to install the software, connecting to America Online is *cake*. (Installing it's not really so rough; there's a marvelous Arthur Murray step-by-step at the end of this chapter. Okay, so there aren't any footprints.)

Note: If you have trouble at any stage, make a "voice" phone call to tech support. Refer to the "Hollering for Help" section later in this chapter.

1. **Double-check to see that your modem is plugged in and turned on.**

2. **Turn on your computer if it's not on already.**

3. **This step varies according to your version of America Online software: Windows, Macintosh, or DOS.**

Using the Windows version? Click the little America Online picture, or *icon*, in Program Manager:

Are you using your Mac to connect with America Online? Click the America Online folder. You'll see the icon shown in Figure 8-2. Click that.

Figure 8-2:
Click the
America
Online folder
if you're
using a Mac.

DOS users should type these three lines at the DOS prompt, each followed by pressing Enter once:

```
C:\> CD\

C:\> CD AOL

C:\> AOL
```

The first command takes you to your computer's *root* directory. The second command moves to your AOL directory — where the America Online software lives. The last command starts America Online's software. (If your computer is set up differently, bribe your nerdy pal to "come over and check out this great new on-line service I just got!")

4. **Click the Sign On button, and then type your password when you're asked to.**

After you type your password, the steps are much the same on all versions: Mac, Windows, or DOS.

Your modem starts dialing and making those Banshee from Hell noises. As the call progresses, the software highlights a picture of a phone pad, little computer cables, and, finally, a picture of a key. Figure 8-3 shows the highlighted phone pad picture.

If you have a Macintosh or a sound card, when the computer connects, a voice booms out to greet you. "Welcome!"

Don't be startled; it's all part of America Online's friendly atmosphere. Just smile and holler "Hello!" back. If you don't have a Macintosh or a sound card, however, you'll just hear your screeching modem finally shut itself up.

Figure 8-3:
Seeing your
progress as
you log on.

America Online at a Glance

When you're finally logged on to America Online, a screen like the one in Figure 8-4 greets you. The little picture of the envelope tells you right away whether you've been popular enough (or argumentative enough) to receive any electronic mail. Clicking the little newspaper on the right side of the screen takes you over to headline news.

The Windows screen at the left differs from the Mac version: Windows lines up a row of tiny pictures along the top. It's called the *Flashbar*, and it zooms you straight to one of America Online's areas when you click that area's icon.

- ✔ Click any buttons that look inviting. If you want to see the bigger picture, click Browse the Service (PCs) or Departments (Macs) — a largish command button on the left (see Figure 8-5). Up comes America Online's entire, grassy range for your grazing pleasure. Moof!

- ✔ The New icon contains a cool command button called What's Hot. Click this button to get other ideas for where to go.

- ✔ Tired of seeing little pictures for everything? You can "put away" these screens through the "close box" in the top-left corner. Double-click the close box in Windows or press Ctrl-F4 (as usual); on a Macintosh, a single-click will do.

- ✔ Don't like buttons? Click the menu bar across the top of the screen, and a little menu flops down. Click the menu item that looks the most fun, and America Online takes you there.

- ✔ To take a quick Tour of America Online, pull down the Go To menu on the horizontal menu bar. Then select Keyword (or press Ctrl-K in Windows). When the little box pops up, type **HIGHLIGHTS** and press Enter. Bon voyage!

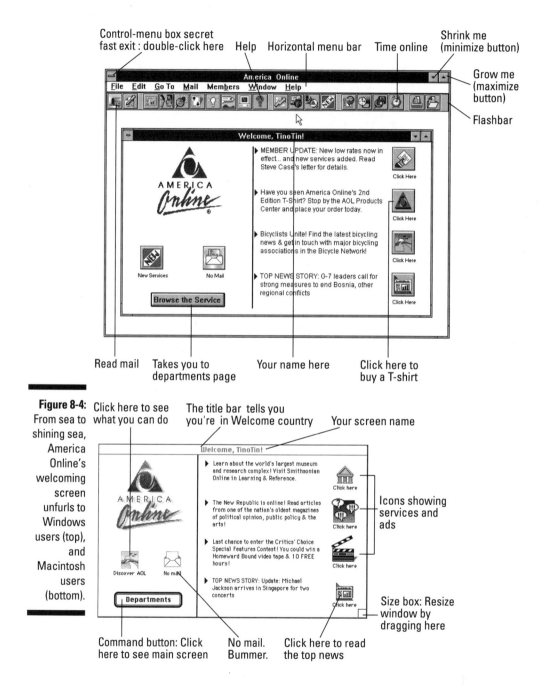

Figure 8-4: From sea to shining sea, America Online's welcoming screen unfurls to Windows users (top), and Macintosh users (bottom).

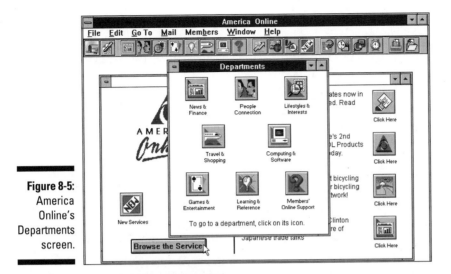

... and Logging Off

Going so soon? You can sign off of America Online in two ways:

1. Do one of the following:

Either, from the horizontal menu bar atop the screen, choose File and select the bottom item: Quit (Macs) or E**x**it (Windows and DOS). Windows users can press Alt-X simultaneously to E**x**it, too. (That's why they underline the letter x in E**x**it.)

Or, pull down the Go To menu, and you'll see another exit sign: Sign Off.

A screen like the one in Figure 8-6 pops up, asking whether you wouldn't rather sign on again maybe pleeease?

> **Goodbye From America Online!**
>
> AMERICA *Online*
>
> Come chat with the Editors of DISNEY ADVENTURES every Wed. & Fri. at 7 pm ET. Bring questions, ideas & a friend. Keyword: DISNEY
>
> Your time for this session was 00:01:57. See you soon!
>
> Hanging up modem, please wait...
>
> **Screen name:** TinoTin
> **Password:**
>
> [Sign On Again] [Setup] [Help]

2. **A third, stealth exit is to double-click the window close box in the top-left corner.**

You can choose between "signing off" but keeping America Online's "access" program open or choosing "Exit the Application" to quit everything entirely.

If you persist in leaving, the disembodied voice returns and bids you, "Goodbye." Make sure that the speaker on your Mac or sound card is turned up all the way for this one.

Hollering for Help

America Online has so many sources for help, it's almost confusing!

🖢 If you're able to connect to America Online, help's just around the corner. Head for one of the "Customer Service Live" chat rooms. Live humans type out "live" advice at keyboards somewhere in Virginia, from noon to 11 p.m. on weekdays; noon to 9 p.m. on weekends (all times Eastern). Use the keyword **CSLIVE** to zoom right there. (Head for the Keyword section later in this chapter to see what this "keyword" stuff's all about.)

America Online sometimes tries a little too hard to be helpful. For example, if you use a Mac at home, but a PC at work, be aware of this: America Online automatically takes note of the type of computer you're using to dial in. So, if you call from your Mac, you'll be stuck with the Macintosh Support Chat rooms and *vice versa*. (So what if *versa*'s got a few bad habits?)

🖢 Customer support areas don't cost anything in connect charges.

🖢 Anytime you're connected to America Online, you can head for the Lobby. (It's under the Horizontal menu's Go To menu.) When you see anyone whose name includes the word "Guide," you can ask them for help in finding your way around.

🖢 Help — free of connect charges — is available for America Online's software commands, whether you're on-line or off-line:

To find help while on-line, Windows users can click the Help button. It's in the Flashbar, and it looks like this:

Bunches of help topics pop up in a stunning array (see Figure 8-7).

🖢 To find help while off-line, click the word **Help** on the menu at the top of the screen. When the menu drops down, double-click **Contents**. (The powerful Microsoft gods told everybody — including America Online — that if they wanted to sell Windows programs, they'd have to include this built-in Help feature.)

`I'd love to, but I've been scheduled for a karma transplant.`

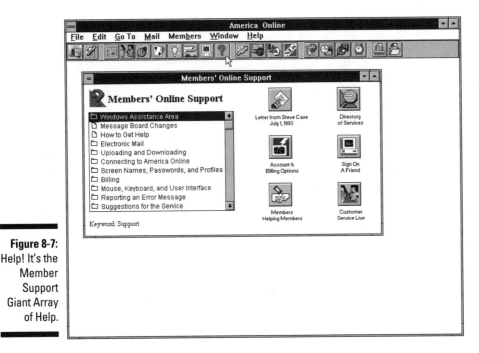

Figure 8-7:
Help! It's the
Member
Support
Giant Array
of Help.

✔ Figure 8-8 shows what happened when I clicked the **S**earch button, typed the term **UPLOADING**, pressed Enter, and clicked the **G**o To button.

✔ See the underlined words? (Actually, they're underlined *and* tinted a sickly green, but you can't tell from the picture.) Clicking one of those words brings up even more information about that particular word. (Nerds have named this clicking-words/jumping-around-to-different-topics feature *hypertext* — but they name their computers, too, so what do they know?)

✔ You can print out any help menus that seem particularly cool.

✔ Help is available anytime anywhere, on-line or off-line. Click the little picture of the apple in the top-left corner. Figure 8-9 shows a useful item on "Using Chat Room Sounds."

✔ In addition, you can find Members' Online Support by clicking the words Go To on the Horizontal Menu. Selecting Members' Online Support brings up a Mac version of the huge Help Screen, like the one in Figure 8-10.

✔ When stumped, look for a button shaped like a question mark or marked "help." Clicking that button usually brings an end to your head-scratching. Can't find a help button? Then click the close window box in a window's upper left corner until you end up in a window that sports a help button. Or go to Help from the Apple Menu.

Marching to a different kettle of fish.

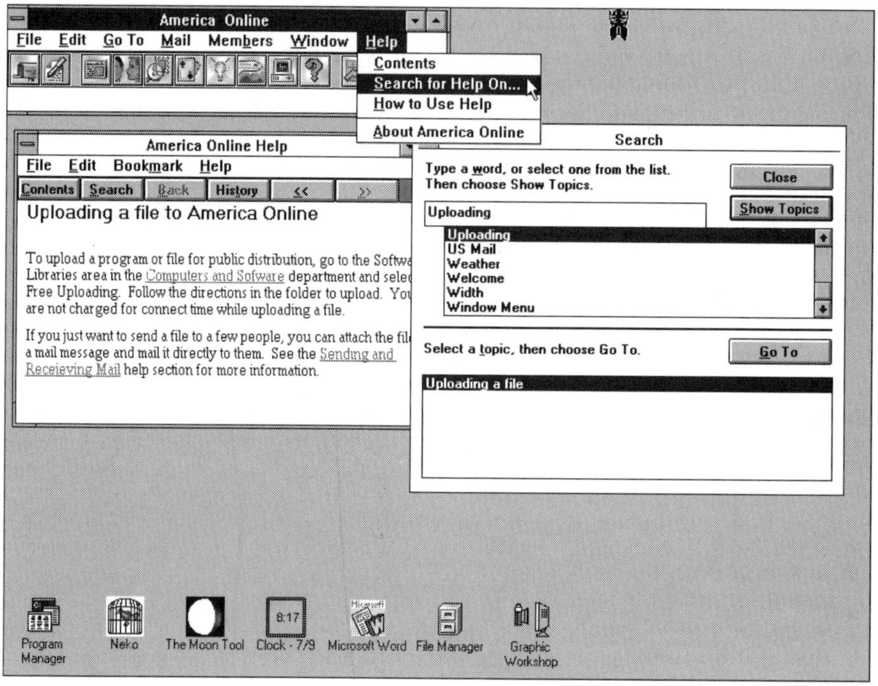

Figure 8-8:
America Online's software offers built-in Help for uploading and other stuff.

✔ A toll-free number connects you with a human voice. You can reach the help staff by "traditional" phone at 1-800-827-6364. (It's usually busy, though, so make it a last resort.) Tech support staffers are available from noon until 9:00 p.m.; weekend and holiday hours may differ.

✔ If you can connect successfully with America Online but you still need help with some other aspect of the service, be a Doo-Bee and log on with your modem instead of calling "voice." Then use the keyword **CSLIVE** to get to the free, live help places. That frees up the toll-free phone line for those poor folks who can't even connect. Head to the "Keyword" section in this chapter for help with using keywords.

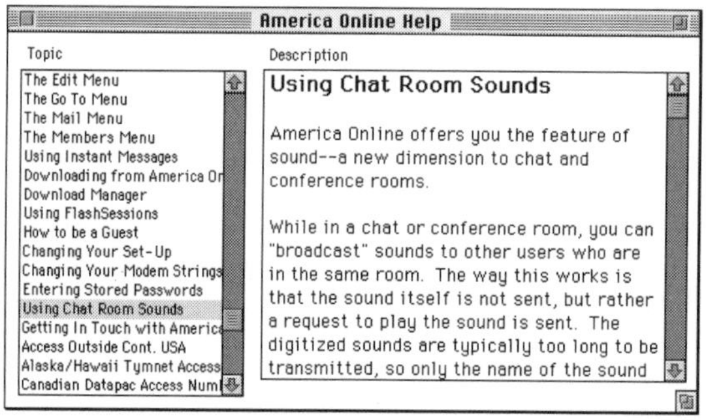

Figure 8-9:
Help is available for critical and not-so-critical features.

Figure 8-10: Members' Online Mac Support — and more!

A toll-free on-line area called Members Helping Members is full of friendly, and, yes, helpful people. Head there with the keyword **MHM** or click the Members Helping Members icon on the Member's Online Support screen.

Ask your questions here about Mac printer cartridges, Windows' memory requirements, America Online, or anything else. Cajole other members to join you in clamoring for an aikido forum on America Online. Or just gripe. If you leave a message in MHM, as old timers like to call it, expect to see a reply in a day or so.

As you gain experience, someday you can H other M.

How Do I Use the Menus?

By now, mouse pointers and menus shouldn't sound quite so exotic. The screen's Horizontal menu shepherds all its submenus in a logically organized herd.

But there's even an easier way: shortcut keys. These zoom you around America Online's menu morass with time to spare. The most popular shortcuts line up in Table 8-1. Press the two keys at the same time for best results.

Table 8-1	America Online Top Ten Shortcut Keys		
Key combination:			*Does this:*
Mac	*Windows*	*DOS*	
⌘ and K	Ctrl-K	Ctrl-K	Opens Go To Keyword window
⌘ and D	Ctrl-D	Ctrl-D	Opens Departments window
⌘ and M	Ctrl-M	Ctrl-M	Compose mail
⌘ and R	Ctrl-R	Ctrl-R	Read mail

(continued)

CAUTION: Incorrigible punster! Don't incorrige.

Table 8-1 *(continued)*

Key combination:			Does this:
Mac	*Windows*	*DOS*	
⌘ and W	Ctrl-F4	Ctrl-F4	Closes a window
⌘ and .	Esc	Ctrl-X	Stops incoming text or action
⌘ and Z	Ctrl-Z	—	Undo (change your mind about a function or command you selected)
⌘ and X	Ctrl-X	Shift-Del	Cuts selected text
⌘ and C	Ctrl-C	Ctrl-Ins	Copies selected text
⌘ and V	Ctrl-V	Shift-Ins	Pastes cut or copied text
⌘ and A	Ctrl-A	—	Select all items
⌘ and T	Ctrl-T	—	Opens Download Manager window
⌘ and I	Ctrl-I	Ctrl-I	Send an Instant Message
⌘ and G	Ctrl-G	Ctrl-G	Get a member's profile
⌘ and F	Ctrl-F	Ctrl-F	Find a member on-line
⌘ and L	Ctrl-L	Ctrl-L	Go to Lobby

Custom Keys for Your Favorite Haunts

Besides the "official" shortcuts in Table 8-1, America Online gives you a way to "record" up to ten of your own key combos. It's the Edit Go To Menu command, and it makes your favorite sections just a keypress away.

✔ Stick a favorite Club on your menu. Or put an essential function within easy reach, like File Search. There's no better way to start feeling at home on the service. And if one of your Go To places grows stale, dump it.

✔ To customize your menu, click the Go To menu and select Edit Go To Menu from the drop-down menu, shown in Figure 8-11. Then replace the software's generic picks with your own Top Ten.

✔ Only places or features that have a keyword can be added to your Go To menu.

Keywords

Keywords are the quickest way to get around America Online. Type a keyword in the Go To Keyword window and press Enter; America Online whisks you there.

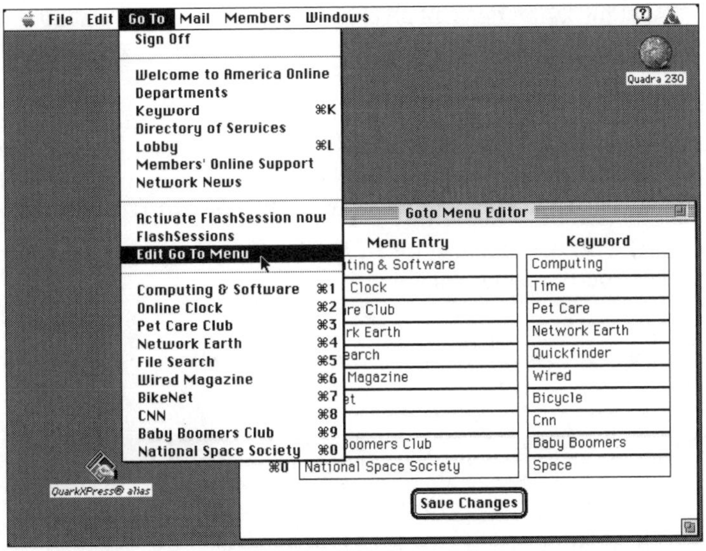

Figure 8-11:
Have fun
shortcutting
to your own
Top Tens
with the Go
To Menu
editor.

✔ Where's the Go To Keyword window? It lives under the Go To menu on the horizontal menu bar. Click Keyword, and the Go To Keyword window pops up.

✔ Or, to be even quicker, press Ctrl-K or ⌘-K. Those shortcut keys open the Go To Keyword window.

✔ So, where are all the keywords? Call up the Go To Keyword window and click the Keyword Help (or Keyword List) button. America Online tosses up a list of departments. Double-click the department you're interested in — News & Finance, for example — and you'll see a list of every keyword relating to that area.

✔ Or, call up the Go To menu and click The Directory of Services. It also lets you call up that same list of keywords.

✔ For more keywords, peruse the Welcome New Member! leaflet that came in the mail when you joined America Online.

Directory of Services

The Directory of Services is America Online's "phone book." It lists the names and keywords of every service on America Online. To browse its little electronic pages, click Go To on the main menu, and then click Directory of Services.

From there, double-click Search the Directory of Services. Up pops a box, where you can type what you're looking for. Type **PETS**, for example, and press Enter. America Online lists all the on-line areas where people talk about retrieving their ferrets from under the couch.

All true wisdom can be found on T-shirts.

- Sometimes, America Online brings up several areas relating to that keyword, letting you choose the most appropriate spot. Other times, it only finds one item, or none at all.

- To find more information about the areas the Directory of Services has dredged up, double-click them. When searching for *Chicago Tribune*, for example, double-click the highlighted entry that comes up, as in Figure 8-12. That brings up a description of what the *Chicago Tribune* area has to offer.

- See the Go button in Figure 8-12? Click the Go button, and America Online takes you there immediately.

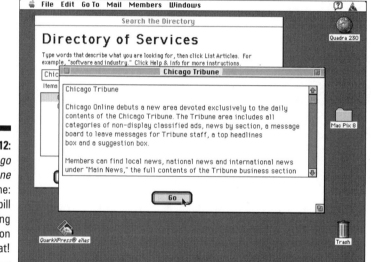

Figure 8-12:
The *Chicago Tribune* on-line: Can't spill my morning coffee on that!

"Clubs," "Forums," or "BBSs"?

Some people refer to on-line "gathering areas" as "forums," "BBSs," or "special interest groups," but America Online's official name is simply "clubs." Unless it's not. Actually, America Online's sorta mixed up that way: It uses a variety of names to describe the forum or club experience.

Whatever they're called, they're the places where people develop their interests or check out new ones. Figure 8-13 shows `Network` as yet another "nickname" for forum.

- Clubs offer messages, programs and other files, and conferences relating to their particular subject. (Conferences are scheduled "events" where members can discuss stuff or "see" guest stars.)

- Some clubs, like the BikeNet in Figure 8-13, carry magazines and newsletters focusing on the members' interest.

Remember the VW Anomaly ... er, Bug?

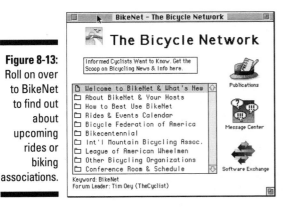

Figure 8-13:
Roll on over
to BikeNet
to find out
about
upcoming
rides or
biking
associations.

🎯 *TIP*

▸ Every club or forum has a leader who'll show you stuff and help you with questions.

▸ Be sure to search the Directory of Services for forums or clubs that relate to your family's hobbies and passions. Then add these areas' keywords with the Edit Go To Menu feature. It makes your sessions more warm and fuzzy.

Message "Boards"

Each forum or club stores members' messages in areas known as *message boards*. (It's the old cork bulletin-board metaphor.) Figure 8-14 shows a typical Club message area.

▸ To read a message, highlight it and then click it. Or click the Read Message icon in the row along the bottom of the screen. Some forums make you "List Categories" and "Find New" messages first.

Figure 8-14:
The
San Jose
Mercury
News'
message
board
incites with
"teasers" on
burning
news topics.

✔ When reading a message, other message tools hop onto your screen. You can add a new message subject, respond to the one you're reading, or move on to the next message just by clicking the darling little icon.

✔ Currently, there's no way to capture or download all the messages in a particular club. You can only scroll from one message to another. America Online is working on organizing the messages by *threads*, or topics, the way the other on-line services and local bulletin boards do. That's good, because right now it's kind of *pokey*.

✔ When posting a message, don't automatically make up a new subject header. Instead, look to see whether people are already talking about your subject; a similar subject header may already exist. The message "threads" on America Online are tangled enough as it is!

Conferences

Clubs, forums — whatever America Online is calling them this week — often schedule conferences where members can discuss issues or meet guests. The members all drop by at the same time and type at each other.

✔ Just like at any other conference, participants all wear the latest funny hats and carry the latest free tote bags. (You just can't see them.)

✔ A forum's welcoming screen usually lists any upcoming conferences. Just jot down the time and place on a sticky note and add it to the collection around your computer's screen.

✔ While you're on-line, the Network News feature automatically alerts you to any currently running conferences and special events. During a recent session, a notice that appeared on my screen in the Customer Support Live area. It worked: I "bit" and headed over for CNN's TV Addiction conference.

✔ For the master list of upcoming forum events, use the keyword **TITF** to bring up a screen called "Tonight in the Forums." The conferences usually cover dull and boring computer-forum stuff, though; nobody ever talks about how to make a million dollars and retire early.

Files, Gimme Files

Files are a main attraction of on-line services. People love those wacky files.

You can find and *download* (copy onto your own computer) shareware programs that are of just as a high a quality as the programs you see in the store. Or grab articles and message collections that relate to your interests.

You can also send, or *upload*, programs you like and feel compelled to share.

- ✔ America Online depends on its members for many of its files, so it doesn't charge you for time you spend uploading a file.

- ✔ Copyright laws being what they are, be sure you only upload shareware programs — not commercial programs.

- ✔ Lots of the files on America Online and other services are articles from magazines and other publications. Be sure you're within copyright guidelines when uploading any articles, too. Let *Cosmopolitan* upload that story on French kissing — not you.

Finding Files

Sure, there are tons of files on America Online. But what's the best way to find them?

- ✔ Browsing is often the best way to come upon new and exciting files. Figure 8-15 shows a "hit" I came across while checking out one of the areas listed in the What's New box. Of course, it's logical that the Hollywood area would go out of its way to assist you in viewing pictures.

- ✔ You, too, can fill up your hard drive with huge graphics files depicting 8-ball spheres or lunar landscapes. Get started today by downloading one of these *GIF viewers*. (No one can agree on how to pronounce GIF, incidentally. You'll hear "Jif," like the peanut butter, and the other way, with a "g" as in "give up and just download it ... ."

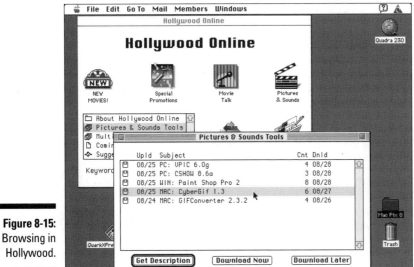

Figure 8-15:
Browsing in
Hollywood.

✔ The File Search tool (keyword: **File Search**) is the quickest way to find a file when you're looking for something specific.

✔ The Windows version of America Online has File Search as one of the little icons in the Flashbar along the top, like this:

✔ Add File Search to your Go To menu. From Go To, choose Edit Go To Menu, and then add File Search to one of the form's blank lines.

Top ten downloadables

Want an idea of the most popular files on America Online? Head for the keyword **Software**. Grab the list called "Top Downloads." Chances are, these are among your best bests for first downloads.

What's new?

Cutting-edge types can download a list of each week's newest and shiniest files by going to the keyword **Software** and selecting "Weekly New Files Listings" from the menu in Figure 8-16. Windows and DOS users, look in "Software Listings" for new files.

Figure 8-16: Software Center for the Mac.

✔ Besides general new files, the Software Center lists business, games, or a dozen other special categories of new files.

✔ Feel free to put the Software Center on your Go To menu. That way, you're just a shortcut key away from File Nirvana.

✔ PC software hunters: You may need to climb through one extra window to get to the new Files list. It's under Specialty Libraries/File Lists.

The Download Manager

Instead of downloading files one at a time, check out the Download Manager. It lets you mark all the files you want to take with you, and download them all in one fell swoop.

✔ Download Manager appears as an option when you select the Download Later option from any download window, as in Figure 8-17 (or anytime you press Ctrl-T or File/Download Manager).

Figure 8-17:
Let's
download
this GIF
viewer later.
The
Download
Manager
button is
always
available.

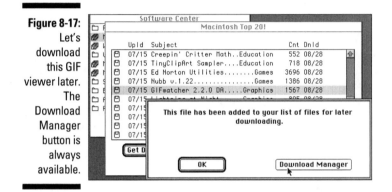

✔ If you choose Download Later, America Online won't let you sign off without reminding you that you've flagged some files for downloading. You can wait some more — maybe until you collect more files to download — or click the Download Manager button to start downloading your files now.

✔ Download Manager has a permanent home under the File menu in the Horizontal Menu Bar.

✔ Downloading your hoard is as easy as clicking the Download Manager's Start Download icon in Figure 8-18.

✔ The Manager's into control. It recaps how many files you've chosen to download, where they'll end up on your computer, and how much space they'll take up on your hard disk. It estimates how long the download's gonna take, too.

✔ Other icons in Download Manager let you View the file's Description to make sure you *really* want it; specify where to put the newcomer; banish a highlighted file from your list; get help; and see a list of the past downloads of your entire career on America Online.

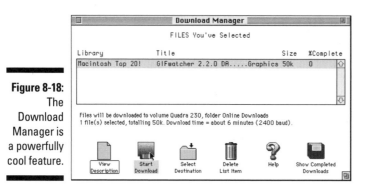

Figure 8-18:
The Download Manager is a powerfully cool feature.

▸ If you get greedy and want to add still more files to your download (or you have second thoughts after seeing how long all this will take), you can bail out at any time by clicking the window's close box. That frees you to head out into the forums (or escape without doing anything at all) to look for additional files to download. Come back anytime you're ready.

▸ If you start a download and still want to bail out, press the Cancel button shown in Figure 8-19.

Figure 8-19:
Click the Cancel button to bail out of the download.

▸ Not sure whether you've already downloaded a file? The America Online software keeps track of the last 50 files (or as many as you specify) you've downloaded. Just click the Show Completed Downloads icon under the Download Manager. You've been busy, eh?

▸ You can tell Download Manager to sign you off the service when your download is finished. Or in Windows, you can do something else — like work — while downloading takes place in the background. That way, you can catch up on your yoga through a Dog Pose while America Online grabs your files and hangs up automatically.

▸ If you want, America Online automatically uncompresses files you download and even deletes the compressed file afterwards. Here's how to set up this cool feature:

1. **Under the Members item in the Horizontal Menu Bar, choose Preferences.**

2. **Select the Download Preferences item from the list, or choose Download Preferences while in Download Manager.**

3. **Highlight Automatically Decompress Files at Sign-Off (or, for a Mac, Auto-Unstuff Files), and then click OK (or, on the Mac, click the Select command button).**

4. **Choose the Delete Files option if you want the original (compressed and unusable) file banished from your hard disk.**

I usually wait and do this "manually," after everything is working okay.

Sending and Receiving E-Mail

Your America Online account gives you unlimited mail privileges. You can send and receive mail with other America Online members, as well as members of CompuServe, MCI Mail, AT&T Mail, AppleLink, Sprint Mail, and the scientific/academic "network of networks," called the Internet.

✔ Your "mail ID" — the "address" other members use to send you e-mail — is simply your log-in name.

✔ Users on the Internet or other services (which usually use the Internet's mail "gateway") can reach you as: **yourusername@aol.com**. If your America Online screen name is Charlie2, that means that your Internet "address" is **Charlie2@aol.com**; whereas your America Online address is simply Charlie2.

✔ A message usually takes less than five minutes to reach its recipient — on AOL or anywhere else. Use the keyword **Internet** to see how it works — if you're in a nerdy mood.

✔ If you have mail waiting when you sign on, the Welcome screen tells you.

✔ If you receive a letter while on-line, a booming, eerie voice announces, "You've got mail." Unless you don't have a Mac or a sound card, in which case you probably need to look for such a message popping up on-screen.

✔ Before you can send someone mail, you need his or her mail address. Ask all your friends and associates for their "e-mail address." Then you can swap e-mail updates on where to meet after work for Happy Hour.

✔ Press the Mail shortcut key or access the mail menu from the menu bar. Then type the recipient's mail address in the To: box, as in Figure 8-20. Refer to the shortcut key table earlier in this chapter if you need a refresher.

✔ The Mac mail screen adds a Send Mail Later icon. That's a good choice for when you've dashed off a huffy note and want to cool off a bit before "really" sending it.

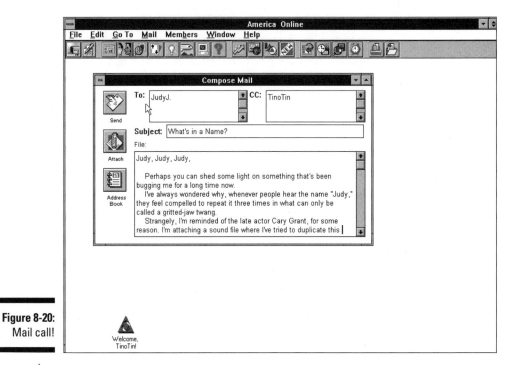

Figure 8-20:
Mail call!

✔ The Mail screen pops up whether you're connected to the service or not. You'll save cash by composing your missives off-line. To compose mail at leisure, load America Online's software and choose the Mail menu's Compose Mail item.

✔ To reply to a message from another member or from some other e-mail system, simply click the Reply icon.

✔ Use the Logging feature under the File menu to "tape" everything you're reading and doing during your on-line session. This gives you a record of e-mail you've read and sent, plus all the places you've visited. You can store this on your hard disk in a permanent file.

At least once while you're on-line, send someone an Instant Message; if only to get one in return so that you can hear the quivering sound or beeps that comes with it.

(Have someone send you a sound card, as well, if your PC doesn't have one yet: You won't hear anything without it.) Macs have built-in sound.

Under the Member's Menu in the Horizontal Menu bar, select Send Instant Message. (See the following section, "Chatting," to learn how to send an Instant Message while you're chatting.)

Chatting

Windows users can reach a chat room by clicking the "people talking" icon.

Chatting is the fine on-line art of typing messages back and forth with a "room-ful" of onlookers. It's different from leaving messages or sending e-mail, because everything happens "live." People come and go in and out of rooms as they please.

Sometimes, chatting seems different on America Online. Maybe it's in the "graphical" look; or the way some of the "chats" are organized around relevant themes rather than "chatter."

- ✔ America Online features several rooms of incessant banter at any time; try it out and experience it yourself.

- ✔ A massive chat calendar of events can be found in the Directory of Services' Calendars area.

- ✔ Role-playing games and other on-line games played by more than one member at a time are called *multiplayer games*; chat opportunities abound in these. Use the keyword **Gaming** to find a list of scheduled gaming "conferences."

- ✔ A legend along the top of a chat screen lists names of people leaving or entering the "room." Of course, most people use a nickname, so you probably won't recognize many friends.

- ✔ Highlight someone's name and then click the resulting dialog box's Get Info button (see Figure 8-21). From here, read more about them from the "member profile."

- ✔ Try sending someone an Instant Message after highlighting his or her name. This sends out a noise that's sure to perk up the ears of any neighboring dogs or cats.

You can easily disable any or all of America Online's sounds by going to the Horizontal Menu bar's Members menu and choosing Preferences.

While you're there, explore a bit and see what other settings you can tweak to improve and customize your sessions.

Fill out your member profile if you want other members to be able to look you up. Keyword: **Profile**. (*Tip:* Time spent in this one is a freebie.)

You can search member profiles for cities, hobbies, and other distinguishing features, too. (*Tip:* When you enter Search Member Profiles through the Member's Online Support, time you spend searching for other members who like Pearl Jam or live in your hometown is charge-free.)

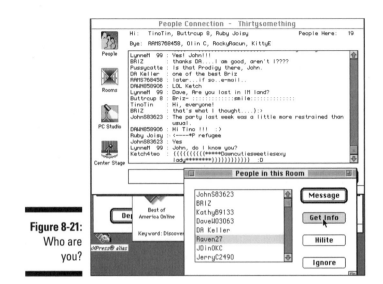

Figure 8-21:
Who are
you?

Network: I Loved That Movie!

On America Online, Network News isn't a Faye Dunaway vehicle. It's an on-line robot that keeps you posted about on-line events about to take place.

Network News is great ... or a great intrusion, depending on your mood. Fortunately, you can "turn off" the robot. Keep a' readin' to see how.

- ✔ Anytime there's an interesting "chat" or event coming up, a Network News window pops up on-screen to let you know.

- ✔ Type the keyword that's listed to go there; or ignore the message by clicking the close window box in the top-left corner. (Even if you don't see a close box at first, click there anyway: it'll appear.)

- ✔ To ignore all incoming Network News briefs, select the Members item from the Horizontal menu bar. Choose System Preferences. Then highlight Ignore Network News and click the Select command button.

- ✔ Even after you disable automatic news updates, you can still get word of the latest happenings anytime you want. From the Go To item on the Horizontal menu bar, select Network News. "And that's the way it is ..." doesn't boom out in a Walter Cronkite voice, though.

How Much Is This Costing Me, Anyways?

Like every other star in the on-line constellation, America Online has recently changed its rates. Table 8-2 shows the new nova.

Table 8-2 America Online: Simplest Rate Structure Ever?		
Base Monthly Fee	*Includes*	*Additional Hours*
$9.95/month	5 hours/month usage,	$3.50/hour 24 hrs; 7 days/week

No translation needed for *this* rate table!

- ✔ No matter how simple America Online's rates are, they're still subject to change. Use the keyword **Billing** to keep current.

- ✔ Like any on-line service, America Online has at least one Premium Service that charges extra for access. These are subject to change; currently, there is AutoVantage and Mercury Center News Library, and possibly others.

- ✔ Callers outside the continental US pay a surcharge to call the service. But Alaska and Hawaii residents are probably used to that, anyway.

How do I find out my current month's charges?

To see how much time you've racked up on-line this month, use the keyword **Billing** and double-click Current Month's Billing Summary. There, you can count how many free hours you have left for that month; if you've used up all the freebies, you can see how much cash you've burned. (The list doesn't include your current session, however.)

What about the charges for my current session?

There are lots of ways to find out how long you've been on-line. You can use the keyword **Time**; select the Online Clock from the Go To menu; or click the little alarm clock icon in the Windows Flashbar.

- ✔ The clock measures your total time on the service, even for time you've spent in charge-free areas like Member Support. Therefore, you don't really get a good indication of your total charges this way.

✔ You can't see the Online Clock item in the Go To menu if you're accessing it from a free area. Why ask why?

Multiple memberships

America Online can be used by four other household members, or you can have four different "screen names." Use the keyword **Names** for more tips on this handy feature.

Dividin' time

America Online old-timers have figured out a way to estimate their "for-pay" hours on-line. They divide their sessions into pay and no-pay. First they visit areas they'll be charged for. Then they look at the clock, to calculate how much they'll pay for their time on-line so far. After that, they head for the free areas like Members Helping Members, Search Member Profiles, and other freebies.

What about 9600 bps access?

America Online was talking about setting up 9600 bps access but hadn't ironed out all the wrinkles (rate stuff, mainly) by press time. So check on-line; type the keyword **9600** to get the latest dirt.

Checking Out PC World Net

One of America Online's strengths is its dozens of magazines and newspapers you can read and search on-line. Last time I checked, the latest additions were the *San Jose Mercury News* (great Silicon Valley coverage), the Dead Sea Scrolls (now *that's* thorough!), and *TIME* magazine.

One of the "paperless" subscriptions that automatically comes with your America Online membership is *PC World* Online. The area's more like a miniature on-line service all to itself. You can download programs mentioned in the magazine, as well as search past issues for news, articles, and product reviews. The idea is similar to CompuServe's ZiffNet and GEnie's MacUser areas, but the easy graphical style is refreshing.

The Mercury News Library in the *San Jose Mercury News* "Mercury Center" is one of America Online's few premium services. When you try to access that area, a notice appears on-screen warning you of extra charges you'll pay to search the library.

Installing America Online's Software on Your Computer

America Online won't "work" without its own special software. That brings advantages and disadvantages.

- ✔ **Minus**: You can't just fire up your general modem software and call America Online; you have to wait for your Welcome New Member! kit to come in the mail.

- ✔ **Plus**: Because America Online comes in DOS, Windows, Mac, and Apple versions, you can call America Online from all the computers you use. Say you usually call from your Mac at home, but you want to be able to check for e-mail on the PC at work. America Online will be happy to send you a free software kit for the work PC.

  After you're connected to America Online, type the keyword **Hotline** to request the other software from the Customer Relations Hotline. Be sure to substitute your current account's screen name and personal password instead of typing in any "new" sign-on passwords and choosing a new screen name.

- ✔ **Plus**: The special software means you don't have to remember complex commands that cajole your modem software to download. It also makes it easy to think up and write your e-mail messages before you dial the service and on-line charges start ticking away.

- ✔ Don't have America Online's software yet? Look in back of this book for America Online's coupon. You'll spot a toll-free number you can call to have them mail you a free copy of the software. (If your dog ate that page, the number's 1-800-827-6364.)

The steps for installing America Online are wonderfully simple. The only thing you really have to remember is to switch on your modem before you start, so that the software can "read" all the gory "init strings" and other unmentionables. Have handy the Registration Certificate they sent you, and your credit card or checking account number. And have fun!

Follow these steps if you use DOS:

1. **Insert the America Online software disk into your floppy drive A or B.**

2. **Type** _A:INSTALL_. **If you put the disk in your B drive, type** _B:INSTALL_ **instead. Press Enter.**

3. **Follow the instructions the installation program gives you.**

Follow these steps if you use Windows:

1. **Insert the America Online software disk into your floppy drive A or B.**

2. **From Windows' Program Manager, select *R*un from the *F*ile menu.**

3. **Type *A:INSTALL*. If you put the disk in your B drive, type *B:INSTALL* instead. Press Enter.**

4. **Follow the instructions the installation program gives you.**

   Install searches for your modem type and files it away. It then copies America Online's files to your hard drive. Then it creates an America Online Program Group, complete with its own cool, eyeball-in-a-pyramid icon.

   To exit the installation program at any time, click the Exit button or press Alt-X (your Alt key and the letter X). America Online's friendliness starts with its installation program — which invites you to run Install later when you're ready.

   You're finished!

5. **If you want to call America Online right away, just click the icon and head to this chapter's "Logging On" section.**

Follow these steps if you use a Macintosh:

1. **Insert the America Online software disk into your floppy drive.**

2. **Double-click the America Online folder.**

3. **Drag the Install icon from the floppy drive window to your hard disk's icon.**

   That copies the Install program over to your hard disk, where it has room to do its stuff. The icon looks like this:

   Install

4. **Close the Install window and drag the Install icon down to the Trash icon.**

   Click the left-corner box to close the window. Dragging the icon to the Trash icon ejects the disk. (Macs make you go through this eject-disk song and dance, but don't really throw your floppy in the trash — you may need it again.)

5. **Find the Install icon on your hard disk window; double-click it.**

6. **Click the Continue button.**

7. **Click the Save button when you see the dialog box asking you** Install software as: "America Online..."

Depending on your Mac's speed, you'll see a quick flash or slow stream as America Online's files uncompress. (My Quadra is a supersonic flash!)

8. Follow those icons! You're there!

The main computer helps you find a local access number. Then it plays operator, disconnecting you from the toll-free number and calling the local number. Finally, enter your certificate number and temporary password, and tell the computer personal stuff like your name and address.

America Online asks you for payment information now; don't worry, you'll still get your first 10 hours on-line free of charge (but you have to use them in the first 30 days). The people in accounting just want to know where to send the bill after those initial hours vanish down your phone lines.

America Online Tens

Does America Online's GUI look really make it easier? It depends on what you're used to.

Finding stuff and knowing what's there is tough on any on-line service, though. Here are some tips and Tens to help make your time on-line a little more fun and productive.

Making America Online easier to use

DOS and Windows version users: Get a mouse.

Add your favorite places to your Go To menu, and use it.

Sign on and check for e-mail from your friend's America Online account. Sign on as Guest, then type your own account information. (Unfortunately, you, and not your friend, will be charged for time on-line.)

Use the Logging feature on the File menu to "record" your sessions so that you can review them later and see where you've been.

Download a file called a *GIF viewer* so that you can download and view photos and computer art on your own computer. Ask other members to recommend the best viewer to get for your computer type.

Buy the America Online Official Guide from a bookstore. At $34.99, it's pricey, but it contains tons of cheerfully written and thorough tips and tours. There's a separate edition for each version of America Online's software; reading it helps minimize the time you spend on-line looking for stuff.

Don't let your eyes rest on this stuff for a minute!

Some of today's advanced modems interfere with America Online's mission to make stuff simpler for you. That's particularly true if your modem has any *error-correction* or *data-compression* capabilities.

Does your modem's box or manual mention "MNP4" or "v.42" anywhere? That's modem shorthand for *error correction*. If you see "MNP5" or "v.42bis" on the box your modem came in or in its manual, that means it has *data compression*. (All these grisly terms get their due over in Appendix A, the buying a modem section.)

Turning off these features improves America Online's performance.

Get your modem pal to crack your modem's manual and look up the "AT commands" that turn off error control and data compression on your modem.

Then get your friendly modem meister to add those AT commands to America Online's modem configuration string. (That's found under the Setup area you see in the Setup & Sign On screen.)

Launch America Online's software, but instead of signing on, click the Setup button. From the Modem and Network Configurations menu, your pal clicks Edit Modem Strings in the bottom half of the screen. Then get him or her to insert the modem's commands from the manual into the top line of gobbledy-gook called the modem configuration string. It should look something like this:

```
AT&....
```

Ugh. I told you not to bother reading this.

Practical Things to Do on America Online

Vexed by problems and questions in running your small business? Save 'em up and fire 'em at grizzled veterans of the fast track. Retired executives who are members of SCORE (Service Corps of Retired Executives) hang out and chat with people just like you in regularly scheduled live sessions over in the Microsoft Small Business Forum (keyword: **MSBC**).

Save a bundle: Travel vicariously through the adventures of people you meet on the Independent Traveler Club. Type the keyword **Traveler** to start your journey. And don't forget to wear comfortable shoes!

Teachers: Swap ideas on using America Online in the classroom with your peers across the nation (keyword: **Electronic Schoolhouse**).

Look up the America Online access number for places you'll be visiting on upcoming trips. Jot these down and keep them in your laptop's carrying case. Don't forget to install your America Online software onto your laptop computer (keyword: **Access**).

Students of all ages: Head for the Academic Assistance Center to find inspiration for term papers, post your toughest homework questions, or "page" a professional teacher who'll tutor you with difficult subjects (keyword: **Homework**).

Use the Search News feature on the Top News screen to search the day's news for specific topics you're tracking.

Figure out how to use that newfangled electronic shaver you got for Father's Day in the Gadget Guru Electronics Forum (keyword: **Electronics**).

Read America Online's list of "25 Reasons to Love America Online" and get more out of your membership (Keyword: **Prodigy** [Huh?]).

Mac people: Set up a FlashSession and have it call America Online while you're asleep. It retrieves your mail and hangs up. There's your e-mail, waiting for you bright and early (when you're still too sleepy to read it).

FlashSessions let you tell your Mac to call the service at a prearranged time and automatically download all those files sitting in your Download Manager stash, too. Check the Apple Help menu under Using FlashSessions to learn the finer points.

Windows and DOS users: America Online is adding FlashSession to the version you use, real soon now.

Cool Things to Do on America Online

Take a look at Ed Curran's Technogadgets column through Chicago Online. (Keyword: **Chicago**; click Chicagoland Lifestyles, and choose Ed Curran's Technogadgets).

SeniorNet Online's Generation to Generation message board offers the strange sight of Students and Seniors actually listening to each other and trading advice and stories about the "good old days." This is a great place to get advice without the attendant "I-told-you-so" when you don't follow it (Keyword: **SeniorNet**).

Download editorial cartoonist Mike Keefe's cartoons and view them on your computer (Keyword: **Keefe**).

Upload your stand-up comedy routines to the Improv Club's Schticks Library. The jokes and one-liners are forwarded to managers of the Improv clubs; fame could be a modem-call away (keyword: **Comedy**).

Consult the Wine & Dine Online Forum's WineBase and do a search for *something* that will go with Cousin Frankie's Triple-Alarm Taos Chili (keyword: **Wine**).

Relive those glorious '60s by turning in, tuning on, and .... oh, whatever it was.... Just go visit the Grateful Dead Forum. Download Dead lyrics, Dead sound files, and GIFs of Jerry at Candlestick Park. Life *is* good (keyword: **Grateful Dead**).

The 5th Wave   By Rich Tennant

"IT HAPPENED AROUND THE TIME WE SUBSCRIBED TO AN ON-LINE SERVICE."

Infinity is just time on an ego trip.

Chapter 9
GEnie

*G*Enie — the General Electric Network for Information Exchange (whew!) — attracts a mellow blend of home users and hobbyists, with a generous dash of professionals and a heaping bushel of computer nerds.

Overall, GEnie recalls days gone by. For one, GEnie offers one of the last refuges for users of very old computers (*dinosaurs*, nerds fondly call 'em). And GEnie's screens still scroll the same texty numbered lists displayed by the earliest on-line services — like CompuServe without CIM.

Even its image harkens back to Ye Olden Days. The acronym "GEnie" evokes magic lamps and three wishes, while GEnie's "make-easy" software is called Aladdin. (Colliding rather jarringly with RoundTables, GEnie's name for its interest groups... sort of a Barbara Eden meets Lady Guinevere.)

Image doesn't always tell the truth, however. That's especially true with GEnie, which offers state-of-the-art multiplayer games; an active mingling with the Internet (a trendy network of computers spanning the globe); and *three* Science Fiction and Fantasy groups, for heaven's sake.

A virtuoso is a musician with real high morals.

Logging On

The following steps show you how to join GEnie and log on. Already signed up and have a GEnie User ID and password? Then substitute your own numbers for the ones in the example. And dial your own city's access number instead of the "joining GEnie" number.

1. **Make sure that your modem is plugged in and turned on.**

   Are your external modem's little lights on? If you have an internal modem, it comes on after you perform Step 2.

2. **Turn on your computer. Hello, little computer.**

3. **Start your modem software.**

   GEnie doesn't need any special software. You can reach it from any general communications software — or *any* brand of computer — even 'dem 'dere dinosaurs.

4. **Adjust your modem software's communication settings.**

   Every on-line service demands some special treatment; GEnie is no different. GEnie doesn't mind you calling at the typical **8,N,1** settings, unlike CompuServe.

   No, GEnie asks you to pamper its little whims in a different way — by turning your software's ECHO setting to ON. (In some software, this means setting your DUPLEX to HALF.) Look in your software's settings — or crack the manual — to see what to tweak in your program.

   Thoroughly confused? Well, you'll be even more confused when you log on and GEnie refuses to show you what you've typed! — which is what will happen if you forget to change your software's ECHO or DUPLEX setting.

5. **Set your software's modem speed setting to 2400 bps, or *baud*.**

   If your modem only goes to 300 or 1200 bps, set your modem software for that. If you have a faster modem, you can rev up your GEnie magic carpet to 9600 bps after you've learned the ropes. (How *did* they steer those flying carpet things anyway?)

   Like most on-line services, GEnie makes members who call at 9600 bps pay a vast surcharge. To find out how much, check the rate info at the end of this chapter.

6. **Tell your modem software to "manually" dial the following number: 1-800-638-8369. (Manual dialing as opposed to adding the number to your dialing directory — you wouldn't want to do that, because you'll only call here this one time.)**

   That number lets you sign up for the very first time. If you already have a GEnie account, dial your normal access number, instead. (To save time later, add your normal access number to your dialing directory.)

Multitask: choke on gum and trip simultaneously.

7. When your modem quits making that *shhhhh* racket, spring into action by typing *HHH* as quickly as possible.

GEnie doesn't care whether you type in uppercase or lowercase letters, as long as you type it within three seconds.

You didn't know modeming involved aerobic activity, did you?

Step 7's mainframe malarkey gives you an irritating example of people jumping to suit the computer's whims, instead of the other way around — the way it should be. The people who designed this sign-on procedure should be forced to watch *2001: A Space Odyssey* 100 times — nonstop (and no fair fast-forwarding through the psychedelic sequence).

GEnie's computer tosses up an incomprehensible series of characters, looking something like this: U#=.

Here's where you type your GEnie User ID and password.

8. Joining GEnie for the first time? Then type the following characters:

```
U#= XJM11701,GENIE
```

That's **XJM11701,GENIE**, with no spaces in the middle, followed by pressing Enter once.

You're on! Hanging on tight? Whoosh! Hold on to your turban!

If you're calling GEnie for the very first time to join the service, GEnie eventually interrogates you about your computer. The following answers usually make GEnie happy:

Q1. How many characters per line does your computer display?

A. 80.

Q2. How many lines can your screen display?

A. 24.

The next two questions ask you to press specific keys on your keyboard:

Q3. What character will you use to delete your typing mistakes?

A. Press the **Backspace** key.

Q4. What character will you use to interrupt GEnie and return to a menu?

A. Press your keyboard's **BREAK** key (you'll have to hunt for this seldom-used key). (Or you can choose the combination Ctrl-C.)

If you decide midway through this procedure not to join up, you can't just type **BYE** or **EXIT** to sign off.

GEnie forces you to commit the unthinkably rude, ultimate on-line gaffe: pressing the key combination that tells your modem software to just hang up. Alas, there's no alternative. (Note that this is not a cool procedure on *any other* on-line service or bulletin board, however — and not on "normal" GEnie, either; just in this one, joining-GEnie area.)

GEnie at a Glance

Figure 9-1 gives a flying-carpet flight plan for the first screen that greets you after callng GEnie.

Typing any of the numbers shown on this menu takes you over to the general interest area.

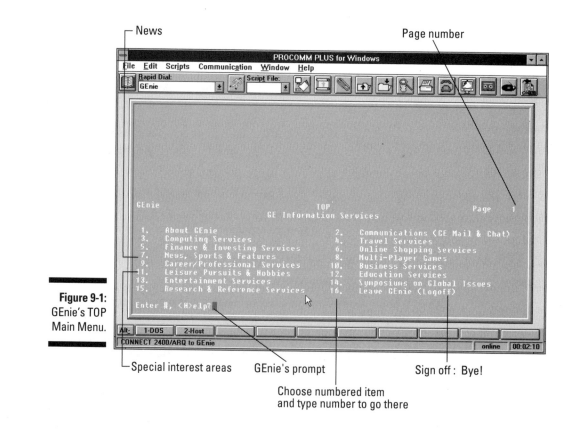

Figure 9-1:
GEnie's TOP
Main Menu.

┌ News   Page number

└ Special interest areas   GEnie's prompt   Sign off : Bye!

Choose numbered item
and type number to go there

... and Logging Off

Type **BYE** and press Enter at any prompt to end your GEnie session. Your wish is GEnie's command.

Two areas on GEnie make you type a special character in front of words you want to be taken as commands. In a conference or chat area, type **/BYE**. In GEnie's Mail editor, when you're typing a message, type ***X** on a numbered line by itself; then you can type **BYE**.

Hollering for Help

When GEnie's got you stumped, type **H** or **HELP** and press Enter at any confusing menu or activity. (Manuals refer to this as *context-sensitive help*, incidentally, and you'll get help on whatever you're doing at the moment.)

Help! I'm stuck in the chat or mail areas

GEnie corrals a whole different set of commands inside its *chat* areas, those on-line versions of CB radio. GEnie's chat areas are "Chat Lines" and "RTCs" ("real-time conferences"). If you're in one of these and you need help, type a slash character (/) in front of the word **HELP** — as in this example:

   /HELP

Otherwise, GEnie thinks **HELP** is part of your chat message. This is true for any other GEnie command you want GEnie to obey while you're chatting up a storm.

Same goes for getting help from GEnie's Mail Editor: When you're typing a message and you need help, type an asterisk symbol (*****) in front of **HELP**, as follows. (If you just type **HELP**, GEnie adds it to your message.)

   *HELP

The word **HELP**, preceded by an asterisk, brings guidance while you're composing an e-mail message, for example.

 ✔ You can get general GEnie help and information by calling the toll-free Client Services number. They're at 1-800-638-9636, Monday through Friday, 8 a.m. to midnight; Saturday, Sunday, and GEnie holidays, noon to 8 p.m. All times Eastern.

 ✔ Or you could try folding your arms over your chest and blinking pertly as you toss your blonde ponytail. (It always seemed to work for Barbara Eden.)

✔ GEnie's on-line manual is available from any prompt by typing the keyword **MANUAL** and pressing Enter. Or you can have a hardcopy manual mailed to you by typing **ORDER** and pressing Enter. You'll pay a few dinar for the privilege, however.

✔ Type **FEEDBACK** and press Enter at the main menu prompt to send GEnie's Client Services a message detailing your problem.

Putting the brakes on GEnie

If you can't keep up with the text scrolling down your screen, these Ctrl key combinations can help. Table 9-1 offers some shortcuts to cut short the display.

Table 9-1	Scrolling, Scrolling, Scrolling — Rawhide!
Press These Keys	*To Make Those 'Lil GEnie Dogies Do This*
Ctrl-S	STOP sending text
Ctrl-Q	RESUME sending text

When you feel lost in a pointless display that rolls endlessly onward, pressing the Break key puts a carrot in GEnie's mouth. GEnie stops sending text and leaves you at the most recent menu or prompt. From there you can move elsewhere or sign off.

(Rejoice! Now you finally get a chance to use the Break key, a keyboard "lurker.")

For example, if reading a long article on the history of artichoke farming on Alcatraz Island makes you hungry for a midnight snack, press Break. The text stops a-scrollin' down your screen, and you're free to log off and graze.

Ignore this bizarre Break stuff

If you're used to another key combination as a Break command, like Ctrl-C, for example, you can change your GEnie account to suit your preferences.

Type SET and press Enter at any prompt, choose GEnie Setup Script, and follow the instructions on your screen.

Finding Stuff

To find stuff on GEnie, you use two shortcuts: *keywords* or *page numbers*, pressing Enter after either one.

✔ Each time you sign on to GEnie, you'll be greeted by an Announcements page. It lists new stuff on GEnie plus any upcoming special events or guests. Each line on the page ends with a *keyword* in all-caps, as shown in Figure 9-2. Typing this keyword at any prompt whizzes you to that area.

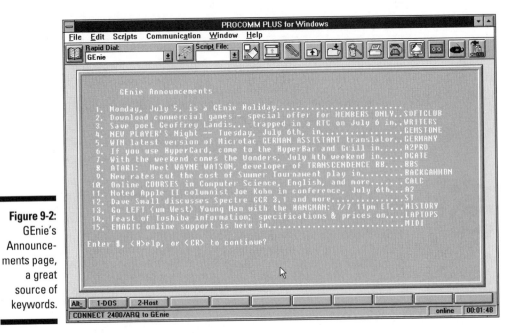

Figure 9-2:
GEnie's
Announce-
ments page,
a great
source of
keywords.

✔ You zoom to other areas of interest by typing keywords, too. If you want to learn about GEnie's prices, for example, you type **RATES** at GEnie's action "prompt":

```
Enter #, <H>elp, or <CR> to continue? RATES
```

✔ To get a massive list of current GEnie keywords that stand ready to jet you into GEnie byways, type **INDEX** at any prompt. Table 9-2 passes you some appetizers. Please note: On-line services change constantly, so these keywords are meant as general samplings.

Table 9-2	Key Keywords on GEnie
Typing This	*Takes You Here*
TOP	GEnie's Main menu
INDEX	Everything on GEnie and the keywords to get to it
MAIL	GEnie's Mail menu
SET	Your settings and billing info on GEnie
TIPS	System tips, GEnie help
GENIEUS	GEnie User Group

✔ Page numbers also stand ready to move you around GEnie. Every page on GEnie has its own number; it's in the top-right corner of each screen.

✔ You use page numbers with a MOVE command. So typing **MOVE1** and pressing Enter takes you to GEnie's top menu, page 1.

✔ Most GEnie members cut even this shortcut, typing the letter **M** and a page number instead. Typing **M1** (or **M 1**) and pressing Enter, for example, brings you to the TOP menu page.

✔ To be really streamlined, you can separate GEnie commands with a semicolon. For example, type a page number, a semicolon [;], and the number of a menu item you would choose after you reached that page. This is most useful when heading for one of GEnie's RoundTables. RoundTables always reserve Menu Item #3 for Libraries, for example. So, typing this line

```
M605;3
```

takes you directly to the Macintosh RoundTable File Libraries. (For more about this RoundTable stuff, see the following section.)

Now you're the GEnie menu-meister.

✔ Keep track of the page numbers listed on the screen above your favorite places on GEnie. By combining the page numbers with the the MOVE command, they banish GEnie's wearying menus.

What's in GEnie's RoundTables?

The word RoundTable has come up a number of times in this chapter. What are they, and what do they do? RoundTables are the special-interest groups on GEnie (much like CompuServe's forums or Prodigy's bulletin boards). Figure 9-3 shows the Internet RoundTable's Welcoming menu.

`Time flies like an arrow; fruit flies like a banana.`

```
Welcome to the Internet RoundTable, a most excellent place on
GEnie!

Your sysops are:
  Andrew Finkenstadt   ANDY        Chief Sysop
  Michael Nolan        MIKE.NOLAN  Assistant Sysop
  Sarah Collier        SARAH       Administrative Assistant
  Gary Smith           GARS        Library Manager
  Janet McNeely        JANS        Bulletin Board Manager


Conference Schedule:
  Tuesday, July 6th, 1993:  10:00pm to 11:30pm Eastern Time
  -> INTERNET EMAIL: Sending Internet email from GEnie.


HELP DESKS in GEnie CHAT LINES:
  Thursdays at 10:00pm Eastern Time, Channel 4
  Sundays   at  8:00pm Eastern Time, Channel 4


GEnie                    INTERNET-RT              Page 1405

                    Internet RoundTable


   1.  Internet Bulletin Board
   2.  Internet Conference Room
   3.  Internet Library of Files
   4.  About the Internet RoundTable
   5.  RoundTable News (930504)
   6.  Send Mail to RoundTable Staff
   7.  Internet Mail
   8.  Download Sysop's Treat File
   9.  Unix RoundTable
  10.  Search the Internet for a file
  11.  Request a file from the Internet
  12.  What is my Internet Address
```

Figure 9-3:
Can't wait to explore this Internet RoundTable!

✔ For a list of RoundTable topics on GEnie, head back to the TOP menu and choose menu item 1, About GEnie. Then use the MOVE command to get to one. Typing **M150** gets you to GENieus, the GEnie users' Roundtable, for starters.

✔ RoundTables are file, message, and conference central on GEnie. If you're looking to meet people who share your interests, look first in the RoundTables.

✔ RoundTable file areas are called *libraries*.

✔ Message areas in RoundTables are *bulletin boards*.

- *Conferences* are also known as *RTCs*, for "real-time conference," in GEnie talk.

- Leaders of RoundTables have a funny name: *sysop* (a contraction of "system" and "operator"). Adopt your local sysop; they're enthusiastic, knowledgeable, and helpful (... loyal, brave, reverent ...). The Sysops' names are listed in a welcoming message that you see as you first enter, as in Figure 9-3. Remember to tell your modem software to capture these important "orientation" messages.

Bulletin Boards: Home of a RoundTable's Messages

You can judge whether a RoundTable will interest to you by looking at its message categories. At the RoundTable's main menu, shown in Figure 9-3, select menu item 1, for the Bulletin Board.

- GEnie's bulletin boards are organized into *Categories*, then *Topics*, and finally, *Messages*. Figure 9-4 shows a typical bulletin board main menu.

- To read messages, for example, choose menu item number 7. To reply to a particularly pithy message, choose number 8. Or, to start your own new topic, go for menu item number 9.

1. CATegories	10. INDex of topics
2. NEW messages	11. SEArch topics
3. SET category	12. DELete message
4. DEScribe CAT	13. IGNore category
5. TOPic list	14. PROmpt setting
6. BROwse new msgs	15. SCRoll setting
7. REAd messages	16. NAMe used in BB
8. REPly to topic	17. EXIt the BB
9. STArt a topic	18. HELp on commands
Enter #, <Command> or <HEL>p	

Figure 9-4: Each RoundTable has a message bulletin board.

✔ When you're first exploring a RoundTable for useful messages, turn on your modem software's capture command. That saves everything you see and places it in a file on your computer. Then you can read the captured text at your leisure — off-line. The next time you log on, you'll know whether you want to return to that particular RoundTable.

✔ You don't need to type entire words on GEnie. If you see a menu with command words in partial ALLCAPS, you can type the capitalized letters to make GEnie do your bidding. For example, typing **SEA** tells GENIE to SEArch.

✔ The bulletin board's messages fall into several broad Categories. After you're in a bulletin board, type **CAT** and press Enter to get a list of every message CATegory in the RoundTable. Up pops a list similar to the one in Figure 9-5.

```
1 Welcome to the Internet RoundTable.  New members "READ
  1" Please.

2 Internet User's Forum.  Welcome and Well Met!

3 GEnie's Internet Mail Gateway. (See page 207 for Internet
  Mail)

4 How to Send Mail to Other Places

5 How to Get USENET News on GEnie

6 How to Get FILES from the Internet on GEnie
```

```
21 Network Information Center

22 Network Administrator's Corner

23 Technical Discussions about the Internet

24 Networking in the Real World
```

Figure 9-5:
The Internet
bulletin
board has 24
Categories.

✔ Within each Category, you see several Topics. You can grab an INDex of every Topic on the bulletin board by typing **IND** and pressing Enter.

✔ Each Topic can hold hundreds of Messages.

✔ Make sure that your modem software is set to capture the display at your end; Category and Topic lists are "keepers" that'll save you time and money by letting you plumb the depths of specific message categories off-line.

✔ Can't be bothered with a certain set of messages? One command lets you IGNore a topic or category permanently.

Uploading/Downloading Files

You can upload and download in two ways on GEnie:

- ✓ Swap files with individual GEnie members through GE Mail (head to "Sending and Receiving E-Mail").

- ✓ Uploading and downloading "public" files in the RoundTables' Libraries, through each Library's menu items.

Files, or Browsing GEnie's Libraries

GEnie Libraries give RoundTable members a place to swap *text files*, shareware, and other files of interest.

- ✓ GEnie has some truly bizarre RoundTables, with some even weirder file libraries tucked inside them. It's worth browsing to see what kinds of files you can find to enjoy on your computer.

- ✓ Like most on-line service biggies, GEnie screens its files for viruses. Should that keep you from screening files you download, on your own computer? Nay. Screen them, lest they screen you!

- ✓ Figure 9-6 shows a Library menu, clearly listing GEnie's commands for browsing, uploading, and downloading the library's files.

- ✓ GEnie supports ZModem, the tip-top file-transfer protocol.

- ✓ For a refresher on telling your modem software to upload and download files, go to Chapter 4. To learn general stuff about files on on-line services and BBSs, turn to Chapter 5. To find out how to handle files you've downloaded, turn to Appendix C, "I've Downloaded a File: Now What?"

```
┌─────────────────────────────────────────────────────────┐
│ ─                    Terminal - GENIE.TRM            ▼ ▲ │
│ File  Edit  Settings  Phone  Transfers  Help            │
│                   Library: ALL Libraries               ▲│
│                                                         ░│
│   1.   Description of this Library                      ░│
│   2.   Directory of Files                               ░│
│   3.   Search File Directory                            ░│
│   4.   Browse through Files                             ░│
│   5.   Upload a New File                                ░│
│   6.   Download a File                                  ░│
│   7.   Delete a File You Own                            ░│
│   8.   Set Software Library                             ░│
│   9.   Save Current Software Library                    ░│
│   10.  Instructions for Software Exchange               ░│
│   11.  Directory of New Files                           ░│
│   12.  Join/Ignore Library Category                     ░│
│                                                         ░│
│ Enter # or <P>revious?█ I                               ▼│
│ ←                                                     → │
└─────────────────────────────────────────────────────────┘
```

Figure 9-6: RoundTable libraries hide their files behind this menu.

✔ GEnie credits you for four hours on-line each month as part of the $8.95 monthly fee. After that, you're charged by the hour. That means you're duty-bound to find and use the quickest methods for finding stuff and getting around. Especially when it comes to files.

✔ Save money by keeping file directories on your computer's hard disk. Here's how.

✔ When a RoundTable captures your interest, capture its file directory with your modem software. Peruse it off-line, jotting down file names that look interesting; *then* call back and download your files.

✔ At the Library menu in Figure 9-6, for example, choosing menu item 2 starts the file directory scrolling down your screen. So, tell your modem software to start the capture, and then choose menu item 2. Figure 9-7 dissects a directory listing.

Submitter's mail address

Year, month, and date file was submitted

```
┌──────────────────────── Terminal - GENIE.TRM ──────────────────▼─▲┐
│ File  Edit  Settings  Phone  Transfers  Help                      │
│     Desc: Segment 5 of Undead from New York                    ▲ │
│  4352 SAMPCRAN.TXT         X K.CRANDELL   930510   22016    5   3 │
│     Desc: sample file for Amy Neal-Crandell                       │
│  4350 SPECPR.MKT           X SPECPRESS    930509    3072   25   6 │
│     Desc: Writers Guidelines/Spectrum Press                      │
│  4347 STRANGE              X J.DECHANCIE  930507   17536   17   3 │
│     Desc: Strange things people say 2 writers                    │
│  4312 WONDERDISK.SEA       X W.GAMMONS    930502   60208    3   5 │
│     Desc: WONDERDISK SF,F,& H Hypercard E-mag                    │
│  4311 SHIFTWOR.K           X R.WILSON78   930502   22016   29   3 │
│     Desc: Science Fiction Short Story                            │
│  4309 CATALOG.TXT          X SPECPRESS    930501   22144   23   5 │
│     Desc: list of electronic books on floppies                   │
│  4304 SI-0393.ZIP          X JUDY.TUCKER  930427   51584   17   5 │
│     Desc: Spilled Ink, Literary E-Mag                            │
│  4302 FIRES...             X K.COOPER1    930425   18368   11   3 ▼ │
│ ◄                                                              ► │
└───────────────────────────────────────────────────────────────────┘
```

Figure 9-7: How to read a library's file directory.

File number   File name   File size,   Number of   Which library
                                   in kilobytes   downloads   file is in
      Description of file   file has had

Sending and Receiving E-Mail

Members can send and receive an unlimited number of electronic mail (e-mail) messages each month through GE Mail, GEnie's "post office."

✔ When mail is waiting for you, GEnie alerts you when you first sign on. Type **MAIL** and press Enter to read new messages right away.

✔ Your GEnie mail address is different from your User ID or password (don't ever give these two out, by the way). GEnie assigns you an address when you first join up; usually it's an unwieldy combination of your name and some numbers.

✔ Unlike most on-line services, GEnie doesn't charge its members to send e-mail to people with accounts on the Internet, a global scientific/educational network of computers. GEnie members can also send mail to members of other on-line services, like CompuServe or America Online. (Use the keyword **MAIL**, press Enter, and browse the instructions for more details on these.)

Type **MAIL** and press Enter to read, create, or send e-mail.

✔ GEnie's Mail Menu pops up, with options you select from depending on what you want to do. Figure 9-8 shows the Mail menu.

✔ To send an e-mail message to another GEnie member, you need to know that member's **To:** address. Just choose menu item 8 to Search GE Mail Directory. It contains the name, state, and GE Mail address of every subscriber. Well, at least the sociable ones (see the following tip).

✔ If, like Garbo, you "vant to be alone," you can choose not to be listed in the Mail Directory by typing **MAIL** and selecting the Change GE Mail Directory Status menu item, #13. Use this same command to reverse your decision and "go public."

✔ After you've got the address, select 6 (Compose and Send GE Mail Online), and then after you've finished typing your message into the GE Mail editor, type ***S** to send it. If you decide against sending the message, end your message by typing ***X** to exit without sending.

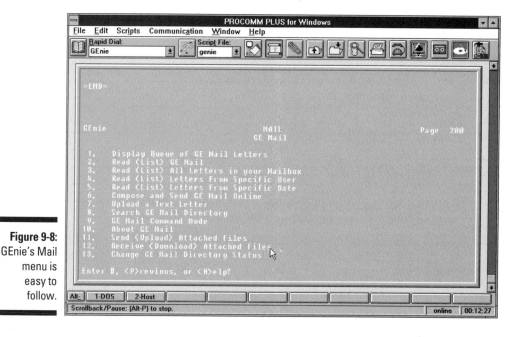

Figure 9-8:
GEnie's Mail menu is easy to follow.

Type ***H** anywhere in your message to summon your Helpful Genie. All commands to GEnie need an asterisk character (*) in front; otherwise, obedient GEnie adds those commands to the message you're composing (and broadcasts your bozo boo-boo to simply everyone).

Just because GEnie commands within Mail use asterisks doesn't mean you can't type ***wow*** in your messages ... GEnie only obeys certain command words — listed when you type ***H**.

Attaching a File to an E-Mail Message

Sending a file to another GEnie member is easy, too. From GEnie's Mail menu, choose item 11, which lets you attach a program file or a longer file of some sort to an e-mail message.

GEnie asks you to name the file you're sending. (Just use the file's regular name if you're not feeling creative.) GEnie then gives you the choice between two common file-transfer *protocols*; choose item 1, XModem (choose ZModem if your modem software supports this). Follow the other simple menu prompts, tell your modem software to upload the file when GEnie asks for it, and your file is on its way. Just fill in the To: area and tell the person getting the message what's in your file. Now's the time when you can add a quick, friendly message, too.

Figure 9-9 shows what it looks like when you first sign on and Mail tells you that someone has sent you a file. After you read a message, it's saved in your mailbox until you delete it. You can read saved letters by choosing Mail menu item 3, for example.

Chatting

To chat with live humans on GEnie, merely type the keyword **CHAT** and press Enter at any menu prompt. (Or you can use the nifty **MOVE** command to go to page 400, the chat page.) From there, follow the simple menus on-screen.

✔ GEnie asks its chat-ees to choose a nickname, also known as a *handle* in on-line parlance. Chat is one of the few places that encourages handles; and that's a Good Thing. (Handles are reviled in most of the on-line world.)

✔ GEnie's manual says that if you make your handle "interesting and distinctive ... it can be a great way to stimulate conversation." So, hey, go wild; bask in the excitement.

- ✔ Chat commands on GEnie are numerous, ranging from basic to advanced. Just remember to type a forward slash (/) character directly in front of any command you want to be "read" by GEnie and not by the chatting masses. In fact, type **/HELP** for a helpful list of chat commands.

- ✔ Strange "punctuation marks" and acronyms like ROFL ("rolling on the floor, laughing") or <g> ("grin") popping up on your screen may confuse you at first. People are only trying to save typing (the acronyms) or show feelings (the punctuation). The Outer Limits announcer was right: "There is nothing wrong with your television set. We control the horizontal. We control the vertical..." Just stay tuned for the Twilight Zone. (Or turn to Chapter 5's "Top Ten Smileys" for a guide to the weirdness, or at least a partial guide — the people chatting I *can't* explain.)

- ✔ Most on-line services offer members an adult chat channel or two; GEnie is no exception. Be warned; it gets rather, er, explicit there.

- ✔ Players dueling to the death in one of GEnie's multiplayer games can chat back and forth while they're killing each other. Type the keyword **MPGAMES** and press Enter to discover these vast (yet enjoyable) drains on your cash and time.

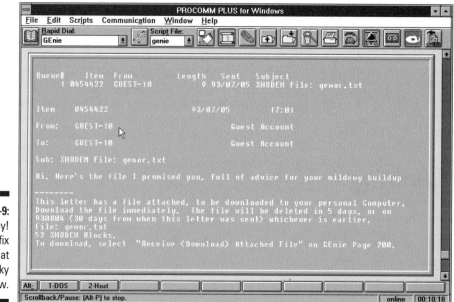

Figure 9-9:
Oh, boy!
Finally a fix
for that
pesky
mildew.

How Much Is This Costing Me, Anyways?

GEnie was known for having the lowest ($4.95) monthly rate of all the services until July 1, 1993, when it dropped the "Basic" membership package and went to more competitive pricing. Table 9-3 puts all this rate stuff into perspective.

Table 9-3	Rates, or Not-Ready-for-Prime-Time on GEnie!			
	U.S.		*Canada*	
Monthly Fee	$8.95		$10.95	
Includes	Four hours in all GEnie services except *Premium*. Hourly rates kick in after 4 hours per month base usage.			
Time of day	*Non-Prime Time*	*Prime Time*	*Non-Prime Time*	*Prime Time*
	6 pm-8 am weekdays, weekends, holidays	Surcharge 8 am-6 pm weekdays	6 pm-8 am weekdays, weekends, holidays	Surcharge 8 am-6 pm weekdays
All callers pay base rate	$3/hour base rate	$9.50/hour surcharge	$4/hour base rate	$12/hour
9600 bps Surcharge	$6/hour	$6/hour	$8/hour	$8/hour

Table 9-3 translation

Basically, after your four hours per month are used up, US members pay the base rate ($3/hour) plus any surcharges that apply.

> ✔ The main thing is to use it after 6 p.m.

> ✔ So, Table 9-3 is saying that if you live in the US and call GEnie after you've used up your standard monthly four-hour credit, and you're calling during a prime-time weekday at 9600 bps, you'll pay the **$3 base rate**, plus a **$9.50 Prime Time rate**, *plus* **$6** for **9600 bps access rate**, totaling **$18.50/hour** to use GEnie. But only if you don't veer off into any Premium services, where there are extra-extra charges too confusing to bother trying to tally.

> ✔ To check out GEnie's current rates, type **RATES** at any prompt and press Enter.

What are Premium Services?

GEnie, like the other on-line services, offers additional, extra-charge Premium Services.

- ✔ Stuff like Charles Schwab Brokerage Services, DowJones News/Retrieval, sending faxes, and Official Airline Guide travel services carry various surcharges.

- ✔ The current lineup of services may vary, and so may the prices... so be sure to check on rates before you creep across the border into these "extra-charge" services.

- ✔ To check out the current Premium Service rates, type **RATES** at any prompt and press Enter.

Phone access charges

Four phone networks currently connect to GEnie: Three of them levy their own surcharges on top of any GEnie rates you would normally expect to pay. The fourth, GEnie's own access network, sometimes charges, depending on whether you live in a smallish city. So, what to do?

Call Member Services at 1-800-638-9636 and ask for the cheapest access number for your location. (Or move.)

GEnie Tens

Even Magic Carpets sometimes need a road map. Here are some tips on making GEnie easier to use, plus some neat places to visit.

Making GEnie easier to use: the TOP command on GEnie

Type **TOP** and press Enter whenever GEnie's got you mesmerized. You'll be whisked right back to familiar turf — GEnie's TOP, home menu.

To escape from Mail, bulletin boards, or chatting, type ***EXIT**, **EXIT**, and **/EXIT**, respectively. *Then* you can type **TOP** and press Enter.

Try Aladdin

Versions: DOS, Amiga, and Atari brand computers

MACINTOSH

According to GEnie, a Mac version is shipping "real soon now" (geologically speaking), which is software-industry snidery for "in our lifetime."

Primary Virtues: Aladdin is the "front end" program for GEnie that helps cut those on-line hourly surcharges. Aladdin lets you read and compose messages off-line and gives you an easier "editor" program in which to do so. Using Aladdin helps you get around GEnie's bulletin boards and software libraries automatically — after you set it up.

Comments: GEnie doesn't charge for downloading Aladdin (depending on the time and modem speed at which you're calling, of course). Aladdin RoundTables for all Aladdin versions exist; each includes a Bulletin Board, Library, and RTC area. Head there for info on downloading Aladdin. Figure 9-10 peeks into the DOS Aladdin RoundTable.

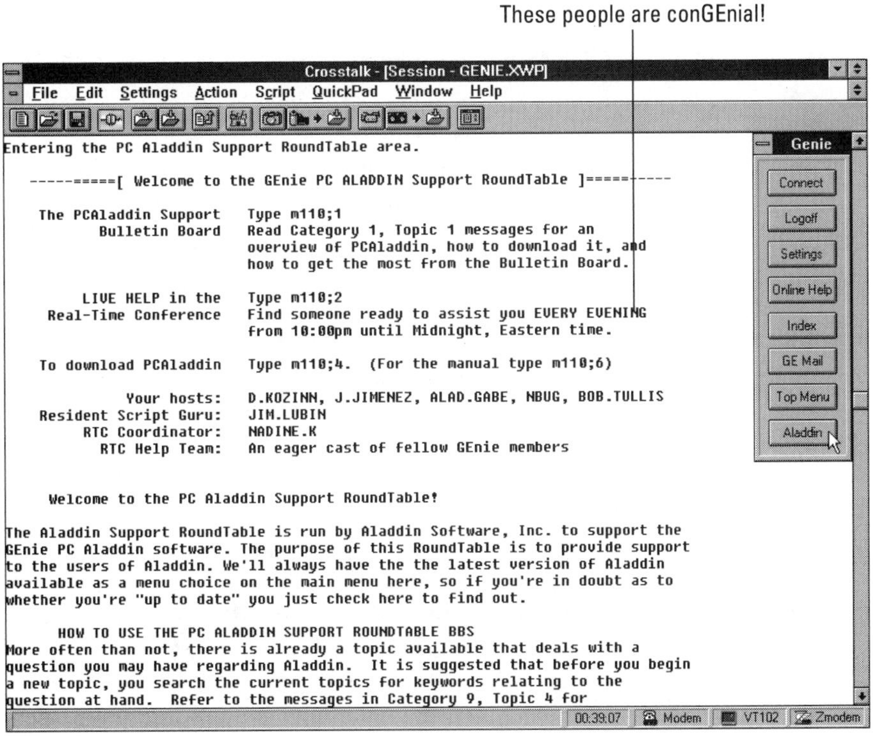

These people are conGEnial!

Figure 9-10: The Aladdin RoundTable is nothing if not helpful.

Ten practical and cool things to do on GEnie

Note: Be sure to press Enter after each word you type.

Type **MONEY** and browse Money Matters and Personal Finance for articles on mutual fund trends.

Type **AUTO** and get other Automotive RoundTable members to diagnose your car's latest *ping* noise.

Type **SFRT** to attend the next Science Fiction & Fantasy RoundTable Conferences — meet the author of *Quivering Quince Blossoms of Mars* or your other favorite visiting sci-fi scribes.

Type **JAPAN** and search the File Libraries for lists of Tokyo eateries that *don't* charge $50 US per steak.

Type **COMPUTING** to download computer sounds, pictures, and screen savers to present to your favorite Modem Guru.

Type **LIVEWIRE** and capture GEnie's bimonthly magazine so that you can look up new features and events from the convenience and thrift of your own hard disk.

Type **MAC** to meet and pick the brains of GEnie's Mac Gurus.

Type **REMINDER** and never feel guilty again for forgetting that special occasion: GEnie mails you automatic birthday, anniversary, and holiday reminders.

Type **WRITERS** and have the members of Writer's Ink RoundTable critique your novel's first chapter, leaving you resolved never to write another word as long as you live.

Type **NOMAD** to read the adventures of an ultra-nerdy guy who bicycles around the world on Behemoth, a giant computing/cycle, dashing off poetic notes about places and people he's met (and uploading them to on-line services via cellular modem — typing recumbently as he rides!).

Type **MEDICAL** and download ASPIRIN.GIF (a photo of the aspirin molecule); download FDA_INFO.ZIP and catch up with all the over-the-counter medicine recalled from drugstore shelves; or just ask the resident meds about that ache you've been having lately... Just kidding about that last, folks — those meds don't want to leave themselves open to any electronic malpractice suits.

Chapter 10
Top Ten Other Guys to Call

. . . .

In This Chapter

▶ DELPHI

▶ MCI Mail

▶ Dow Jones News/Retrieval

▶ The ImagiNation Network

▶ ZiffNet

▶ BIX

▶ Lexis

▶ OAG Electronic Edition

▶ Nexis, DIALOG, and Knowledge Index

▶ The WELL

▶ Top Ten Cows

. . . .

*T*he big on-line services described in the previous chapters are like the department stores anchoring a typical shopping mall: You're likely to try shopping at one of the "Big Guys" first. They're easy to find and they carry a little of everything.

You know you'll find just about the same stuff in each "store" — at nearly the same prices. (Okay — let's forget about Neiman-Marcus, for illustration's sake.)

Sometimes, though, you need pesky specialty items you'd never find at the mall. For hand-tied fly-fishing lures, for example, you'd shun Kmart and trot over to Captain Hank's Fishy Flies instead. Or, if your prized Sicilian Cauliflower calls for a grating of stinky caciocavallo cheese, you'll head for Luigi's Deli, not Food Hulk. And when your kids clamor for the grittiest, zittiest band's new T-shirt, you'll find yourself treading the paint-splotched floorboards of the local Heads-R-Us, not the carpeted aisles of Bloomingdale's.

The on-line world works pretty much the same way. This chapter describes some of the more "boutique" on-line services and what they have to offer. When you've fine-tuned your information needs to a pinpoint, try one of these specialty "shops."

DELPHI

DELPHI offers its smaller community a slightly "tighter" range of the same basic stuff you get with other on-line services.

- ✔ DELPHI's special interest groups aren't called "oracles" or anything cool: They're plain old forums.

- ✔ Callers probe the burning issues of the day in bulletin board messages and e-mail; they trade files relating to their interests. (Refer to Chapter 5 for *e-mail* and *files*.)

- ✔ Members converge in electronic "rooms" and type at each other in conferences, chat sessions, and multiplayer games — just like people on the biggie on-line services. (These chat "rooms" aren't named after the Acropolis or other famous Grecian ruins, either.) Refer to Chapter 5 for the scoop on conferencing, chatting, and games.

- ✔ DELPHI's smaller size may have something to do with its being one of the most intimate and least expensive of the on-line services.

- ✔ So, what's different about it? DELPHI's biggest distinguishing feature is a low-cost link to the Internet, a global web linking millions of computers — even ones at the North Pole.

- ✔ At this writing, DELPHI's members pay a pittance ($3/month) for complete Internet access during evening and weekend (non-prime time) hours. (For a translation of *complete Internet access*, read the following sidebar. To learn more about the Internet, troop over to Chapter 12.)

- ✔ Head for the Internet chapter to find out more about *mailing lists*, *newsgroups*, and other renegade terms laying siege to the preceding sidebar.

- ✔ Check the coupon in back of this book for rate information and other stuff on connecting to DELPHI. If someone beat you to it, you can call 1-800-695-4005.

Mail Call! MCI Mail

If sending and receiving electronic mail (e-mail) is your sole goal, check into a mail specialty service called MCI Mail. MCI Mail subscribers receive an electronic *mailbox*, similar to the one you get when you join an on-line service. When you have mail, MCI tells you so the moment you sign on.

INNER SANCTUM

They should call it the "In"-ternet

The Internet: Everyone wants it but no one knows what it is.

Whether people really *need* access to the millions of connected, "internetworked" computers that are the Internet is beside the point; what's certain is that it's hot and growing hotter.

Although most on-line services offer a way to send and receive e-mail, DELPHI was one of only two general services at this writing to provide its members with "complete" Internet access. (Lots of the other services were talking about it, though.)

What exactly does "complete Internet access" mean?

It means you get:

- ✔ The Internet e-mail connection that many of the other services provide

- ✔ *Plus* the capability to send files to and receive files from Internet computers

- ✔ To "sign on" to Internet computers through your DELPHI connection (and search for information or just look around)

- ✔ To obtain electronic "mailings" of information relating to your special interests by asking people on the Internet to let you join their *mailing lists*

- ✔ To follow discussions and contribute your own nuggets of wisdom in special interest areas called Usenet *newsgroups*, which are sort of comparable to the forums on other on-line services (not all Internet accounts come with news "feeds," however).

- ✔ Muddying further these turbulent waters, the newsgroups just mentioned are actually part of a different computer network, called *Usenet.* Among other differences, the Usenet network includes more personal computers than are usually counted among the Internet's academic- and scientific-dominant computer population.

- ✔ No matter how "incorrect" it may be, most everyone thinks of the Internet and Usenet as being just about the same thing — mostly because Internet access *usually* includes a *feed*, or *stream*, of Usenet news. (Some Internet connections don't offer newsgroups.)

That's why you'll see people who admit to these things referring to themselves as *being on The Net* when they're connected to either one.

- ✔ MCI members swap messages worldwide with other MCI subscribers — as well as with CompuServe, GEnie, SprintMail, AT&T Mail, and Internet subscribers.

- ✔ MCI Mail calls the "letters" you send and receive *Instant Messages*; these cost $.50 for the first 500 characters and go up in 10-cents-per-thousand-word increments. Additionally, you pay a $35 per year "base" fee to join MCI Mail. Because a person can easily spend that much per month on a general on-line service, this is a good deal for those who just want mail.

✔ For an extra fee, MCI can send faxes for you or print out and mail "real" letters through the U.S. Postal Service. If you ask real nice (and pay a real lot), they'll even have a courier hand-deliver a letter somewhere and impress the bejeezus out of someone.

✔ Help's available at any point on the service by typing **HELP**.

✔ You can buy MCI Mail Express, a software "front end" that makes composing and sending messages easier. Besides creating messages while you're off-line, MCI Mail Express lets you check messages for spelling errors, as well as store e-mail addresses in a built-in address book. It dials the service on command after you configure it to work with your modem, too — similarly to the way a general communications program works.

✔ AT&T Mail, SprintMail, and other services compete with MCI Mail. These guys offer about the same services for somewhat similar prices.

Contact MCI Mail at 1-800-444-6245.

SprintMail can be reached at 1-800-736-1130.

AT&T Mail's number is 1-800-624-5672.

✔ As a member of MCI, you can access portions of the serious big-time investor-oriented Dow Jones News/Retrieval financial service — for approximately $2 per minute. For more on Dow Jones, see the next section.

The Business-Meister: Dow Jones News/Retrieval

If you need the same corporate and financial information as the big Wall Street brokers, why not go to the source they use? (No, you don't have to endure a power lunch or a jail visit with one of those junk bonds guys.) For up-to-the-minute, on-line information, you may want to check into a Dow Jones News/Retrieval membership.

✔ Dow Jones contains the full text of articles from the *Wall Street Journal*, *Barron's*, *Business & Finance Report*, and all those other publications the business majors always hoarded during finals week.

✔ You can search for information tucked away in articles from 1,300-plus publications by using any of Dow Jones' 60 specialized publication collections, or *databases*. (For more on databases on-line, turn to Chapter 5.)

✔ Dow Jones has news pouring in through five different news wires. And stock quotes, too, providing subscribers with late-breaking market reports of national and international exchanges. The service offers MCI Mail, the Academic American Encyclopedia, and the OAG Travel Service, among others.

✔ Dow Jones subscribers should seriously consider the extent of their *need* for any information they gather on-line. It's possible to rack up astronomical charges of more than $200 an hour on this service.

✔ In an effort to attract more personal investors and home users, Dow Jones developed the "After-hours/Flat-fee pricing plan." This may be a good way to break in, considering the intimidation potential of sky-high rates multiplied by user-unfriendly menus and prompts.

✔ Several of the general on-line services offer "lite," slightly less comprehensive versions of Dow Jones News/Retrieval.

✔ Contact Dow Jones at 1-800-522-3567 to learn more.

Just Playing Around: The ImagiNation Network

If you've played Space Quest, Leisure Suit Larry, King's Quest, or any of Sierra Online's computer games, you've already set foot in The ImagiNation Network.

Don't look for gruesome text menus. Instead, ImagiNation sprinkles colorful graphics across your screen, turning it into a carnival. Logging on to The ImagiNation Network is like turning your computer, modem, and phone line into a pack of Fun-Zone ride tickets.

✔ The big difference between buying a game for your computer and logging on to ImagiNation comes in finding teammates, opponents, and companionship in the other members who call ImagiNation. It's like playing a computer game, but having dozens of other people across the world playing along as well.

✔ In one section, for example, you'll find LarryLand. Named after one of Sierra's more popular adult-oriented games, it draws adults from all over to get together and swap bar talk as they compete against an on-line Blackjack dealer or spin Roulette wheels for "LarryBucks."

✔ Adults pursuing tamer pleasures can head for the Clubhouse and find on-line opponents to challenge in Go, chess, cribbage, and the other traditional card and board games.

✔ The kids aren't left out, though. They can visit the Clubhouse and whomp their elders in chess, or hike over to SierraLand's arcade and action games. They can pick Boogers, for instance: a Reversi-style strategy game. Or they can choose Paintball and MiniGolf.

✔ Even the flight simulator crowd will enjoy SierraLand's Red Baron, where they become World War I aces as they zoom their planes around the screen and try to blow each other up. SierraLand's Red Baron offers speed and graphics comparable to the stand-alone version you buy in the software store.

✔ Fans of fantasy and role-playing games should try MedievaLand's Shadow of Yserbius, where you create your own character to explore the misty volcano — complete with membership in a guild, a grisly mask, and attributes like strength, agility, or initiative.

✔ ImagiNation compounds its overall fun-and-games aura by letting you choose and design different personas for each "Land" you visit. People can burn up hours of connect fees playing with their on-line hair color, clothing, facial expression, nose size, and other visual imagery. Be a boy, a girl, a hot mama, or a leisure-suit dude ... even a wizard. No one will suspect it's really just you!

✔ To enjoy all these pictures (and some sounds, if you have a sound card), you need to install ImagiNation's special free software. Don't try to run ImagiNation's software on anything less than a 386SX with 640K of memory, a hard disk with 9M of free space, a 2400 bps modem, and an EGA color monitor. (Shadow of Yserbius requires a VGA color monitor.) At this writing, ImagiNation doesn't come in a Mac version.

✔ Prodigy has announced that it will offer a gateway to some of ImagiNation's Lands.

✔ There's a coupon in back of this book for a free trial membership to ImagiNation.

✔ You can reach ImagiNation at 1-800-SIERRA-1.

ZiffNet and ZiffNet/Mac

Zounds! It's the ZiffNets — a dynamic duo on-line service package created by Ziff-Davis, publishers of *MacUser*, *PC Magazine*, *PC Week*, and thick bunches of other popular computer mags. Like most on-line services, the ZiffNets offer files, messages, forums, and conferences. The ZiffNet offerings focus pretty much exclusively on computer software, hardware, news, and opinion, however. After all, what could be more appealing than computers?

✔ While on-line, members can search through magazine back issues or swap messages with the editors of ZD's many publications. Sadly, Zsa Zsa Gabor doesn't have a Zsa Zsa's ZiffNet weekly column on-line.

✔ ZiffNet houses a giant collection of shareware programs, demo software, and other files you can download and use on your computer.

✔ You can access ZiffNet through two popular on-line services for an extra fee on top of the service's base fee. On CompuServe, use the GO word **GO ZNT**. Prodigy members can use the JUMP word **ZIFFNET** to access Prodigy's version.

✔ Call 1-800-848-8990, and a pleasant CompuServe representative will give you rates and access info for ZiffNet and ZiffNet/Mac.

BIX

In the unlikely event that any hardened computer nerds are reading this, they'd chastise me heavily if I left out BIX, the BYTE Information Exchange. Founded by McGraw-Hill as the on-line version of its *BYTE Magazine*, BIX is one of the oldest and most beloved of the on-line services.

✔ BIX aims straight for the computer-nerd jugular, abandoning all the EAASY SABREs, Grolier's Encyclopedias, and weather maps to the more general on-line services.

✔ On BIX, you can read back issues of *BYTE*, an extremely technical computer magazine that nerds drool over. You can swap e-mail, attend on-line conferences and forums, download software, and read news about the computer industry. BIX's offerings are almost completely devoted to computer stuff.

✔ BIX recently started offering full Internet access.

✔ Next to CompuServe, BIX may be one of the best on-line services to call when you need help with computer problems.

✔ Recently, DELPHI's owners acquired BIX, but it's still run as a separate service. Contact General Videotex Corporation at 1-800-695-4005 to learn more about BIX.

Lexis: It's the Law!

If you need to get your hands on legal statutes, codes, and regulations, you're either in deep trouble or you're a lawyer. Either way, if you need this type of information often, consider getting your hands on a Lexis account.

✔ It's pronounced just like the luxury car but it's all Law — 300 databases bristling with briefs and rulings and other aracana of interest only to the legal professional (or the career criminal of a strategic bent).

✔ Lexis offers sizable searching grounds, matched only by its sizable fees and a daunting command system.

✔ This extremely specialized information-retrieval service is geared toward the legal professional. If that's you, call 1-800-227-4908 to learn more. The rest of you can continue peacefully browsing the remaining sections.

Traveler's Aid: OAG Electronic Edition

One sure sign of our hands-on times is how the OAG has evolved from a travel agent's computer tool to computerized travel agent. People seem to enjoy the autonomous feeling brought about by searching for cut-rate air fares and booking their own rent-a-cars.

✔ OAG is geared toward frequent flyers, but members can also spot cruise bargains, find out what's going on in far-away lands, and browse the government's travel advisories warning against vacationing in war-torn and strife-ridden nations.

✔ You can access OAG "lite" through many of the general on-line services.

✔ You can call OAG with general modem software; PC users can call it with OAG's special "front-end" program, Auto Logon.

✔ Subscribers pay a one-time sign-up fee and a per-minute rate, reduced for access during evenings and weekends.

✔ Contact OAG at 1-800-DIAL-OAG for current rates and fees.

Nexis, DIALOG, and Knowledge Index

Nexis and DIALOG would both make great challengers on "Jeopardy". Both services shelter hundreds of separate information databases. Unlike Lexis — the legal trivia specialist — these databases dabble in every conceivable interest and field.

If you seek information — no matter how obscure — one of these services is likely to harbor it.

✔ To give you an idea of its immensity, DIALOG keeps tabs on more than 300 million summaries, articles, and abstracts in fields ranging from agriculture through the social sciences.

✔ For a better idea of its complexity, DIALOG includes *training classes* or a how-to *video* as part of its $295 (!) sign-up fee, which includes $100 worth of on-line time. Renewing your DIALOG membership costs $75 per year.

✔ An average search will take you from 4 to 8 minutes, and cost anywhere from 40 cents to $5 per minute, depending on the database you're flipping through.

✔ Nexis and DIALOG serve as vast "database clearinghouses" — huge switchboards, if you will. These *gateways* let you access thousands of separate databases without making thousands of separate modem connections. And unlike other popular clearinghouses, there's not nearly as much junk mail.

✔ Knowledge Index is the name for DIALOG's after-hours version. It offers fewer databases at a far lower cost.

✔ You can access Knowledge Index through CompuServe during evenings and on weekends. Knowledge Index stands out for having one of the easiest menu systems. To find out more about Knowledge Index, call CompuServe at 1-800-848-8199.

✔ Contact DIALOG at 1-800-334-2564.

✔ Call Nexis at 1-800-227-4908.

The WELL

Seeking fellowship and forums rather than file downloads? The WELL may be the service for you. The Whole Earth 'Lectronic Link offers members a progressive atmosphere full of topical debate and interchange.

✔ The service offers over 200 interest groups, or *forums*, touching on just about any interest imaginable, from cooking to virtual reality.

✔ Complete access to the Internet and Usenet newsgroups comes with membership in The WELL.

✔ If you're a student (or rocket scientist) or you already have Internet access for some other reason, you can jump in The WELL by using the Internet telnet command: **telnet The WELL**. Refer to Chapter 12 for more about the telnet program.

✔ You can find out more about current rates or your local access number for The WELL by calling 1-415-332-4335.

Top Ten Cows

```
        (_)

       (oo)  <----Cow

    /-------\/

   / |    | |

  * ||----||

    ^^    ^^
```

Smileys left too long on the range? Beef by-product of bored programmers? Or simple cows?

Sooner or later, everyone en-cow-nters one of these creations on-line. They cavort and gambol through on-line messages, newsgroup "articles," and anywhere else the electronic bits and the antelopes play.

What are they, and what do they mean? They're the Cows of the Net. Where did they come from? Well, the point is "moo-t" ... but legend has it that one day, programmer Steve Schultz was showing another programmer how to make characters blink and stuff. Steve needed something to draw. Voilà, a Cow was born. Someone picked up Steve's doodlings and "steered" them onto another computer. Because that computer was connected to the Internet, soon hundreds, then thousands of people were having a cow, man.

The whole thing stampeded. Everyone was adding new cows to the herd. In the spirit of the Net, Cows are *public domain* (they're free-ranging cows) and belong to everyone.

You too can contribute new cows: Send them to David Bader at the Interent address in the box that follows. "When drawing cows, please DO NOT use tabs — use only "real" spaces," says David.

While you're at it, send a word of thanks to David for "editing" and posting new cow-reations to the Net, to Eric for being in the credits, and to Stephen for creating the first calf for everyone to cowment on and enjoy. Thanks, guys!

Stephen L. Schultz	(SLS4255@RITVAX.isc.rit.edu)
Eric W. Tilenius	(etileniu@Oracle.COM)
David A. Bader	(dbader@scarecrow.csee.lehigh.edu)

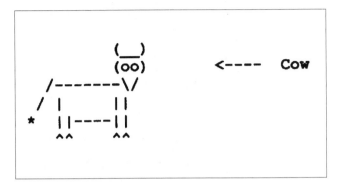

```
       (__)
       (oo)            <----   Cow
   /-------\/
  /  |    ||
 *   ||----||
     ^^     ^^
```

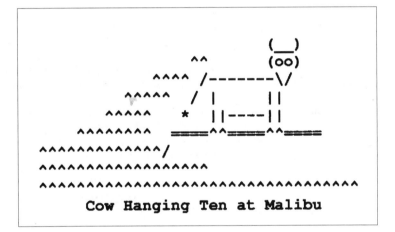

```
                          (__)
                 ^^       (oo)
             ^^^^ /-------\/
          ^^^^^  /  |    ||
         ^^^^^  *   ||----||
      ^^^^^^^^    ===^^===^^====
  ^^^^^^^^^^^^^/
 ^^^^^^^^^^^^^^^^^^^^
^^^^^^^^^^^^^^^^^^^^^^^^^^^^^^^^^^^^^^^^^^
```

Cow Hanging Ten at Malibu

```
       o___o
       (oo)
   /-------\/
  /  |    ||
 *   ||----||
     ^^     ^^
```

Cow at Disneyland

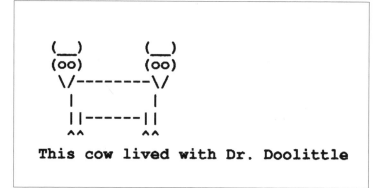

This cow lived with Dr. Doolittle

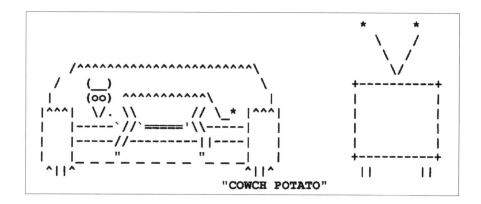

"COWCH POTATO"

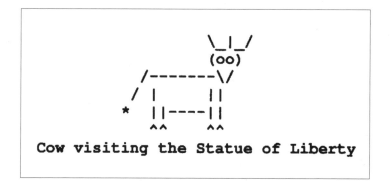

Cow visiting the Statue of Liberty

```
              (__)
             (oo)
    /-------vv
   /  |     | |
  *   ||----||
   ~~        ~~
    bela lugosi cow
```

```
                (__)
               ($$)
      /-------\/
     /  |=====||
    *   ||----||
     ^^        ^^
   This cow is a Yuppie
```

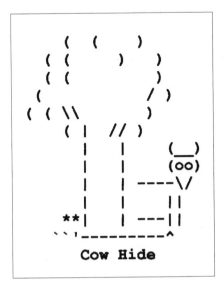

```
      (  (    )
     ( (    )   )
     ( (        )
    (          / )
   ( ( \\        )
    ( |  // )
      |   |      (__)
      |   |     (oo)
      |   | ----\/
      |   |     ||
    **|   | ---||
   `` `|_____^
        Cow Hide
```

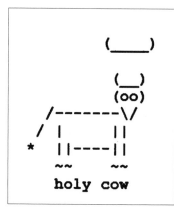

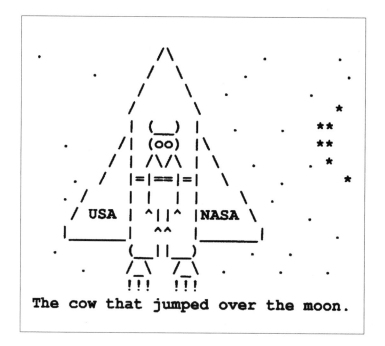

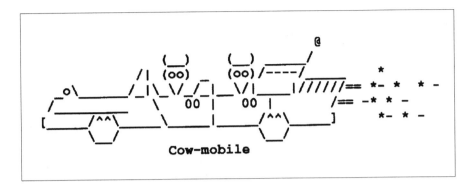

Cow-mobile

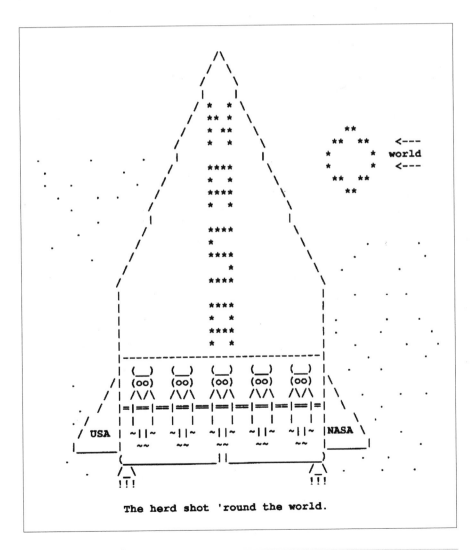

The herd shot 'round the world.

BEWARE -- Tagline thief in this echo.

Chapter 11

Electronic Bulletin Board Systems: Where Do All the Pushpins Go?

● ●

In This Chapter

▶ Understanding what a BBS is

▶ Logging on and logging off

▶ Finding stuff on the main menu

▶ Calling a company's customer support BBS

▶ Calling the computer at work

▶ Calling a hobbyist BBS

▶ Top Ten Mondo Bizarro BBSs

● ●

*E*lectronic bulletin board systems, or BBSs, offer varying mixes of the same things people look for anywhere else on-line: files, chatting, e-mail and messages, special interest groups, and other members. Just on a smaller scale.

Calling a local BBS usually doesn't cost anything. That's why BBSs are a natural choice for many modemers taking their first teetering ride on-line.

BBSs aren't just for the novice. Some folks never call anywhere else, preferring the BBS scene over the commercial on-line services they migrate from. And many business offices — especially those with people "in the field" — create BBSs just for employees to call.

Creating a BBS doesn't necessarily cost a lot either. Anyone with a computer, modem, and phone line can start a BBS and become its *sysop*, or system operator. That explains another thing about BBSs: like their creators, each one is different. Some are weird, some are cool, some are power-trips, and some are purely business.

This chapter tugs at the veil of that most mysterious on-line entity, the BBS.

What's a BBS?

A BBS is a computer that someone has set up with a modem, phone line, and special BBS software so that it can take calls from other modem users.

BBSs have their analog, "real-world" equivalent in the cork bulletin boards found in Laundromats. There you can pin up your index cards saying, "Darling fluffy kittens, Absolutely *FREE!* to Good Home," and read messages left by others.

✔ Although they lack pushpins, BBSs offer much more than their cork counterparts. BBS software, the special software that BBSs run on, provides five main areas of interest: file downloads/uploads, forums, chatting, on-line games, and e-mail. Whether a particular BBS offers some or all of these things is up to its sysops — the people who run it.

✔ More than 13 million North Americans access some 53,000 electronic bulletin boards on a regular basis, looking for fluffy kittens.

Who Runs These BBSs?

Computer hobbyists run most of the BBSs you'll see. You can call Percy-Down-the-Street's BBS and find cool games to download and play on your computer. You might dial up the Au Naturel BBS and leave messages debating the best suntan lotion for nude sunbathers. Or you can discover BBSs targeted to entrepreneurs, veterinarians, or other professionals.

✔ Some cutting-edge sysops run 24-hour, 7-days a week super BBSs on advanced computers, using high-speed modems and BBS software tuned to within an inch of its life. Hundreds of modems operate at once, granting access to hundreds (naturally) of callers and vast file accumulations. These BBSs usually charge subscription fees.

✔ Other BBSs might run on alternate Saturdays — out of someone's garage — on ancient dinosaur computers linked to creakingly slow modems. There's probably a BBS fired up and taking callers on every type of computer that exists.

✔ A BBS reflects the goals and personalities of its sysop owner — for better or for worse. Hobbyist BBSs can specialize in freewheeling message boards and discussion groups linked by BBS networks to the entire universe. Others concentrate on making the latest, hottest shareware programs available to its callers. Still others specialize in X-rated "cheesecake" or "beefcake" photo-images and adult-oriented chatting — even matchmaking.

✔ Is it just me, or does *beefcake* sound incredibly unappetizing?

✔ **Sysop joke:** Sysop (sis'op), n. : the person laughing at your typing.

✔ Most major computer companies run BBSs devoted to customer support and public relations. Software "fixes" lurk in the file areas for customers to download. Customers can head for the message boards to discuss their experiences and suggest improvements. The company can welcome callers with announcements touting the latest products or milestones.

✔ Your modem's manufacturer probably runs a BBS. There, you can call and leave e-mail questions for tech support. Or you can download a "modem driver" program that prods your modem software into cooperating with your modem.

✔ Calling the computer at work can come under the broad BBS heading, because the software is likely to be similar. Employees can transfer e-mail or access files off the network, for example. Your office may have such a BBS set up.

✔ Most BBSs are small, no-charge affairs run by hobbyists seeking the shivers of delight found only in running a BBS. No one can explain why, but sysops get *big whoops* out of devoting an entire computer and lots of spare time (and no small amount of spare cash) to setting up a BBS and watching perfect strangers call up and complain about "what a grouch the sysop is" to other perfect strangers.

✔ BBS friendships have been known to run on for years without the parties ever meeting face to face! (This *may* be a Good Thing.)

Preparing to Call a BBS

Each BBS is different from all the others. Besides the obvious differences in content and philosophy, a BBS can be running one of 20 or so different BBS software programs, on any of a dozen types of computers.

Each brand of BBS program greets you with a unique "look" and a different set of commands. (You were thinking this was going to be easy or something?)

✔ Calling a BBS is like making a "voice" telephone call, except that you're using your modem, communications software, and computer.

✔ If the BBS's number is a long-distance phone call for you, the phone company will charge you long-distance rates while you're connected to the BBS.

✔ When you're using your modem, you're hogging the phone line. Anyone who calls you will hear a busy signal. Unless you have "call waiting," that is. Then, your caller will get through — and your modem will be tossed off-line quicker than you can say, "Uh oh."

✔ Do you have call waiting as part of your phone service? Then make your modem turn it off automatically each time it makes a call. Chapter 4 gives the scoop.

✔ Brandish a rotten rutabaga or do whatever else is necessary to keep other household members from picking up any extension phones while you're on-line — and knocking *you* off-line.

✔ People seriously into "BBSing" end up installing a second, "data" phone line for any or all of these reasons.

Matching the BBS's settings

You can dial most BBSs with your general modem software. One or two require fancy front-end software, à la Prodigy. If so, the BBS greets you with screens telling how to download this software, set it up on your computer, and call back.

But for the most part, just make sure that your modem software's settings match the BBS's settings. That means you need to find out the following things about the BBS:

✔ Phone number (this goes in your software's dialing directory)

✔ Modem speed (try calling at the fastest speed your modem supports)

✔ Communication parameters: data bits (8); parity (None); stop bits (1)

✔ Terminal emulation (try PC/ANSI first, then VT-100/102, and finally TTY, if nothing else seems to work)

When calling another computer, you're dealing with two sets of software: The *modem software* on your computer drives your modem. The *BBS software* runs on the computer you're trying to call — the *host* computer.

Different BBSs often use different settings. How do you keep track? By creating a new setting or dialing directory entry in your modem software for each BBS or on-line service you call. Choose "new setting" or something similar from your software's menu, and then give it the BBS's name.

Then record the BBS's phone number and any alternate phone numbers. Specify the best terminal emulation, the data parameters, modem speed, and any other particulars for that BBS. Finally, save the setting. Then when you want to dial the BBS again, everything will already be set up correctly.

Settings are handy for people who call several services that require different terminal settings or modem speeds. Head to Chapter 4 to learn more about modem software styles of the rich and famous.

`I'd love to, but I have to answer all my "occupant" mail.`

Trying to connect

After saving the settings for a BBS, press the keys or click the spot that tells your modem to start dialing. Keep an eye on the computer screen to watch the action.

- First, you hear (and see) your software dial the number, prefacing it with the ATDT command:

```
ATDT287-8616
```

- Pay close attention with your eyes and ears, especially when the screeching modems fall silent.

- Hear a busy signal? See the word Busy appear on your screen? Somebody else is already using that BBS. That happens pretty often in BBS-land, especially with BBSs that only offer one or two phone lines. Set your modem software to redial — or just try again later.

- Peak BBS hours commence right after school/work lets out. If you *must* connect with a busy board, try calling before 5 p.m. or after midnight.

- Hear a human voice answer, "Hello"? Hang up right away and don't call back to "make sure you've got the right number." Nothing irritates someone more than picking up the phone to hear an eerie silence on the other end. BBSs come and go, and you may have an old number.

- I once worked at a computer magazine that published a list of regional BBSs. In one issue, we accidentally listed a guy's home phone number next to a BBS's name. After receiving incessant calls at all hours from hundreds of modems, the poor guy called us and complained. We promised to correct The List in time for the next issue. In the next issue, to our dismay, the guy's phone number had crept in *again*. Fie!

  The publisher threatened to dock the editors' pay for the cost of the phone company installing the new phone number the guy insisted upon — and got.

  If you come across an irate human in your BBS travels, try to notify the people who listed that person's phone number by mistake.

- If all goes well, you should see the word Connect from your modem software. Then a greeting or some sign of life usually manifests itself from the BBS. If fortune is smiling upon you in this manner, head for "Logging On and Logging Off," later in this chapter.

Can't Get No (Modem) Satisfaction

Alas, instantaneous, happy connections like the one in the preceding bulleted item are rare. Because BBSs present so many variables, it's more likely you'll see one of the following:

✔ See nothing after Connect? Press Enter a few times if you see nothing but a blank screen and your flashing cursor. This should nudge the other computer into action.

✔ If a bizarre stream of hieroglyphics dances across your screen, your settings don't match those of the BBS computer. You need to press the keys that make your modem software **hang up**. Now look at Figure 11-1. What kind of garbage are you getting?

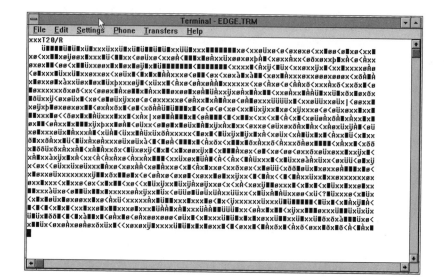

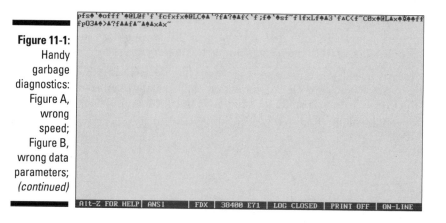

Figure 11-1:
Handy
garbage
diagnostics:
Figure A,
wrong
speed;
Figure B,
wrong data
parameters;
(continued)

Alt-Z FOR HELP | ANSI | FDX | 38400 E71 | LOG CLOSED | PRINT OFF | ON-LINE

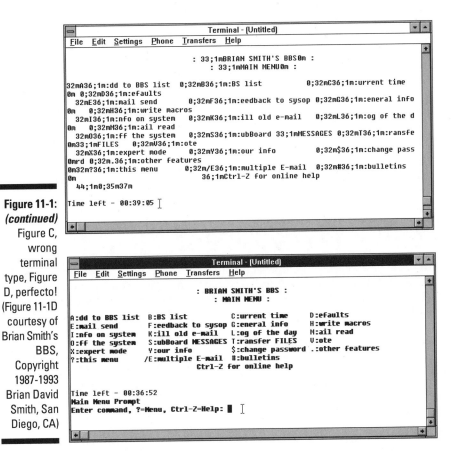

Figure 11-1:
(continued)
Figure C,
wrong
terminal
type, Figure
D, perfecto!
(Figure 11-1D
courtesy of
Brian Smith's
BBS,
Copyright
1987-1993
Brian David
Smith, San
Diego, CA)

Now it's time to try the adjustments to your modem software's settings shown in Table 11-1. (Just change *one* setting before redialing the BBS; if it still doesn't work, try changing another one. Altering everything at once will promote confusion and that *lost* feeling.)

Table 11-1 Garbage on the Screen? Change One of These

Fig.#	Setting	Try This	Example
A	Modem Speed	Drop your speed; your modem is not "negotiating" automatically	Go from 9600 bps to 2400 bps
B	Parameters	8,N,1	Go from 7,E,1 to 8,N,1
C	Terminal Type	ANSI	Go from TTY to PC/ANSI
D	Perfecto!	Clap your hands in delight!	

Keep your words soft and sweet, in case you have to eat them.

- ✔ Don't just turn off your computer or modem at the first signs of trouble connecting. This can mess up the BBS computer temporarily and does nothing to improve your future relations with the BBS's sysop.

- ✔ If you can see part of what you're typing but gibberish appears randomly out of nowhere (and you've had only one mint julep), your settings are okay; you're just experiencing a noisy phone line, as in the following logon. The gibberish has the wildly imaginative name of *line noise*. Just hang up and call back later.

```
User name: TinÇ◆Σathbone ÇÇÇÇÇ◆◆
```

Logging On and Logging Off

Now that you've jolted the BBS's computer into action, it wants to know who's stumbling around on its front porch. Telling the computer about yourself is known as "registering" on the BBS. Type your name, address, and other information at its promptings.

- ✔ You're giving out your real name, address, and phone number to a stranger each time you register on a new BBS. (Unless it's the BBS at work, of course.) Most sysops strive to reassure callers that their private information will stay private. Other sysops may openly sell your information as a mailing list. A few scum may secretly sell your name, and so on. If this "trust factor" bothers you, calling BBSs may not be for you.

- ✔ Some BBSs ask new callers to type **NEW** as a user name. Then you go on to register.

- ✔ Sensitive BBS types give you the choice of logging on as "guest" and looking around before you register right away. That lets you avoid answering all those pesky questions until you're sure you want to join. (Especially for the harried residents of the Southwest, who can bypass typing out their ubiquitous Spanish street names, like "Caminito San Luis Obispo de las Estrellas," each time they're prompted for their street address.)

- ✔ Calling the office BBS? Your office guru may have given you a special user name and perhaps a password. Try not to write these down on a sticky note and affix it to your monitor at work. You don't want coworkers broadcasting randy e-mail limericks under your name.

- ✔ Unless you're told to use a nickname — or *handle*, in BBS parlance — don't. Always give your correct name and phone number. Some BBSs will log you off shortly after you do so; the sysop will then call you to verify that you gave your real phone number.

✔ Some fancy BBSs automate the "call-back verification" routine: Their modem automatically calls you back the first time you sign on, to verify your phone number. The BBS software asks you to disconnect, but before you do, it tells you how to set up something called "manual-answer" to leave your modem "door" open. When you see RING on your terminal screen, type **ATA**, and then press Enter. Your modem answers the phone and the BBS keeps asking new questions.

✔ Eventually, the BBS asks you to dream up a *password*. Think up something that you can remember but that no one else will be able to guess. Unlikely combinations of food groups, punctuated by some offbeat punctuation marks and at least one uppercase character, will do.

✔ ***Be very, very, very careful*** about giving out your credit card number on a BBS, even on pay BBSs that charge members' credit card accounts as a normal payment method. If you join a pay-BBS, consider paying by traveler's check, even if it means delaying your log-on gratification for a few weeks until your check arrives and clears.

✔ After distributing illicit copies of commercial software (*software piracy*), credit card/calling card fraud is the most common crime committed by on-line "phreaks" (*phone freaks*) and "crackers." These crooks use their modems to crack other computer systems the way some criminals crack safes. They dig up account numbers and charge vast monetary sums against them, before distributing them on-line — on "underground" BBSs — for even more crooks to access.

These people think they're cool, but they're merely common criminals. There's no need for you to be unduly alarmed. Despite media hype, these incidents happen relatively infrequently and it is unlikely to happen to you. Just keep all your account numbers — "plastic" and otherwise — to yourself.

Telling "all" about your computer

You've replied to interminable questions about your name, age, birth sign ... and now the BBS is prompting you to cough up yet another round of confessions. Common questions and answers about your computer and display appear in Table 11-2.

Table 11-2 Computer Settings to Tell the BBS Computer

Question	Answer to Try First
Number of characters per line? (also seen as "Terminal Width")	80
Number of lines on screen? (sometimes called "Terminal Height")	24

(continued)

Table 11-2 *(continued)*

Question	Answer to Try First
Terminal type?	ANSI
Can your computer display upper/lowercase letters?	Yes
Video adapter, or video display?	Tell what type of video adapter your computer has; answer "VGA" if you have a VGA monitor, for example
Page pause	Yes
Number of nulls	0
Line feeds?	Yes (most Mac users can say No)
Default file-transfer protocol	ZModem, if your software supports it and the BBS offers it
	If you choose None, you can select a protocol prior to each transfer — kinda nerdy, though.
When in doubt, accept the default.	

Other fine points of display dispute

The settings in Table 11-2 should get you started. Most BBSs provide a way to make changes to your settings in future calls, should you confuse everyone by switching computers or something.

- ✔ If you're at a "dumb" terminal that usually connects with a mainframe computer, use "dumb terminal" for your terminal type. (If that's not listed, try TTY, or sometimes "glass terminal," that most generic of terminal settings.)

- ✔ If you see a Hot Key or similar option, choose it. You can type a letter that performs the menu item's command without having to press your Enter key after every command.

- ✔ SEEIINNGG DDOOUUBBLLEE of everything you type after you're on-line? Your modem software is valiantly trying to "echo" all your immortal prose. Toggle your software's Echo On setting to Echo Off. Some software labels this Half Duplex — if your software's menus say Duplex, try changing this to Full Duplex.

- ✔ Can't see anything you're typing? Then toggle your software's Echo Off setting to Echo On. (Sometimes it's a case of putting an X in a box next to the word Echo — no X means no Echo.) If your software calls this feature Duplex, change it to Half Duplex. GEnie is one of the few on-line services that requires this adjustment, incidentally.

The person who is all wrapped up in himself is overdressed.

✔ For gory definitions of Echo and other intimate interludes with your modem software, head to Chapter 4's section "Calvacade of Software Settings."

✔ If you're seeing double-spaced or run-together text, or having other bothersome BBS troubles, head over to Chapter 13, "Common Modem Mysteries."

✔ You can change Echo On/Off and a couple other of your software's settings "on the fly," while you're on-line. But trying to change some settings while you're on-line, particularly modem speed and data parameters (8,N,1), can possibly dump you off-line. Unless you're feeling particularly adventurous, wait until you log off to perform these adjustments.

Logging off: Now it's time to say goodbye (or was that exit?)

Typing **G** for Goodbye usually logs you off a BBS. That's *usually*. If that doesn't work, try **X** or **E** for Exit, **O** for Off, **B** for Bye, or **H** for Hang Up (or try **I** for I Give Up).

Figure 11-2 shows a typical display of line noise, or *garbage*, you may see just before you successfully log off the BBS. The NO CARRIER message is the one sure sign that you're off, *finito*, done, history, outta here.

Figure 11-2:
The NO
CARRIER
message in
this garbage
means you're
logged off
(courtesy of
Nashville
Exchange
BBS).

```
              Tina, thanks for calling INE!! Call again soon.

      Have you checked the INE CLASSIFIED ad section on the main menu ?
                  You can always find some great BARGAINS.

        >>>>>>> Your Subscription will Expire on: 00/00/00 <<<<<<<

   <*> Got something you want to say to the Mayor? Press [L] on the main menu
   -
   <*> Joining the Mayor: Jim Travis Channel 4 & John Paul WNQM
   -
   <*> [J] The DOCTOR is in! Dr. Gary Smith is here to answer your questions

Logged on at 15:55:14
Logged off at 15:59:45
```

You don't always need to see garbage to know that it's over, as you can see by the sleek NO CARRIER displayed in Figure 11-3.

Always log off the host system properly; don't just hang up (unless you can't possibly avoid it).

Figure 11-3:
Saying
goodbye
doesn't
always
involve
gibberish
(Courtesy of
N.A.C.D.
BBS,
Gainsville,
Florida).

```
Logged on at 15:52:52
Logged off at 15:58:44

Thanks for calling the N.A.C.D. BBS
Please Hang Up Now

The following message brought to you by Bell Telephone:

NO CARRIER
```

ANSI   Offline 38400 8N1 [Alt+Z]-Menu   FDX 8 LF X ♪ ♫ CP LG ↑ PR   14:07:46

Before you try to log off, wait until you've finished logging on. (Be patient. This can take a while.) See the BBS's main menu? An Exit Sign should be clearly marked there. Commonly seen main menus proffered by BBS software stack up in the section "It's My *Main* Man! — the Main Menu," later in this chapter.

Getting ANSI?

Some BBSs offer dancing graphical screens sprinkled with colorful pictures, blinking boldface text, and other delightful manifestations of *ANSI Graphics*. (ANSI rhymes with "antsy," which is how many modemers respond to ANSI graphics.) It's not unusual to find BBSs with ANSI art galleries on-line. Figure 11-4 shows a finely wrought ANSI bald eagle from San Diego's Tradewinds BBS.

Figure 11-4:
Colorful
ANSI
graphics
enliven life
on-line
(courtesy of
Tradewinds
BBS, San
Diego).

Dedicated to those who serve and especially those who have given all.

Alt-Z FOR HELP| ANSI | FDX | 19200 N81 | LOG CLOSED | PRINT OFF | ON-LINE

I'd love to, but it's my parakeet's bowling night.

Changing your display

You can change the way the BBS displays on your computer screen in future calls to that BBS. Just head for the likeliest looking spot on the BBS's main menu. Usually it's User Utilities or User Profile, or something similar.

For example, sometimes I use my ancient Tandy Model 200 laptop to call BBSs. It only displays 40 characters per line and 16 lines per screen. So, right after I connect, I high-tail it over to the Change User Settings menu (so everything doesn't scroll right off my screen).

People with Atari Portfolios and other palmtop models will also want to change their BBS settings as they navigate from these tiny screens.

✔ Your computer needs a color monitor and video card before you can see the pretty pictures. Your modem software's terminal type must be set to PC/ANSI, IBM/ANSI, ANSI/BBS, or something similar. Also, you need to tell the BBS that you want things displayed in ANSI, by changing the settings in your User Profile or somewhere similar on the BBS.

✔ ANSI is fun the first couple of times you call the BBS. Most on-line old-timers turn off the BBS's ANSI setting for subsequent calls because ANSI graphics slow things down. (ANSI uses special codes that take time to send. Every color change is a code.) A few curmudgeons seriously roll their eyes at the faintest mention of ANSI graphics. (And if the BBS is a long-distance call for you, those colors and flashing words take longer — and cost more — to arrive.)

Now That You're Here ...

You've dog-paddled through raging rivers of pesky questions. Now you're finally on-line to the BBS. You still may need to wade through a trickle of "ads" and announcements — whatever the sysops feel a new caller should know.

✔ You may find yourself scrolling through more credits and acknowledgments than an Academy Awards ceremony.

✔ Other opening screens list new phone numbers to access the BBS or news of a total system crash that has wiped out every message on the board since day one.

✔ The BBS at work may contain bulletins on Irv-and-Sue-in-Accounting's sizzling office romance.

✔ Go ahead and read the bulletins when it's your first time on the BBS. Rules, policies, and membership fees (if any) hide here.

Oh dear! I think you'll find reality's on the blink again.

✔ On fairly large BBSs running more than one modem, the bulletins list phone numbers that callers with particular modem brands or speeds should use to access the BBS. Table 11-3 shows what one of these lists might look like, along with a translation. (Note that modem fads come and go, so the BBS's phone lists may feature new brands of "in" modems.)

Table 11-3 What Do All Those Letters and Numbers Mean?

Example Access Number	Modem Types	Calling This Number Will
123-4567	Courier HST 16.8	Connect with U.S. Robotics Courier HST modems at 16,800 bps
123-4568	ZyXEL 16.8 or 19.2	Connect at 16,800 bps or 19,200 bps, depending on your ZyXEL model
123-4569	V.32*bis*	Connect at 14,400 bps with 14,400 bps modems
123-4570	V.32	Connect at 9600 bps with 9600 bps modems

✔ As you can see by Table 11-3, a few modem brands work at higher speeds only when calling a kindred modem. Super high-speed modems sometimes go even faster than "advertised" maximum speeds when connected with one of their own kind. Like identical twins, these modems devise quick and efficient ways to communicate.

✔ Table 11-3 is designed only to decipher some of the more common funny numbers and initials you may see on a BBS's access number list. You should know that just having a capable modem isn't enough to connect at these high speeds: You also need to set your software correctly and perform Big Magic on a half-moon night in the ocean surf. Head to Appendix A for more on configuring high-speed modems and other agonizing rituals.

✔ Turn on your modem software's Capture or Log command each time you call a new BBS. You can save the captured session to a file on your hard disk. Name the file something that relates to the BBS's name. Then you can peruse all the announcements and even keep a "copy" of the access numbers and the Main Menu on your hard disk. (Capture files and other unseemly phrases are deconstructed in Chapter 4, "Making Your Software Talk Nice to Your Modem.")

✔ You may not want to plod through every announcement each time you call a BBS. Table 11-4 shows some key combinations that should pause or end text displays, file listings, and other unwanted actions on a BBS. Of course, each BBS runs different software (this couldn't be straightforward or anything), but these key combinations are about as standard as you're likely to find.

Table 11-4	Controlling the Scrolling
Pressing These Keys	*Should Do This*
Ctrl-S	Pause incoming text
Ctrl-C	Quit incoming text or action
spacebar	Stop incoming text
- *or* P *or* Esc	Return to the preceding menu

It's My Main Man! — the Main Menu

All BBSs differ in their purpose, look, and personality, but most of them share at least one characteristic: a main menu. From there, you can find your way to files, games, e-mail, messages, or chatting.

Typing the menu option's first letter should get you where you want to go. The letter you should press is usually in brackets or highlighted in some other way.

Macs, PCs, and most all the other types of personal computers can call the same BBSs. The only reason you may not want to call a board from the "other camp" is if it focuses only on files and support for that one type of computer. An IBM caller may not find much of interest on the "All-Mac, stay-out-you-icky-PCs" BBS, for example.

You will encounter dozens of BBS programs. The following sections present some of the most common main menus you'll see.

PCBoard BBSs

PCBoard is a popular program running multiline, often pay BBSs. Figure 11-5 shows a main menu on a BBS running PCBoard bulletin board software. Here, I've dialed a "superboard" called Aquila in the Chicago area.

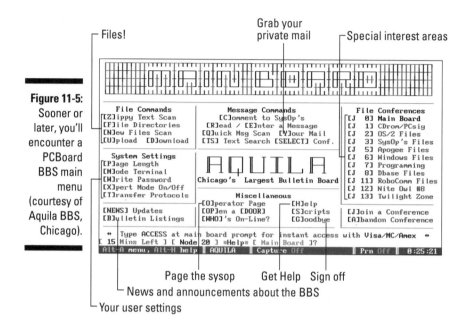

Grab your
private mail

Files!

Special interest areas

Figure 11-5:
Sooner or
later, you'll
encounter a
PCBoard
BBS main
menu
(courtesy of
Aquila BBS,
Chicago).

Page the sysop   Get Help   Sign off
News and announcements about the BBS
Your user settings

TBBS BBSs (try saying that three times quickly)

TBBS is another top contender in the BBS software lineup. Its main menu usually looks like the one in Figure 11-6A. A sysop can customize a given BBS program to look like anything, however. Figure 11-6B shows a highly personalized TBBS menu. Incidentally, "TBBS" is short for "The Bread Board System."

Wildcat! BBSs

Wildcat! is yet another common board program. This software offers several main menus nested within a *main* main menu that looks similar to the one in Figure 11-7A. Choosing M for Message Menu summons a submenu that looks almost exactly like its "parent" screen, as shown in Figure 11-7B.

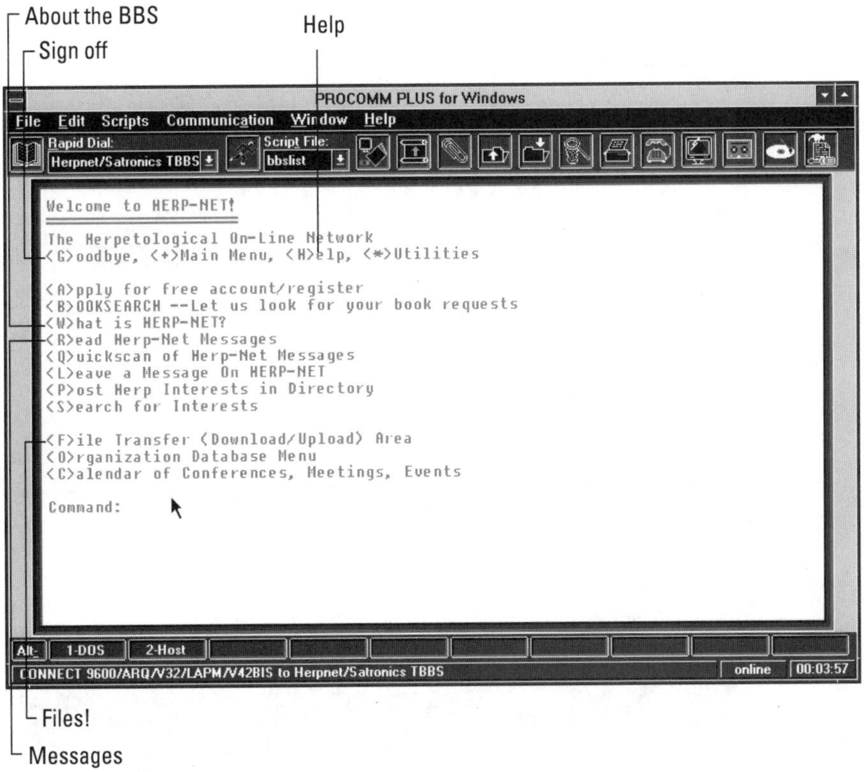

About the BBS

Sign off

Help

```
                              PROCOMM PLUS for Windows
 File  Edit  Scripts  Communication  Window  Help
 Rapid Dial:                 Script File:
 Herpnet/Satronics TBBS      bbslist

 Welcome to HERP-NET!

 The Herpetological On-Line Network
 <G>oodbye, <+>Main Menu, <H>elp, <*>Utilities

 <A>pply for free account/register
 <B>OOKSEARCH --Let us look for your book requests
 <W>hat is HERP-NET?
 <R>ead Herp-Net Messages
 <Q>uickscan of Herp-Net Messages
 <L>eave a Message On HERP-NET
 <P>ost Herp Interests in Directory
 <S>earch for Interests

 <F>ile Transfer (Download/Upload) Area
 <O>rganization Database Menu
 <C>alendar of Conferences, Meetings, Events

 Command:


 Alt-   1-DOS   2-Host
 CONNECT 9600/ARQ/V32/LAPM/V42BIS to Herpnet/Satronics TBBS        online   00:03:57
```

Files!

Messages

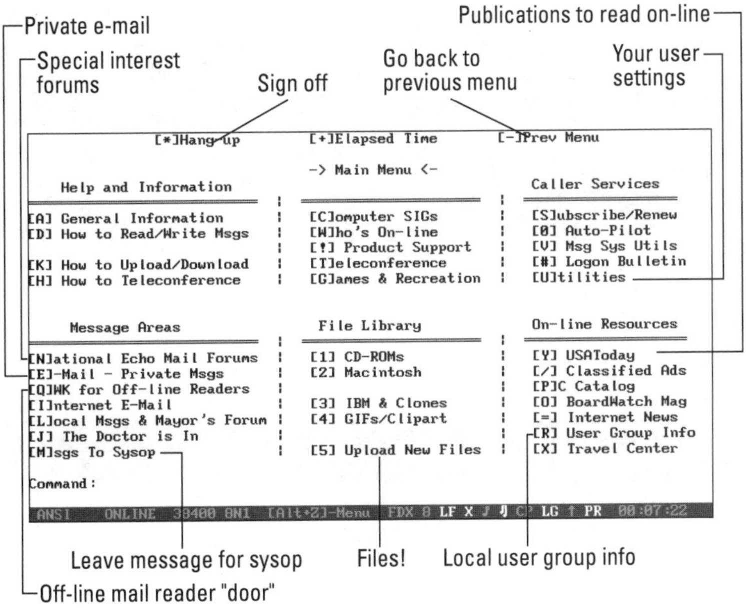

Figure 11-6:
You're
bound to run
across a
typical
TBBS menu
(top).
Bottom, a
variation on
the TBBS
theme
(Figure 11-6A
courtesy of
Herp-Net;
Figure 11-6B
courtesy of
Nashville
Exchange
BBS).

Private e-mail

Special interest forums

Sign off

Go back to previous menu

Publications to read on-line

Your user settings

```
         [*]Hang-up        [+]Elapsed Time      [-]Prev Menu

    Help and Information        -> Main Menu <-        Caller Services

 [A] General Information      [C]omputer SIGs      [S]ubscribe/Renew
 [D] How to Read/Write Msgs   [W]ho's On-line      [0] Auto-Pilot
                              [!] Product Support  [V] Msg Sys Utils
 [K] How to Upload/Download   [T]eleconference     [#] Logon Bulletin
 [H] How to Teleconference    [G]ames & Recreation [U]tilities

    Message Areas              File Library         On-line Resources

 [N]ational Echo Mail Forums  [1] CD-ROMs          [Y] USAToday
 [E]-Mail - Private Msgs      [2] Macintosh        [/] Classified Ads
 [Q]WK for Off-line Readers                        [P]C Catalog
 [I]nternet E-Mail            [3] IBM & Clones     [O] BoardWatch Mag
 [L]ocal Msgs & Mayor's Forum [4] GIFs/Clipart     [=] Internet News
 [J] The Doctor is In                              [R] User Group Info
 [M]sgs To Sysop              [5] Upload New Files [X] Travel Center

 Command :
 ANSI   ONLINE  38400 8N1  [Alt+Z]-Menu  FDX 8 LF X J J C  LG ↑ PR  00:07:22
```

Leave message for sysop

Files!

Local user group info

Off-line mail reader "door"

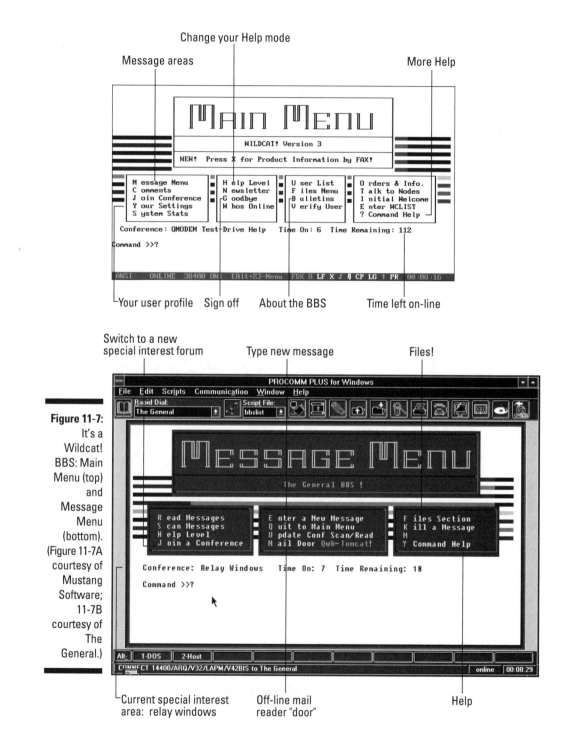

Change your Help mode

Message areas

More Help

Your user profile   Sign off   About the BBS   Time left on-line

Switch to a new
special interest forum

Type new message

Files!

Figure 11-7:
It's a
Wildcat!
BBS: Main
Menu (top)
and
Message
Menu
(bottom).
(Figure 11-7A
courtesy of
Mustang
Software;
11-7B
courtesy of
The
General.)

Current special interest   Off-line mail   Help
area: relay windows   reader "door"

If you see an onion ring, answer it.

Major BBSs

Another BBS program that pops up pretty often is the Major BBS. Figure 11-8 shows a Major main menu. What's that line of letters the arrow's pointing to? That's it, Major's main menu.

Top menu:
Long-winded version

You've got mail!

```
              NO WINNER YET!

Numbers drawn were:  1   5  19
=====>> Today's pot : 88060      <<=====


There is new mail in your mailbox!

Please select one of the following:
(N)onstop, (Q)uit, or (C)ontinue?
   T ... Teleconference
   I ... Information Center
   F ... Forums (Public Message Bases)
   E ... Electronic Mail
   L ... Library of Files
   A ... Account Display/Edit
   C ... Chatlink
   R ... Registry of Users
   G ... Games/Entertainment
   Q ... QWK-MAIL Services
   X ... Exit System (Logoff)

(TOP)
Make your selection (I,I,F,E,L,A,C,R,G,Q,? for help, or X to exit): ail
  Alt-A menu, Alt-H help  CELEBkAT  Capture Off        Prn Off   0:04:10
```

Figure 11-8: Major's menu sometimes looks a bit terse (courtesy of Wayne BBS).

Your command options

Major BBSs assume you usually want to navigate in "expert" mode, with only abbreviations to guide your actions. Typing a question mark should spell things out, summoning a long-winded menu similar to the one above the menu line in Figure 11-8.

Going First Class

Increasing numbers of Mac BBSs run the First Class BBS program. First Class offers a graphical user interface (GUI) that looks great, especially on color Macs. Using First Class, you drag folders, click icons, and generally revel in the more Mac-like environment. There's a Windows version for PCs, too.

To access a BBS's First Class mode, you need to install special software on your computer. That means you have to do two things:

1. Make sure that the BBS offers First Class.

2. Call the BBS in the usual, boring text way and download the special First Class modem software for your computer.

3. *(Optional)* Bribe your favorite Mac modem god (or goddess) to come over to "see" (i.e., help you set up) "a cool new GUI modem program."

Figure 11-9 compares the Boston Computer Society Mac BBS's main menu in graphics and text modes.

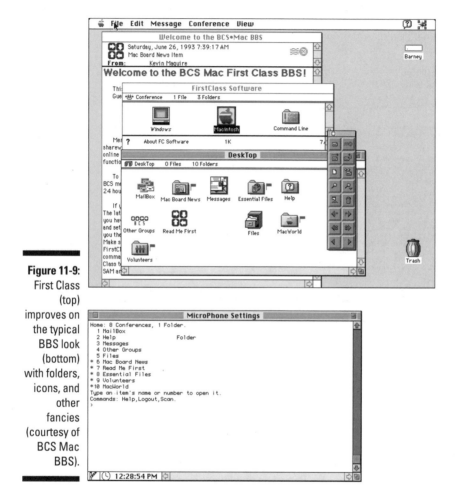

Figure 11-9:
First Class
(top)
improves on
the typical
BBS look
(bottom)
with folders,
icons, and
other
fancies
(courtesy of
BCS Mac
BBS).

GUIs for the PC Bunch

This GUI movement has spread like wildfire from the Mac to Windows and now to BBS programs. Some folks think GUIs are the (sticky) wave of the future for on-line services.

- Clicking buttons is easier than typing a lot of forgettable commands.

- You need a mouse to take full advantage of a GUI.

- Even CompuServe, the venerable on-line service, has seen the GUI light, offering DOS, Windows, and Mac graphical "front-end" programs.

✔ Two graphical BBSs, Roboterm, top, and RIP, bottom, show up in Figure 11-10. Both run on PCs, and RIP also can be found on Mac BBSs. Ask a BBS friend about local boards running these way-cool programs.

✔ Newer versions of leading modem software like Qmodem Pro and Telix include RIP as one of the terminal-emulation settings; setting RIP "On" and calling into a RIP-enabled BBS generally drops you into graphics mode, although you may need to tell the BBS software you want to RIP.

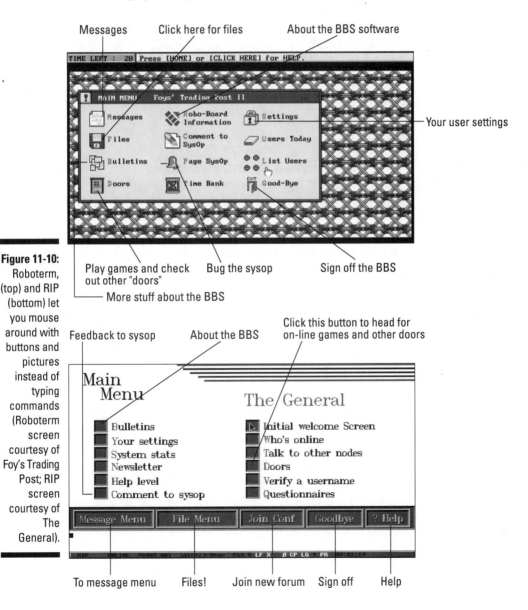

Messages   Click here for files   About the BBS software

Your user settings

Figure 11-10:
Roboterm,
(top) and RIP
(bottom) let
you mouse
around with
buttons and
pictures
instead of
typing
commands
(Roboterm
screen
courtesy of
Foy's Trading
Post; RIP
screen
courtesy of
The
General).

Play games and check   Bug the sysop   Sign off the BBS
out other "doors"
— More stuff about the BBS

Feedback to sysop   About the BBS   Click this button to head for
on-line games and other doors

To message menu   Files!   Join new forum   Sign off   Help

BELL announces: Free Call waitiXn$g#@0(/

Calling a Customer Support BBS

It may mean modem-dialing clear across the nation, but calling a customer support BBS is the most efficient way to obtain service and support for the computer products you own. Figure 11-11 shows the array of support areas to be found on the WordPerfect BBS. Head to Chapter 14's section "Top Ten Computer Company BBSs" to see many leading industry services.

Figure 11-11:
A call to a customer support BBS can yield product upgrades, announcements, and support! (Courtesy of WordPerfect Corporation).

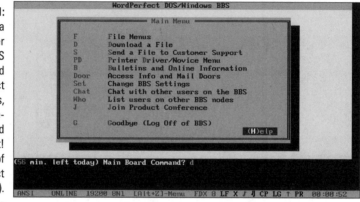

```
                    WordPerfect DOS/Windows BBS
                  ╔══════════ Main Menu ══════════╗
        F         File Menus
        D         Download a File
        S         Send a File to Customer Support
        PD        Printer Driver/Novice Menu
        B         Bulletins and Online Information
        Door      Access Info and Mail Doors
        Set       Change BBS Settings
        Chat      Chat with other users on the BBS
        Who       List users on other BBS nodes
        J         Join Product Conference

        G         Goodbye (Log Off of BBS)
                                          (H)elp

(56 min. left today) Main Board Command? d

ANSI   ONLINE  19200 8N1  [Alt+Z]-Menu  FDX 8 LF X ♪ ♫ CP LG ↑ PR  00:00:52
```

✔ Nearly every computer software and hardware company runs a BBS. I've never seen a charge for joining such a board.

✔ Tech support folks usually respond quickly to questions left on the company's BBS.

✔ An occasional call to a company's BBS — even when nothing's wrong — can reap rewards such as free software updates you can download, or a first crack at special offers.

✔ Fun message areas and even games liven up the business atmosphere on many company BBSs.

✔ Some companies ask you for your product registration number as part of logging on.

✔ The company BBSs run on the same selection of BBS software as hobbyist or office BBSs. If anything, their menus may be clearer than most. Typing the first letter of a menu option should get you there.

✔ After you check out the file areas for "patches" and other product improvements you can download, head over to the message areas to exchange opinions and findings with other users.

✔ Messages you leave on a customer support BBS — or *any* BBS — are public and visible to all the other callers. Conduct yourself as you would in any other place of business. Private e-mail, if offered on the BBS, can always be viewed by the sysop, and possibly others. Never post a message you wouldn't want to be made public on *any* BBS, even in "private" e-mail.

✔ Company presidents and other employees usually welcome the chance to keep in touch with customers through their BBSs. You can leave e-mail thanking the president for a job well done, or complaining about shoddy documentation. Remember, the company *wants* your feedback.

✔ You can find a huge list of customer-support BBS phone numbers on the DAK BBS; the DAK number is at the end of this chapter.

Some computer vendors include "remote access" programs on their computers' hard disks. Diagnosing and fixing a customer's problem can be as easy as a modem call from technical support and a quick probe into the customer's computer.

Modem troubleshooting can shave by half the time it takes to complete a tech-support call. Basically, a customer calls tech support by voice and quickly describes the problem. The tech staffer makes sure that the customer's remote access program is loaded and modem is turned on. A few modem screeches later, and the staffer can be roaming the customer's system, avoiding the laborious process of "Type this" "This?" "No, this" so typical of most voice phone calls to tech support.

Calling the Computer at Work

If your company has a BBS computer for e-mail and file transfers, use it. They're great for touching base with the office while you're at home.

✔ Some offices have modems connected to the office network. You can call into the computer from your business trip and print an urgent document on the office's nice laser printer, for example, and have it waiting for your coworkers the next morning.

✔ Ask your office computer guru about all the options the BBS can open up for you.

✔ Some software lets you dial up and run programs on your work computer — but from the keyboard of your home computer. This *remote access software* works sort of like a BBS program. Instead of navigating a BBS with your modem, you're commandeering Lotus 1-2-3 or Microsoft Word on the computer at *work*, from afar. Chapter 4 has more about remote access software.

Calling a Hobbyist BBS

Calling BBSs for fun is just that: You're there to have fun. So, don't call while in a bad mood or looking for a fight. Other than that prime rule, these other long-standing BBS guidelines should get you through the basics.

Files by the armloads

You don't have to call a commercial on-line service to download files for your computer. Shareware authors *want* you to see their wares — on as many BBSs and on-line services as possible.

The minute the last bit of programming code is in place, shareware houses zap their latest versions over to the larger bulletin boards. It's just a matter of hours before they spread to BBSs of all sizes, shapes, and colors.

✔ Downloading a file from a BBS is usually free — but that doesn't mean the software you've downloaded is free. Files marked *freeware* and *public domain* are truly free of charge. Most of the programs distributed by BBSs, however, are *shareware*. People who like and use a shareware program are obliged to register with the author (and pay the requested fee or be doomed to a lifetime of "guilt screens" each time they use the program).

✔ *Leeches:* The derogatory name on BBSs for people who call and download tons of files but never contribute a message or a file in return. Contribute messages and files to support your favorite BBSs, which wither away — they really do — without active, contributing members.

✔ To discourage leeches, many BBSs impose an *upload/download ratio.* Under the ratio system, you're allotted a specific number of downloads for every file you upload to the BBS. Uploads can also increase your daily time limit on-line. Before uploading a file, check the BBS's file areas to make sure that it's not there already.

✔ Files to upload include shareware you've downloaded from other BBSs and enjoy, plus your text files collecting your thoughts on the state of particle physics research today.

✔ When you upload a file, add a concise, appealing description saying what it is and why you like it.

✔ The sysop will (should) screen all uploaded files, but you should scan any files for viruses anyway.

✔ Another antileech practice (or an official sanction — I'm not sure which) is to charge callers a fee that allows unlimited downloads but exempts them from having to upload anything.

- ✔ If you've paid a fee, download all you want; you've paid to do just that.

- ✔ Most BBSs impose a daily time limit. Usually there's a command that tells you how much time you have left on-line. Check your remaining time before you launch into a lengthy download or upload.

- ✔ BBSs can offer some sophisticated file-searching capabilities. Depending on the BBS software, you can look for a file by name, keyword, or date.

- ✔ Like message areas, files are generally organized into categories on-line. Mac users can head for the Mac file areas without wading through Windows files, for example.

- ✔ Follow the steps in Appendix C, "I've Downloaded a File: Now What?" after you nab a file. And don't leave out the virus-scanning tips.

E-mail ethics

Electronic mail is a godsend for cheapskates. You can keep in touch with friends, family, and business associates — for free, if the BBS is a local call. If the BBS participates in one of the many BBS networks, your reach can extend nationwide ... or across the globe.

BBSs have two varieties of e-mail:

1. Public messages you read and leave in forums, conferences — or whatever that particular BBS calls it.

2. "Private" mail you send to another user or to the sysop. Be warned, however that the sysop can read anything you leave on the BBS. "Private" mail is *not* completely private.

When you first log on to the BBS, a note pops up to indicate whether you have any mail. Some BBS software lists all your messages together, whether they be from someone on the public "Coffee Shop" forum or private e-mail. Others separate private mail and public forum messages to you.

Posting public messages

Leaving your first public message, or *posting*, in on-line lingo, is rife with dread and angst for most new callers. "Will they like me?"

Relax and join in the fray, after reading a few messages and getting a sense of the atmosphere, of course. Forums, conferences ... or whatever ... are the lifeblood of a BBS.

✔ From the main menu, choosing Join or Conferences or something similar should take you to the public message area. Usually there's a way to bring up a list of the BBS's forums. You "Join" a topic of interest and start reading the messages.

✔ Leaving a forum message can be a simple as choosing to Reply to one that sparks your interest. Leaving a message as a response will put the original sender's name in the To: line, but your reply is still visible to everyone.

✔ You can respond to any public message, regardless of the person it's addressed to.

✔ If you want to send a "fresh" message on some new aspect of the forum's topic, you select New from the message menu.

✔ Don't post in ALLCAPS; see Chapter 5 for other e-mail pointers.

✔ To spread messages farther and more efficiently, more and more BBSs are hooking up to other BBSs in a scheme called *networking*. Like runners in a relay race, one BBS calls another, uploading mail and forum messages from its users and downloading that BBS's responses and new messages. Such forums are known as *echoed* forums. Figure 11-12 shows two BBSs' lists of forums — echoed and local.

✔ Don't go begging for mail. Leave mail and you'll get mail.

✔ Turn on your modem software's Capture or Log command for your first message-reading sessions. You'll "take home" a list of forums, instructions on commands, and other invaluable helpers.

✔ It's a nice idea to send a message to All when you first join a forum. Introduce yourself and say something about your interests and experience in the forum's subject area. This is the best way to make friends (and get mail!) on a BBS.

Fido-Net's SHAREWRE echo offers the Hack Report, a monthly publication that lists viruses and other malicious programs masquerading as common shareware or commercial software.

Before you download a file, ask a veteran BBSer whether the file's version is for real, or wait a week or so after the file first appears on-line. It may be one of these bogus programs listed in publications like the Hack Report.

Leaving private mail

Sending private mail is tricky. Some BBSs discourage private messages if they're on topics of interest to the entire community. If I sent Joe a private message asking for a good source of floppy disks, for example, the sysop might ask me to make it public, because my question (and Joe's answer) will help everyone.

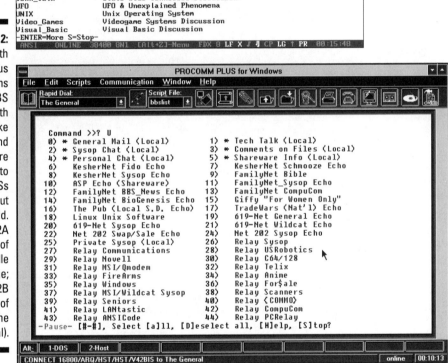

Figure 11-12: Forums with a local focus (top); forums on this BBS (bottom) with names like *Fido* and *Relay*, are echoed to BBSs throughout the world. (Figure 12A courtesy of Nashville Exchange; Figure 12B courtesy of The General).

On the other hand, choosing the Mac Software forum as the place to send Marie a long message detailing my sister's wedding might be frowned upon. The sysop has every right to barge in and ask that Marie and I make our nuptial tête-à-tête private mail — or take it to one of the "social" forums.

Smart mail: Using an off-line mail reader

Off-line reader programs like SLMR (Silly Little Mail Reader) or Off-Line Xpress make reading and responding to mail easy. You just collect your mail when you're on-line, and save reading and responding to it for when you're logged off the BBS. Not every BBS offers an off-line reader. If you're lucky enough to be on a BBS that does, you'll find that the reader gives you time to participate in many more forums and conferences.

Briefly, here's how it works: The first time you use the reader, you call the BBS and enter the mail reader *doorway* that starts the reader. It then asks you to select the forums you want to participate in. The reader scurries off and collects all the messages in each of your selected conferences into one file, which you download into your com-

puter. You log off the BBS, enter the mail reader at your end, and then read and respond to the messages at your leisure, off-line.

When you're ready to call back, you enter the mail doorway again and choose the command that uploads all your responses and new messages onto the BBS. The reader automatically deposits each message in the correct forum, with private mail going to the private e-mail section.

You need to set up the reader to work with your modem software. (Qmodem Pro has Off-Line Xpress built-in.) Your BBS buddies will be happy to help you through setting it up. Incidentally, like everything else in the on-line world, each reader program has devoted followers who heatedly defend its superiority.

Fido-Net? Woof!

Modems excel at file transfers, right? Well, modern sysops put the many modems driving their operations to good use. Late at night (or whenever they can get through), networked BBSs send out packets of e-mail to BBSs all over the nation. These BBSs are members of one or more networks: Fido-Net, PCRelay, USENET, RIME, RaceNet, and many others.

Your local BBS may participate in one or several networks. Some BBSs set themselves up as hubs, collecting messages from hub peers nationwide

and distributing the packets to smaller BBSs in the region.

Faster computers and more sophisticated modems and software have transformed the world of BBSs from isolated, quirky domains to speedy, efficient communications gateways.

This capability has benefits. Your questions are seen by more people. The scope of your new friendships can extend beyond the next block. And you can call a local-phone-numbered BBS yet have your stuff sent long distance for free!

Doors and Other Mysteries

A *door* is nothing more than a gateway into another program running within the BBS software. Games like Trade Wars, publications like *USA Today*, adult GIF files, or mail readers are some of the things you can find behind doors. Figure 11-13 shows one BBS's available doors, top, plus a close-up peek behind the Trade Wars door, bottom.

Pretty Pictures

Many BBSs offer GIFs, which are pictures you can download and view on your computer. Depending on your monitor and video card, some GIFs look just like a photo on your screen. Many GIFs are scenic landscapes or computer art; some are way cool.

Adult Matters

A number of the GIFs on-line seem to be scanned photos of nude models — mostly female. Steamy body shots and other sexually oriented subjects abound. Chat areas, too, can center on adult topics.

This sort of stuff is just a fact of on-line life. Boards attempt to screen out underage callers, and usually you can't stumble across adult stuff without a bit of effort. If you're squeamish about adult matters, you can screen adult boards from your dialing directory. Because adult BBSs, like adult movie theaters, advertise their contents, you can simply ignore them if they don't appeal to you.

Gurus, Co-Sysops, and Other People Not to Annoy

Each forum generally has a guru or co-sysop. Don't upset them; they can be very helpful.

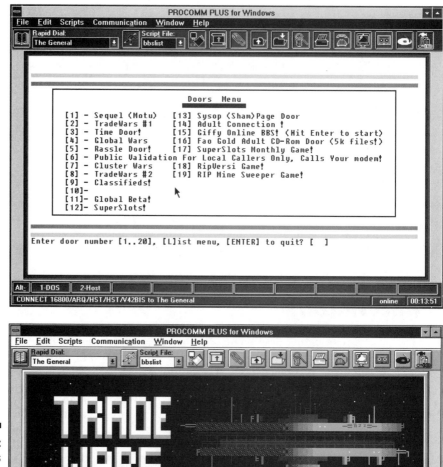

Figure 11-13:
What's
behind door
number 1? A
listing (top)
and games
like Trade
Wars
(bottom)
(courtesy of
the The
General).

When I want your opinion, I'll read your entrails.

Paging the sysop

Most BBS menus have a "Page the Sysop" command, beeping at the sysop's end and summoning him or her to the keyboard. You can chat or ask a question. Try to avoid this unless absolutely necessary — and *never page the sysop late at night!*

If you do, you may unknowingly encounter Sham Page, an artificial intelligence program that looks and talks just like a real, rudely awakened, and irritable sysop.

Lurkers and Flamers

Lurkers are people who never post any messages. *Flamers* are obnoxious types who send angry e-mail to everyone they meet. Try to fall somewhere in the middle.

PC-Pursuit: Pursuing Low Cost BBSing

If you find yourself regularly calling long-distance bulletin boards, you've been bitten hard by the BBS bug. You should probably look into getting an account on PC-Pursuit, a service of the SprintNet phone network.

If your city offers a PC-Pursuit connection, you can connect to BBSs in more than 30 major cities by making a local call. You pay a monthly fee that allows you a fixed number of non-prime-time hours on-line. Any hours above the monthly allowance are charged at a higher hourly rate. Contact SprintNet at 1-800-736-1130 to learn more about PC Pursuit.

How Do I Find BBS Numbers?

Where to find BBSs to call? That's the first question most people ask when they find a modem in their possession.

  ✔ Modem software may come with one or more BBS lists on files. You can call the software's BBS and find many more — and more current — numbers.

  ✔ Local computer stores often distribute freebie BBS listings. Just ask the sales clerk for one the next time you pop in for a new mouse.

  ✔ *Boardwatch Magazine* lists BBSs worldwide; frequently, an issue zeros in on a specific region or theme in addition to its regular list.

- ✔ *Computer Shopper* offers a sizable listing; wear wrist protectors if you're lugging this one home from the magazine rack. (It's hefty but well worthwhile.)

- ✔ When you do make a connection on-line, be sure to ask other callers you meet about BBSs related to your interests. BBSs usually store one or two files full of national, regional, and local boards to call.

Top Ten Mondo Bizarro BBSs

These BBSs aren't really so bizarre — they're just highly focused — but they give you a glimpse into the wide range of interests covered by BBSs. Here are ten (or so) BBS numbers to get you started.

A new car!

... or maybe a clunker. If you need new wheels, take a spin on Automobile Consumer Services BBS, 513-624-0552, which specializes in keeping its callers updated on new and used car prices. (Toll call outside of Cincinnati, OH.)

The Mac track

Macs can call 'most any BBS, but it's fun to dial one completely dedicated to Mac talk, Mac files, and Mac fellowship.

East Coast Mac users will want to try BCS Mac BBS, 617-864-0712, brought to you by the dedicated Mac wing of the Boston Computer Society. (It's toll call outside of Boston, MA.) West Coasters can try the equally Mac-centric BMUG BBS, 510-849-2684, from the Berkeley Macintosh Users Group. (It's a toll call outside Oakland, CA.) Both boards offer USENET Newsgroups, Internet mail, and the First Class GUI BBS software, as well as plenty of technical support, camaraderie, and pointers to other Mac specialty BBSs.

Time travel

Turn back the clock to the very first BBS: CBBS, 312-545-8086, founded by modem-world luminaries and BBS pioneers Ward Christensen and Randy Suess in the primeval ooze (for computers) days of 1978. The sysops warn casual tourists away with the message that dedicated computer hobbyists or "wannabes" need only apply. And CBBS contains only messages (no files or chat). It's of great historical interest to budding BBSmeisters, however. (Toll call outside of Chicago.)

Key to the city

Cleveland Freenet is one of a wonderful breed of BBSs offering users a free Internet e-mail box, along with city info and a library catalog, at 216-368-3888. Call, look around, then get one started in your city. Freenets forever! (Toll call outside of Cleveland, OH.)

Buried treasure

Metal detecting enthusiasts taking a break from finding watches, rings, and the other stuff we lost over the weekend can call the Computer Garden BBS, 301-546-1508. Dig through the on-line catalog here or just boast about your finds to other treasure hunters. You need to type **X** at the Computer Mall to get to the treasure stuff. (Toll call outside of Salisbury, MD.)

Breadmakers ... and so much more

Seemingly everyone's got a breadmaking machine cookbook on the bestseller lists. But do you know anyone with a breadmaker? Where are all these machines hiding? Probably in the pages of the legendary DAK Catalog. If you're curious about breadmakers, computer toys, or other wonder gizmos, call DAK's Online Resource Center BBS, 818-715-7153. This BBS contains a world-class customer-support BBS file. (Toll call outside of Canoga Park, CA.)

The greatest show ...

Easy menus; 500,000 messages; 5,000 callers per day; and 280 phone lines: It's the world's largest BBS, Exec-PC, out of Elm Grove, Wisconsin, at 414-789-4210. (Makes you wonder what else there is to do in Elm Grove, Wisconsin...) Local access numbers are available after you log on. (Toll call outside Elm Grove, WI.)

FedWorld, FedWorld...

Calling FedWorld is like opening a door onto all the BBSs run by the federal government, as you can see in Figure 11-14.

Besides being a gateway into other BBSs, all of 'em open to anyone, the FedWorld board offers a lot all by itself. You can learn about cryptography and other secret practices, or find out about federally sponsored toxic-waste cleanup business opportunities (!). FedWorld is run by the U.S. Department of Commerce's National Technical Information System (NTIS), at 703-321-8020. (Toll call outside of Washington, D.C.)

`What does *this* button do? +++ATA+++ CARRIER LOST`

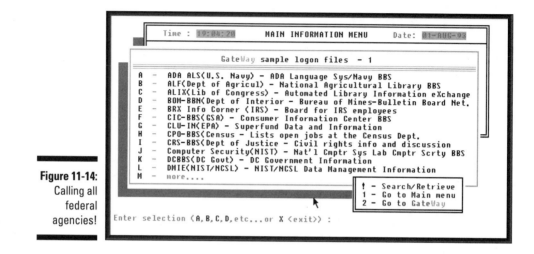

Figure 11-14:
Calling all
federal
agencies!

Not-so creepy crawlies

Some people like the idea of snakes, lizards, and poison toads. If you're one of these herpophiles, slither on over to the Herpnet/Satronics BBS, 215-698-1905. (Toll call outside of Philadelphia, PA.)

NASA Spacelink

It's NASA's official bulletin board. Rave, rave. Teachers, students, and everyone on Earth should call NASA Spacelink, at 205-895-0028, out of Marshall Space Flight Center. Download GIFs of Saturn and Jupiter to view on your computer. Read flight logs and mission announcements. Teachers can find loads of suggestions for classroom activities. See you on Spacelink! (Toll call outside of Huntsville, AL.)

Chapter 12

Enter the Internet

*P*atron: "Waiter, what's this chapter on the Internet doing in my beginning modem book?"

Waiter: "The backstroke!"

To give you an idea of how technoidal, advanced, and completely unrelated to beginning modeming the Internet is, hardly *anyone* on CompuServe, Prodigy, and the rest of the on-line services *and* BBSs on earth ... *combined* ... (and all the king's horses and all the king's men) knows much about the Internet.

Even so, increasing numbers of people harbor a burning interest in what the Internet is and how they can get "on" it — or at least try it out. This chapter inches you a little closer to the flame with a brief outline of what the Internet is and what it offers.

Because it's impossible to use the Internet without understanding a little bit about its beginnings, you'll find historical tidbits sprinkled here and there. For thorough Internet coverage, including step-by-step commands, check out (or, better yet, buy) *The Internet For Dummies*. And if you're really going to get into the Internet, pick up *UNIX For Dummies* while you're at it.

Disclaimer

You, gentle reader, may not give a whit about the Internet. And that's perfectly okay. Millions of people lead fulfilling lives without the Internet ever casting a shadow across their brow. To protect you, I've posted our Technical Geek at the beginning of this chapter as an early warning system against any of the following seeping unsolicited into your cerebrum.

If the handy Geek sentry didn't warn you away, how's this ... "UNIX." (Whisper this last word if you're reading aloud and don't want to be mis-taken for a serious computer nerd.)

UNIX is a computer operating system similar to DOS but about a zillion times more cryptic. Because most of the computers tied to the Internet use inscrutable UNIX, most Internet sessions require you to know and use at least a few UNIX commands. That is, if you want to do anything more than stare at a flashing cursor.

Feeling brave? Stick around.

What's an Internet?

The Internet is like an interstate highway system. Except it's global.

And, instead of hefty 18-wheelers lugging watermelons to and fro (and keeping a wary eye out for Smokey), this highway shuttles *data* to and from thousands, nay, hundreds of thousands of *internetworked* computers. And *millions* of users — from Tierra del Fuego to Greenland.

The Internet is a thoroughfare for an almost infinite number of files, programs, articles, art, treatises, poems, pictures, medical diagnoses, sounds, opinions, scientific visualizations, and sheer bull — and anything else that can be squeezed down into computer-readable data and shipped over a phone line or cable.

Its big advantage: right now, it's by far the cheapest solution for shipping large amounts of data — especially long distances.

Is that something like the politicians' data highway?

The U.S. strip of this computer highway is called the National Research and Education Network, or NREN for short. And yes, it's the same data highway you see in the papers. The reason it's news is that Washington is trying to sell us taxpayers on beefing it up so that more computers can shuttle even more data around on it even faster.

Breeding rabbits is a hare-raising experience.

The politicians, phone companies, and other interested parties endlessly point to how Little Janey will be able sit down in her second grade classroom in Imogene, Iowa, to tap into endless streams of information at the press of an Enter key ... after they get enough money to "widen" the NREN highway ... and allow for even faster and faster "information retrieval."

Who's on first?

Although NREN is the acronym *du jour*, in this chapter it's still just "the Internet." Like the interstate highway system, the Internet is subsidized by Federal funds — partially. Like other Federal entities you've experienced, the Internet is just a bit disorganized — it's loosely administered and the federation's member networks basically do what they please.

No one is really "in charge," so new computers link up to smaller networks on this giant computer network every day. And there's no central "directory assistance" to keep track of these new "sites." Or what type of information is on them.

Like shifting desert sand dunes, the Internet is never the same "place" any two days in a row. This constant growth and change is the second reason (UNIX is the first) that the Internet is so baffling.

So the Internet's free?

The ARPAnet, the Internet's extinct ancestor, started out with the mission of linking military and research computers. It transferred some data around, but its primary purpose was as a sort of network "lab" to teach military types and researchers how to build a bomb-proof computer network. (So the Internet's decentralized, distributed setup was originally intended as a defense measure.)

Soon, university computers hooked up. Then other networks joined in the fray, notably, the National Science Foundation's network of supercomputer centers, NSFnet.

Today, in a move toward self-sufficiency and even profitability, some of these sites have become Internet *providers*, selling Internet access to businesses and individuals. Other, commercial businesses that focus on providing Internet access are cropping up weekly. In fact, if the government "chunks" went away, some people say the Internet would run just fine.

If you're in a university setting or work in a governmental branch or research organization, chances are you have a "free" Internet account. Many employees of large businesses find themselves "on the Net" as well. If you're without any of these contacts, you have to access the Internet through a provider, or some other way — usually for a fee. (Several pointers await you later in this chapter.)

After all is said and done, usually more is said than done.

The Internet's elusive "Who's on first (and who's not)?" quality is the third reason for the tradition of confusion and befuddlement surrounding the Internet.

Is my computer at home on the Internet?

Any discussion of the Internet in such a general, PG-rated book is bound to be oversimplified. It may sound as if just about every computer on earth is connected to the Internet. In truth, certain restrictions apply even to this seemingly boundless Mother of All Networks.

- ✔ Users of computers with _dedicated_, wired-in Internet connections don't bother with modems when they want to connect to "the Net." Instead, they can _log on_, or connect, to another Internet computer from their computers simply by typing a command.

- ✔ Instead of modems, true Internet computers all "wear" the same brand of "glove," which makes computer "fingers" of various sizes and shapes all look and act the same way to the data being handed around. (Budding computer scientists will enjoy learning that the glove's real name is the _TCP/IP protocol_, which is itself being challenged by a new, more international glove on the block called _OSI_, for Open Systems Interconnect.)

- ✔ Computer users who don't have this true-blue, TCP/IP connection must use a modem to "dial up" an access provider. After you make a modem connection to the provider's computer, you can use all the normal Internet commands on it through your computer (sort of the way you would navigate CompuServe or Prodigy from your own computer by typing commands on your own keyboard).

- ✔ You can get this type of connection by asking an access provider to sell you a _dial-up_ account. Dial-up accounts pose an attractive solution for small businesses and individuals who need Internet access. (Pointers to regional access providers await you at the end of this chapter.)

- ✔ Larger businesses and nonprofit corporations can buy a _dedicated_ Internet connection from the access provider. It costs more money than a dial-up account, but it provides a direct, _leased-line_ connection that's faster and therefore handles more data.

- ✔ If you call a BBS or commercial on-line service with your modem, you may already be "partially" connected to the Internet. Sorta. Some of them already let you send and receive e-mail through the Internet — often for no extra cost. Some BBSs also offer USENET newsgroups, the Internet's unique form of message boards dedicated to special interests. Refer to the e-mail and USENET sections of this chapter.

- ✔ Some people mistake e-mail or USENET newsgroups for a true Internet connection. (It doesn't really matter how "fully connected" you are, though, as long as you're getting what you need.) But a true link brings much more power.

A stitch in time would have confused Einstein.

REMEMBER

What, exactly, is "being on the Internet"?

A full Internet connection, whether dial-up or dedicated, doesn't necessarily provide access to USENET newsgroups. It's up to each provider to decide when and if to add news — as well as how many newsgroups to carry (there are more than 2,500 of them).

A full Internet connection _does_ mean the capability to send and receive e-mail; _plus_ being able to

log on to other, _remote_ Internet computers from your computer — through a program called _telnet_. Having an Internet account also means you can transfer files to and from your own computer by using a program called _ftp_. (Refer to the section "Roaming Around on Other Places" later in the chapter for more on telnet. Refer to "Getting Files from Other Places" to find out more about ftp.)

This "What's an Internet connection?" mystique is the fourth reason that the Internet is not for everyone.

The Internet's Dirty Little Secret

INNER SANCTUM

As the ultimate network of networks, the Internet boasts some amazing resources. Unfortunately, it also boasts a few bigots who enjoy intimidating and "shutting out" newcomers.

Most of the people on the Internet are helpful — welcoming and instructing the beginners they meet. But a visible and vocal minority of Internet users act like waterfront mansion owners — despising any hapless vacationers who'd dare lay towels on the sand outside _their_ homes.

The roots for this elitist behavior probably stem from so many of the Internet's sites being college campuses. Because just about any student can get an Internet account, maybe the threats and intimidation are the inner circle's "freshman" initiation into the "right" behavior on "the Net." Proof for this theory can be found in the many cautions not to "waste" the Net's resources with needless messages.

As more individuals and businesses link up to the Internet, a gradual "democratization" is taking place. Until that's completed, just smile and remember that _you_ have a life.

This snooty attitude is the fifth reason that the Internet is tough on beginners.

So Why Bother?

So many explanations of the Internet end with breathless gushing on how you can tap into this awesome universe of information. Actually, the Internet poses a number of disadvantages — especially for people like you who just want to get some work done.

To quickly summarize:

- ✔ The Internet usually asks you to memorize and use arcane UNIX commands.

- ✔ Finding a hookup can be tricky.

- ✔ You can't place a phone call to "Internet Central" to get help or answers. No one's in charge!

- ✔ It's almost impossible to know who's on the Net and how to reach them.

- ✔ Finding files, articles, catalogs, and other stuff is even more difficult.

- ✔ And the Internet still suffers from an overbearing, elitist culture that discourages newcomers.

So why bother with the Internet? In truth, there's not much in the Internet to recommend it for pure beginners. However, after you've become more familiar with being on-line, the sheer amount of information available may be irresistible.

Ways "Around" Connecting to the Internet

In search of a few shareware programs or common text files? You can probably find them on and download them from a local BBS or a commercial on-line service much more easily than off the Internet. That's doubly true for getting computer help and support.

After you've been on-line a while on BBSs and on-line services, you start hearing lots of Internet pointers. (You can hardly avoid them!) Eventually, you may even be nudged along into "cyberspace" — that wide-open on-line realm — by a friend or colleague. Just take your time, don't go before you're ready — and don't let anyone intimidate you.

Until then:

- ✔ You probably can find a plain old local BBS that lets you swap e-mail with Internet users — for free.

- ✔ Several BBSs and on-line services offer USENET newsgroups — often at little or no cost to you.

- ✔ On-line services offer varying degrees of Internet access. DELPHI and The WELL offer full access. GEnie users can request files and newsgroup archives through the Internet forum there. And CompuServe and America Online offer e-mail connections to Internet users. (Prodigy is still talking about opening an e-mail Internet "gateway" as I write this.)

- ✔ Some of these on-line services offer helpful menus that make it all much easier. And each one has a lively and informative Internet forum full of discussions, tips, useful articles, and even some software. Before logging on to the Internet, browse one of the many Internet forums first.

- ✔ There's nothing really *gross* about the Internet; it's just needlessly difficult for most people. Maybe your boss wants you to access the Internet for some reason; or you need some important information you can't get anywhere else. In this case, you'll end up on the Internet.

Your Internet Address

Each person on the Internet has a unique *address*. You use this address when you want to send an e-mail message to someone.

- ✔ Internet addresses usually consist of a person's last name and the computer's name *plus* some weird symbols that look as if the cat climbed onto the keyboard.

- ✔ You have an address, too: it's the name of your "host" computer on the Internet, combined with your name. My current address looks like this:

```
rathbone@cerf.net
```

You would pronounce this: "Rathbone at surf dot net."

The host with the most

There are two main computers in any Internet connection. You connect to the *host* computer through a dedicated phone line or a dial-up account. And then there's the *remote* computer, the one you access through special commands like *ftp* or *telnet* in order to look around, run a program, search a database, or download a file.

Actually, you can add a third computer to the fray: the home or office computer where you're tapping your fingers. That's the *local* computer. Figure 12-1 shows what all this stuff looks like in real life!

(Ftp and telnet get more coverage later in the chapter.)

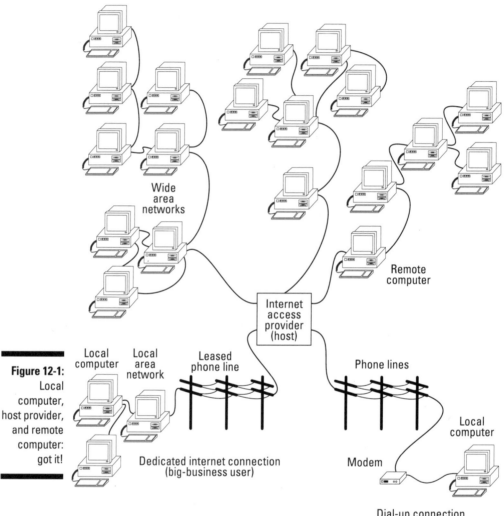

The Internet

Wide area networks

Remote computer

Internet access provider (host)

Figure 12-1: Local computer, host provider, and remote computer: got it!

Local computer

Local area network

Leased phone line

Phone lines

Local computer

Dedicated internet connection (big-business user)

Modem

Dial-up connection (home user)

TECHNICAL STUFF

> ✔ The little group of letters at the very end of an Internet address is the *domain*, telling what type of host you're calling. When there are three letters, you can tell whether the addressee's host computer is at a business, school, or somewhere else. Two letters after the last dot tell you whether your e-mail pal's computer is in a different country. Head to Table 12-1 for the most common domain names.

> ✔ Normally, I'd wave you around such drivel, but understanding Internet computer names can help you find people. Refer to Table 12-1 for the some of the more popular domain names.

Apathy Error: Don't bother striking any key.

Table 12-1	The Seven Most Common Internet Domains, Plus a Country Code	
Domain	*Translation*	*Example*
com	"commercial," businesses	compuserve.com
edu	"educational"	ucsd.edu
gov	"government," nonmilitary	nasa.gov
mil	"military"	navy.mil
org	"organization," other places	fidonet.org
net	"network" backbone systems	cerf.net
int	"international" organizations	nato.int
it	country code; here, "Italy"	iunet.it

Using the information in Table 12-1, you can try sending e-mail to a friend who attends the University of California at San Diego by trying her last name, the "at" symbol (@), ucsd, and the .edu domain, which looks like this when you put it all together:

```
monalisa@ucsd.edu
```

Searchin,'searchin'

There's no "411" on the Internet. A few "directory assistance" programs help people find out each other's Internet address, but they're incomplete and work only sporadically, depending on the state of pork belly futures or something.

- ✔ Simply *everybody* has thought of compiling a vast Internet directory; soon reason seeps in and the idea is abandoned.

- ✔ **Netfind** is a directory-assistance searcher program; it works better than most at finding someone you think is on the Internet. Figure 12-2 shows Netfind finding me (when I was feeling a little lost, I guess). Head down to this chapter's section "Ten Internet Tidbits" for more tips on using Netfind.

- ✔ **WHOIS** is like a big Internet white pages; it lists some Internet host machines, users, and networks — but not all.

- ✔ **Finger** is another finder command you can use.

- ✔ When you receive an e-mail message, stash the sender's Internet address somewhere. Soon you'll have your own white pages.

- ✔ The best way to get someone's Interent address is still to make a (low-tech) voice phone call to them and ask. (Or check out the bottom of their business card.)

Figure 12-2:
Netfind
results: real-
life identity,
most
promising
e-mail
address, and
the last
machine
used by its
target.

```
▢▢▢              MicroPhone Settings              ▢▢
SYSTEM: nic.cerf.net
     Login name: rathbone              In real life: Tina Rathbone
     Directory: /users/cerfnet/dnc/rathbone  Shell: /bin/csh
     Last login Sat Jul 17 10:30 on ttyp6 from dial-sdsc.cerf.n
     No unread mail
     No Plan.
( 4) do_connect: Finger service not available on host sdsc-ucr.cerf.net -> canno
t do user lookup
( 5) connect timed out
( 5) Attempting finger to current indication of most recent "Last login" machine
  dial-sdsc.cerf.net
( 5) check_name: checking host dial-sdsc.cerf.net.  Level = 1

SUMMARY:
- Among the machines searched, the machine from which user
  "rathbone" logged in most recently was dial-sdsc.cerf.net,
  on Sat Jul 17 10:30.
- The most promising email address for "rathbone"
  based on the above search is
  rathbone@dial-sdsc.cerf.net.

Continue the search ([n]/y) ? --> ▮
▯ ◷ 3:31:34 PM  👁▯
```

Swapping E-Mail through the Internet

E-mail is one of the best things the Internet has to offer. You can collaborate on projects with persons living halfway around the globe or send a message to coworkers down the hall, asking where to have lunch.

✔ E-mail messages, like other Internet stuff, can look complicated. They really boil down to two main components: a header — giving you all sorts of profound insight into the date, time, location, and subject of the message — and a message body. In Figure 12-3, the message body starts about three-quarters down the page, with `CERFnet has ...`

✔ Some thoughtful Internet providers offer easier menu systems and mail programs. Figure 12-4 shows an e-mail message with content similar to the one in Figure 12-3 — but much friendlier looking, thanks to the nice-and-easy mail program called PINE that's offered by my access provider, CERFnet. (So if you discover that someone is pining for you, they may mean it literally.) Notice the little menus at the bottom of the screen. (For more information on making the Internet easier, head to this chapter's section "Ten Internet Tidbits.")

Figure 12-3:
A typical
Internet
e-mail
message.
Scary.

```
▢▢▢              MicroPhone Settings              ▢▢
Message  4:
From dnc@cerf.net Wed Jul 14 10:54:56 1993
Received: from  (localhost.cerf.net) by nic.cerf.net (4.1/CERFnet-1.0)  id AA17
Date: Wed, 14 Jul 93 10:53:03 PDT
Message-Id: <9307131757.AA22515@nic.cerf.net>
Comment:  Dial n' CERF Updates
Originator: dnc@cerf.net
Errors-To: matzked@cerf.net
Reply-To: <dnc@cerf.net>
Sender: dnc@cerf.net
Version: 5.5 -- Copyright (c) 1991/92, Anastasios Kotsikonas
From: Dan Matzke <matzked>
To: Multiple recipients of list <dnc@cerf.net>
Subject: [CERFnet] Scheduled Maintenance at UCI/IRVINE On  [Mon Jul 19 07:30 PD
Status: RO
X-Status:

CERFnet has scheduled maintenance at UCI/IRVINE with the following
particulars:

TIME START:      [Mon Jul 19 07:30 PDT 1993]
--More--
▯ ◷ 8:13:34 PM  ▯
```

Figure 12-4:
Internet
provider
CERFnet
offers easy
PINE mail.

> ✔ You don't have to worry about composing your own header: it's added automatically when you create a new message. Figure 12-5 compares how composing mail looks in the nice, friendly PINE composing screen and in regular mail.

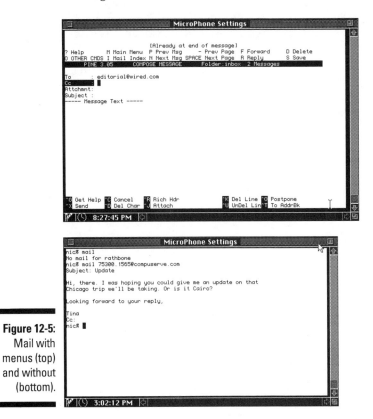

Figure 12-5:
Mail with
menus (top)
and without
(bottom).

✔ Unless your provider offers PINE or another "easy" mail program, you'll see a numbered list of mail messages that await you in mail, similar to the following:

```
>N 1 75300.1565@CompuServe.COM Tue Jul 20 15:06 12/535 Test
 N 2 75300.1565@CompuServe.COM Tue Jul 20 15:29 13/454 Foo!
```

Typing **1** brings up the first one to read. If you want to reply after reading it, typing **re** or something similar brings up the mail editor, with your recipient's address already filled in.

✔ Just to prove how nerdy this system is — the Backspace key doesn't even work dependably. Talk to your provider or Internet guru about how to set up your account's Backspace key. Mail editors differ, but there's nothing more frustrating than trying to backspace over a typing mistake ... and trying, and trying ...

✔ Besides sending e-mail to people you already know, you can forge connections through e-mail with numerous academic and professional users in scads of fields.

You're not just limited to Internet users (if that can be called a limitation). You can send mail to people on other computer networks, like CompuServe, for example, even though you don't subscribe to CompuServe. Table 12-2 shows how to send mail to and from various "cyberplaces."

Table 12-2	Sending Mail to Barney from Everywhere	
	Note: Barney's Internet address is barney@cerf.net.	
Network	**To Network from Internet**	**From Network to Internet:**
CompuServe (Barney's account # is 12345,543.	12345.543@compuserve.com	>internet:barney@cerf.net
DELPHI	barney@delphi.com	barney@cerf.net
America Online	barney@aol.com	barney@cerf.net
GEnie	barney@.genie.geis.com	barney@cerf.net
MCI Mail Barney's account # is **23477**.	23477@mcimail.com	To: Barney EMS: INTERNET MBX: barney@cerf.net
UUCP	barney%host.UUCP@uuunet.uu.net (You must know the host computer's name.)	cerf!net!bjones

The Barney e-mail examples in Table 12-2 are fake. Please don't scramble onto the Internet and ask some poor innocent person lots of questions about Baby Bop.

Getting on Mailing Lists

The mainstays of on-line services — discussion groups, forums, and special interest groups — all have a counterpart on the Internet, which are known as USENET newsgroups (discussed at the end of this chapter). But sometimes it's inconvenient or impossible to read newsgroups in your areas of interest.

So, Internet aficionados often set up mailing lists as a way to distribute all messages in an interest area to hundreds or thousands of "subscribers." More than 1,000 mailing lists keep subscribers posted on topics ranging from home-brewed beer to computer networking to potential cures for AIDS.

- ✔ You can subscribe to a mailing list by sending an e-mail request to a person or computer program that maintains the list — called a *listkeeper*.

- ✔ Methods for subscribing vary ... this couldn't be easy or anything, nooooo. But usually the following will get you on the list: Send an e-mail message to the listkeeper by using the following syntax:

```
beerbrew-request@bubble.suds.edu
```

  This message, for example, would register your *request* to join a hypothetical "beerbrew" mailing list that focuses on brewing beer. Refer to "Where Do I Find These Mailing Lists?" in the next section for more on finding lists.

- ✔ *Caution:* You may get more mail than you can handle — much of it meaningless. Some of the more active mailing lists can generate more than a hundred messages per day, many of them mere electronic "noise" instead of meaningful "signal." Be prepared for lots and lots of mail.

- ✔ Sending a message to the same listkeeper will let you "unsubscribe" to the mailing list. Usually any message that includes the word *unsubscribe* somewhere in it should do the trick.

Where Do I Find These Mailing Lists?

There are always at least three ways to do something on the Internet. For example, you can find a giant listing of mailing lists just by reading the "news.announce.newusers" USENET newsgroup.

Figure 12-6 shows a few out of dozens of screens full of mailing list names and descriptions. As you can see, lists contain discussions on everything from bagpipes to Bonsai.

. . . .<------ Stealth Tagline

```
┌─────────────────────────────── MicroPhone Settings ───────────────────────────────┐
│ Last-change: 19 July 1993 by arielle@taronga.com (Stephanie da Silva)              │
│                                                                                     │
│ [This is the first of five articles on mailing lists.]                             │
│                                                                                     │
│ Quick Summary of Changes                                                            │
│ ------------------------                                                            │
│   Added since last list:                                                            │
│     Ars Magica            The Chaosium Digest      dg-users                         │
│     IAMS                  informix_nca_sig         Middle-Eastern Music             │
│     Miracles              MLB Scores and Standings Novice MZT                       │
│     ShadowTalk            Sri Lanka Net            tbi-sprt                          │
│     UK-DANCE              VEGGIE                                                     │
│                                                                                     │
│   Deleted since last list:                                                          │
│     cyberpunkRPG          stormbringer             UK-RAVE                          │
│                                                                                     │
│   Changed since last list:                                                          │
│     AR-News               AR-Talk                  Catholic Doctrine                │
│     dark-shadows          mmos2-l                  Tolkien                          │
│     torg                                                                            │
│                                                                                     │
│ Intro                                                                               │
│ -----                                                                               │
│ This is a list of mailing lists available primarily through the Internet           │
│ and the UUCP network.  A mailing list is different from a newsgroup                 │
│ because you do not receive anything unless you specifically request it.            │
│ To be added to a mailing list, please mail a note to the contact for that          │
│ list, listed below.                                                                 │
│ --MORE--(4%)                                                                        │
│  [ ] 3:45:25 PM                                                                     │
└─────────────────────────────────────────────────────────────────────────────────┘
```

```
┌─────────────────────────────── MicroPhone Settings ───────────────────────────────┐
│ 12step                    30something              386users                         │
│ 3d                        90210                    a2-bbs                            │
│ ABC                       accordion                act-up                            │
│ add-parents               adoption                 aeronautics                       │
│ agenda-users              aids                     airplane-clubs                    │
│ alife                     allman                   alpha-osf-managers                │
│ Alspa                     alternates               AltInst                           │
│ Amazons International      America                  America                           │
│ American Hockey League     AM/FM                    AMOS                              │
│ anneal                    apc-open                 apE-info                          │
│ argentina                 AR-News                  Ars Magica                        │
│ AR-Talk                   artist-users             Art of Noise                      │
│ att-pc+                   auc-tex                  autox                             │
│ Aviator                   Ayurveda                 backstreets                       │
│ bagpipe                   bahai-faith              balloon                           │
│ ballroom                  ba-poker-list            barbershop                        │
│ Basic programming         bbones                   bears                             │
│ Bel Canto                 BETA                     Between the Lines                 │
│ big-DB                    bikecommute              bikepeople                        │
│ biodiv-1                  Biomch-L                 biosym                            │
│ Birthmother               BMW                      Bonsai                            │
│ BosNet                    brasil                   brass                             │
│ british-cars              BTHS-ENews-L             bugs-386bsd                       │
│ bx-talk                   cabot                    ca-firearms/ba-firearms           │
│ ca-liberty/ba-liberty     Cards                    catalunya                         │
│ Catholic                  Catholic-action          Catholic Doctrine                │
│ cavers                    cd-forum                 CDPub                             │
│ chalkhills                Chaosium Digest          chem-eng                          │
│ chem-talk                 chessnews                chorus                            │
│ christian                 C-IBM-370                cisco                             │
│ clarissa                  Class of '96             Cleveland Sports                  │
│  [ ] 3:47:36 PM                                                                     │
└─────────────────────────────────────────────────────────────────────────────────┘
```

```
┌─────────────────────────────── MicroPhone Settings ───────────────────────────────┐
│   handicapped.  Topics include, but are not limited to: medical,                   │
│   education, legal, technological aids and the handicapped in                       │
│   society.                                                                          │
│                                                                                     │
│   Note: The articles from the Handicap Digest are also posted in the               │
│   Usenet Newsgroup, "misc.handicap".                                                │
│                                                                                     │
│ hang-gliding                                                                        │
│   Contact: hang-gliding-request@virginia.edu (Galen Hekhuis)                       │
│                                                                                     │
│   Purpose: Topics covering all aspects of hang-gliding and ballooning,             │
│   for ultra-light and lighter-than-air enthusiasts.                                │
│                                                                                     │
│ Harleys                                                                             │
│   Contact: harley-request@thinkage.on.ca (Ken Dykes)                               │
│            harley-request@thinkage.com                                              │
│            uunet!thinkage!harley-request                                            │
│                                                                                     │
│   Purpose: Discussion about the bikes, politics, lifestyles, and                   │
│   anything else of interest to Harley-Davidson motorcycle lovers.                  │
│   The is list is an automated digest format scheduled for twice a                  │
│   day Monday through Friday.  Members may access an email archive                   │
│   server for back-issues and other items of interest.                              │
│                                                                                     │
│ hey-joe                                                                             │
│   Contact: hey-joe-request@ms.uky.edu (Joel Abbott)                                │
│                                                                                     │
│   Purpose: Discussion and worship of Jimi Hendrix and his music.                   │
│   Although Jimi has been dead for about 2 decades, we feel that                    │
│   his music is still worthy to be recognized.  Prerequisite to                     │
│   joining: appreciation for his music.                                              │
│ --MORE--(17%)                                                                       │
│  [ ] 3:50:07 PM                                                                     │
└─────────────────────────────────────────────────────────────────────────────────┘
```

Figure 12-6:
A partial list
of mailing
lists, along
with some
descriptions.
Jimi Lives!

A person without a navel lives within all of us.

✔ A book called *Internet: Mailing Lists*, sold in bookstores, offers lists, although it's probably out of date by now.

✔ That book's same "list of lists" also exists as a text (ASCII) file on the Internet. You need to use the **ftp** command to log in to the computer at **nisc.sre.com**. The file is in a directory called *netinfo*. Use the **get** command to grab the *interest-groups* file. (Head to the ftp section of this chapter for more stuff on ASCII files and on getting files for your computer.)

✔ You can get the list through e-mail by sending a message to mailserver@nisc.sri.com with the words **send netinfo/interest-groups** somewhere in the body of the message.

✔ Your colleagues and people you meet on local BBSs and commercial on-line services can point out mailing lists of interest to you.

✔ CompuServe's Electronic Frontier Foundation (GO EFF) forum is a goldmine of information on hot mailing lists, files, sources, and general lore about the Internet.

Roaming Around on Other Places

The telnet command lets you log on to other host computers from your access provider. It's as if you're working *at* the other computer; your provider's computer is completely invisible.

But why would anyone want to be on two (or three, with a dial-up account) computers *at once*? Oooh, scary!

✔ Run programs like Netfind to search for a long-lost love.

✔ Search for a rare "classic" dirty book in a library's catalog.

✔ Browse menus to look for pictures of exotic ferns.

✔ Read a text file and discover how to build a wine cellar.

✔ In Internet IRC Chat, talk about what frogs are *really* saying. (There's gotta be more to this "ribbit" stuff, eh?)

✔ Play bizarre "role-playing" games with people in Yugoslavia and Kansas. (Maybe Dorothy really met the Wizard of Dubrovnik.)

Many people have authorized accounts and *real* reasons for using telnet to get onto other computers and get some work done. A medical researcher can log on to a supercomputer and crunch numerical data to create visualizations of her work, for example, stuff that would take a personal computer months to crank out. Even average users can get on a big computer — with an authorized account, a very good reason (and a lot of smarts).

The thronging masses like to use telnet to search for ways to grow ferns in wine cellars.

How Can I Find Anything Cool?

It's all too much, as the Beatles sang. Happily, you're not the only person to have been struck by that thought. And some of the folks who found the Internet too vast to be meaningful set to work ordering the chaos.

Following are some Internet helpers: programs people have devised to get around more productively. You can ask your site administrator or Internet guru where to find some or all of these useful programs.

- **Archie** is a bold attempt to index a majority of the files on the Internet so that people can search for stuff by name. When you achieve a *hit*, Archie points to one or more *file holding tanks*: special Internet *server* computers called *anonymous ftp archives*. (Head to the next section for more details on anonymous ftp.)

- Several Archie-dedicated computers, called *Archie servers*, are sprinkled throughout the world. To use Archie, you use the telnet command and log on to the server closest to you. Some Internet providers offer Archie clients special programs that allow more sophisticated Archie searches.

- Those persons without actual Internet connections will be heartened to hear that they can perform Archie searches through e-mail.

- **WWW**, or World Wide Web Project, aka The Web ... What happens when a group of particle physicists at CERN in Geneva get together and decide it's time to catalog all the information in the Internet? The Web is a program that organizes the Internet's resources and presents the results as a boundless *hypertext* document. Far from complete, it's constantly being updated.

Hypertext is like a normal document you can read on your computer screen, with one big difference. Some of its words look highlighted, underlined, bold-faced, or just plain *different*. When you position the cursor over one of these words and press Enter (or click it with your mouse), a new screen of information appears, devoted to explaining that word. That screen contains yet another round of *hypertext-linked* words — enabling you to go "text" spelunking on whichever words stimulate your urge to learn more or satisfies your search in some other way.

Do you use Microsoft Windows? If you've pressed F1 for help, you've seen an example of hypertext — those little underlined (usually green) words that jump to other pages when you click them.

✔ **WAIS** searches the Internet for keywords that match a sentence or phrase in your quest. It's based on the efforts of volunteers to index the Internet's resources; WAIS doesn't search the actual Internet as much as it searches indexes.

✔ **The Internet Gopher** is aptly named: This finder program tunnels through the Internet, looking for resources you specify by using key words and menus. When Gopher finds a likely suspect, it'll "go fer" it by grabbing and downloading the file or program you need without you bothering to look up the remote computer's address and telnet to it.

✔ All of these search methods have their devotees, mailing lists, and other sources of help. Find out more from the books mentioned at the beginning of this chapter — or by talking to your access provider or UNIX guru.

Getting Files from Other Places

Files of every description await you on the Internet. Shareware, articles, catalogs, and information databases mean more programs for your computer and more resources for your professional and personal enrichment. And more padding for your term papers.

Many Internet computers allow perfect strangers like you and me to hop on board (log on) and grab (download) a surprising amount of "freebies." This strange process goes by the equally strange name "anonymous ftp."

✔ Anonymous ftp was set up on various Internet computers to provide a place for file archives that the general public can access.

✔ Anonymous ftp requires two things. First, you need a full Internet connection. And second, you need to be able to type the word **anonymous** flawlessly upon command.

✔ It may be that you have an actual account on the remote system you're logging on to. If so, you use the telnet command and your user name and real password in the examples that follow.

So, instead of typing **ftp, anonymous** to log on to the host computer, you type **telnet** and use your real user name and real password.

✔ UNIX is case-sensitive. UNIX views the file names **Net-god**, **NeT-god**, and **net-God** as completely different. If you're a competent typist, UNIX can almost sort of be fun.

✔ Jumping onto a distant computer is as easy as typing

```
ftp computer.name.foo
```

and pressing Enter from your host computer. The "computer.name.foo" part of the command is something *you* provide, depending on the name of the computer you're trying climb onto.

✔ Each computer on the Internet has a unique name, which covers up an incredibly intricate number that the system uses, but we'll stick with stuff humans can remember.

✔ When you connect successfully, the remote computer asks you for a "username" or, sometimes, just "name." Here's where you type the magic ticket:

anonymous

✔ The computer then asks for a password. Typing your Internet address usually works fine. For example, I would type

rathbone@cerf.net

(For a refresher on e-mail addresses, head back to the section "Swapping E-mail through the Internet" earlier in this chapter.)

✔ Figure 12-7 shows some of the prompts you would see while **ftp**ing to a popular archive computer.

✔ After you're on the remote computer, you need to know how to get to the file's location, or *directory*, on the remote computer.

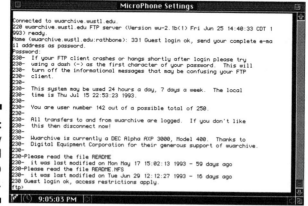

Figure 12-7: The hairy look and feel of an ftp login.

A file's name plus its directory name are known as the file's *pathname*. Printed catalogs, Internet directories, and other guides often mention the file's full pathname in the same breath as the file. That way, you know just where to find it. When someone points you to an interesting-looking file that's on the Internet, try to jot down the pathname as well as the file's name.

✔ A common command for changing directories in UNIX is

cd new_directory_name

✔ When you're ready to see a list of the files that await you in a particular directory, type either the *dir* (directory) or *ls* (UNIX's list) command. Figure 12-8 shows the extremely cryptic results of looking in a directory.

Figure 12-8:
A rewarding
peek into
the remote
computer's
directory.

```
                    inet ftp, directory
                   MicroPhone Settings
579 bytes received in 0.14 seconds (4.1 Kbytes/s)
200 PORT command successful.
150 Opening ASCII mode data connection for /bin/ls.
total 9
-rw-rw-r--   1 root     archive        0 Mar 24  1992 .notar
drwxrwxr-x   4 root     archive     1024 Jun 30  1992 EFF
drwxrwxr-x   5 root     archive      512 Jun 30  1992 SJG
drwxrwxr-x  12 root     archive     1536 Jul  1  1992 academic
drwxrwxr-x   2 root     archive      512 Jun 30  1992 bcs
drwxrwxr-x   2 root     archive     1024 Jun 30  1992 cpsr
drwxrwxr-x   3 root     archive     1024 Apr  8 18:21 internet-info
drwxrwxr-x   9 root     archive      512 Jun 30  1992 journals
drwxrwxr-x   2 root     archive      512 Jun 30  1992 pub-infra
226 Transfer complete.
579 bytes received in 1.6 seconds (0.35 Kbytes/s)
We only support non-print format, sorry.
```

✔ At any time, you can type *help* and press Enter to see a list of commands available to you. And that's a Good Thing. To see a brief explanation of any command, type **help commandname** and press Enter.

✔ Typing **get** (UNIX's command to *copy* a file) and the "target" file's name starts the file's transfer from the remote computer to your host. For example, typing

```
get cookie.nuke
```

starts the cookie.nuke file on its happy journey to your workspace on the host computer.

✔ If you want to rename the file for its new existence on your host computer, you use the **get filename newfilename** variation. Typing

```
get cookie.nuke cookie.txt
```

brings you a file named cookie.txt.

If you have a dial-up account, you need perform one last, crucial step. You have to *download* the file from your host's workspace to your own home or office computer.

Host systems vary, but they generally work with the typical file-transfer protocols found in your communications software. To start an XModem transfer of cookie.txt to your home computer, for example, you type

```
sx cookie.txt
```

and then press the key or keys (typically PgDn) or click the buttons that start a file transfer in your own communications software.

All things are possible except skiing through a revolving door.

Table 12-3	Transferring Files from the Host to Your Local PC
Command	*Translation*
sx	Typically, "send XModem"
sy	Typically, "send YModem"
sz	Typically, "send ZModem"

✔ Some Internet hosts offer menus to ease the downloading process. Figure 12-9 shows a menu offered in CERFnet's Compass program. Your local provider may offer a similar cushion to soften these hard, cold metal seats.

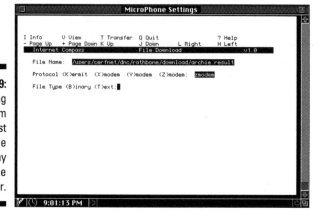

Figure 12.9:
Downloading a file from my host workspace onto my home computer.

What File Is This?

Files on the Internet, as elsewhere, are of two types: they're either *ASCII*, or text files; or *binary*, or program files.

✔ ASCII files are flexible creatures. They'll happily import themselves into your word processor or text editor for your reading pleasure — even when created by an "incompatible" computer type.

✔ Binary files are just the opposite. They're almost always programs, and they're mighty fussy about the type of computer they'll execute themselves upon. Other binary files may be program data from a specific spreadsheet, word processor, or database program. (Refer to Chapter 5 for more on file transfers.)

✔ On the Internet, ftp transfers come in ASCII or binary, too. You specify one or the other — depending on what type of file you're trying to grab — before you start the transfer.

✔ Usually articles, papers, or other things you want to read are ASCII files. Before you engage the ftp command, type **ascii** to make sure that the file transfers in text mode.

✔ If you have reason to suspect that your target file is a program or some sort of program data file, type **binary** before you bring the file onto your host computer. Forgetting this step doesn't make your computer explode or anything. It's just handier when it works right the first time.

✔ Generally, Internet files are compressed before they're stored on an archive computer. Compression saves both space and transfer time. Refer to Appendix C for instructions on uncompressing files you download.

✔ UNIX users have an additional file-compression utility to worry about: any file ending in .Z has been squished down with a program called *compress*. You need another program, imaginatively named *uncompress*, to uncompress it before you can use it on your computer.

✔ Users who find themselves with e-mail connections but not the capability to ftp or telnet can still get files, using a special "gateway" program called *ftp mail*. You can find out how to use ftp mail by sending an e-mail message with the lone word **help** in it to this address: ftpmail@decwrl.dec.com. An e-mail reply will show up in your mailbox, complete with instructions for using this nifty service.

IRC and MUD: R & R on the Internet

The Internet: pinnacle of stolid academia, province of stuffy researchers. Not! The folks who use the Internet take a break every once in a while. And when they do, things can get pretty wild.

✔ The chat system on the Internet is known as IRC, for "Internet Relay Chat." Chat is where people collect in electronic "rooms" and type "Howdy" back and forth. (For background on chatting, head over to Chapter 5.)

✔ On the cyber-playgrounds of the Internet, chat rooms are called *channels*. Some are public and some are private. Some are PG-13, and some are rated R. And then there's the X-rated chat sessions, held purely in the interest of scientific investigation, of course.

✔ During the Persian Gulf War, people gathered in one channel to hear the updates someone over there was typing across the Internet.

✔ Join the IRCHAT mailing list to find out more about IRC. Send an e-mail message to irchat-request@cc.tut.fi.

✔ Figure 12-10 shows a handy chat session, from San Diego to San Diego, routed through a chat server in Taiwan (!) — not that we'd want to use up any of the Internet's valuable resources or anything! Oh, well ...

✔ If someone on IRC instructs you to type some gibberish words and characters, don't do it. (This isn't a bad policy in general, unless it's your UNIX guru doing the talking!)

IRC has a bug, and evil strangers can enter and mess with your computer — with the right entrée (that you glibly provide).

✔ Plenty of Internet fun and games await your competitive urges. Multiplayer role-playing games like MUD (for "Multi-User Dungeons") are big right now. There are word games and logic puzzles, and even a Go server, for devotees of the revered Japanese strategy game.

✔ Check for mailing lists and newsgroups on your favorite games of skill and mastery.

Figure 12-10: Got my chat nickname, time for some chatting!

```
                                    MicroPhone Settings

*** 487 channels have been formed
*** This server has 13 clients and 1 servers connected
*** - IRC.NSYSU.edu.tw Message of the Day -
***              I N T E R N E T   R E L A Y   C H A T
*** -                  -D●nSSS=§je«««₂N∫□□□SU▪┐SSS□▪Sн●□□ð●tsΣ
*** -
*** -      This is a Chinese multi-chatting program.  Sincerely hope
*** - you have a nice conversation here.  Please use BIG-5 Chinese to
*** - read following Chinese messages.  You have to use Chinese
*** - system and 8-bit telnet before using Chinese IRC.
*** - < anonymous ftp 140.117.11.33 to get /pub/nsysu/Chinese-IRC/* >
*** -
*** -      ₂o˝090aWs f§●▪˝ðbe«₂N∫□□□SW˝┐SSS□▪SнSH●□□ð●ts£, ●➟□●□SSS═§je«
*** - ▪q┌□SSS□∫□□□S˝¶XS□═●β□●˝S□˝9IRC¶▪˝¶ð, ●9µ(¶°˝∫=□¶□¿,ð□□Тмð●9£˝▪q┌□
*** - SSS□▪∫ anonymous ftp server IRC.NSYSU.EDU.TW(140.117.11.33) §W˝∫
*** - /pub/nsysu/Chinese-IRC/*.  ¶p˝σм□SSS□IRC¶₂▪□¶□=□/D3£˝ÿf₂, Ω- e-mail
*** - anqu●mis.nsysu.edu.tw  hung●cc.nsysu.edu.tw  lmj●cc.nsysu.edu.tw
*** -
*** - Ω-§jℇa /join ●chinese or ●888□, ●9¿Wмð▪w˝□§jℇaℝ●▪£SSS§□●□□ð°D
*** -                    Happy Travel!!  ;-)
*** -
*** - ==== CC of National Sun Yat-sen University, Taiwan, R.O.C. ====
[1] 03:04 chuffers * type /help for help  <NSYSU-Chinese-Guest-IRC>
```
12:16:32 PM

Catching Up on the News: USENET Newsgroups

Where can you go to gab, schmooze, gawk, and generally burn up zillions of hours leaning over the biggest "back fence" on the planet? It's USENET: the world's largest electronic bulletin board.

✔ USENET fills a gap for many Internet users: its *newsgroups* are the Net equivalent to the discussion groups, forums, special interest groups, and message bulletin boards of the rest of the on-line world.

✔ Each day, hundreds of messages created by thousands of users appear on more than 3,500 newsgroups.

✔ USENET has its own terms: *newsgroups*, instead of forums; *articles*, instead of messages. And you don't *write* an article, you *post* it. (But before you do that, read the following Inner Sanctum sidebar.)

"Daddy, what does FORMATTING DRIVE C mean?"

INNER SANCTUM

USENET ins and outs

Like so many aspects of the Internet, USENET grew from an ad hoc solution to a widely used (and wildly successful) service.

USENET started as a way for UNIX machines to share information. Today, USENET circulates inconceivably heavy message traffic — linking "netnews" browsers on machines ranging from research supercomputers on the Internet to tiny bulletin board systems run on a PC in someone's back bedroom.

Because the traffic's so heavy, newcomers are seriously discouraged from posting any articles of their own until after they've read the *entire contents* of two newsgroups: news.newusers.questions (questions and answers for newcomers) and

news.announce.newusers (announcements and other important tidbits). News.answers is another good place to find information.

Even after you do that, still, before you dare to post a question or article of your own in a newsgroup, you're advised to find and read even more stuff. Locate and read the newsgroup's archived messages, going back into that newsgroup's prehistoric mists of time. Also, get hold of the newsgroup's set of Frequently Asked Questions (FAQ).

FAQ rhymes with *flak*, something you'll get major doses of in the off-chance that you (gasp!) post a question or comment that has ever been seen before in the history of that particular newsgroup.

What is this thing called a newsgroup?

Newsgroups are difficult to pin down. People can hardly agree on whether newsgroup is one word or two, for example; you'll see it both ways.

A few ground rules about naming newsgroups have been set down, however. Newsgroups' names go from broad category to ever smaller categories as you read from left to right. Table 12-4 dissects a newsgroup called comp.sys.ibm.pc.hardware.

Table 12-4   What's in a (Newsgroup's) Name?

Category	Subcategory	Subcategory	Subcategory	Subcategory
comp.	sys.	ibm.	pc.	hardware
computers	computer systems	IBM types of computer systems	personal computers of IBM/ compatible types	hardware topics of personal computers of IBM/ compatible types
Ad infinitum.				

As you can see, newsgroup names say a lot. As you can also see, comp.sys.ibm. pc.hardware would be a great place to look for hard disk recommendations for your IBM or compatible computer. It's not a great place to find out about Macintosh software, however.

Fortunately, hundreds of comp newsgroups focus on every type of computer in existence. Table 12-5 displays some other primary categories.

Table 12-5	Selected Newsgroup Categories	
Name	*Means This*	*Actual, Sample Newsgroups*
alt	alternative, often "trial" groups on timely subjects; not subject to normal newsgroup conventions; weird	alt.tv.infomercials alt.death-of-superman
bionet	biology focus	bionet.molbio.genome-program bionet.women-in-bio
biz	business (press releases and other commercial stuff nixed elsewhere)	biz.comp.software biz.tadpole.sparcbook
comp	computer groups	comp.publish.cdrom.multimedia comp.software.testing
k12	education, K-12th grade/ teachers	k12.chat.elementary k12.ed.life-skills
misc	miscellany	misc.jobs.offered misc.consumers.house
news	USENET groups, news reader programs, and so on	news.groups news.newusers.questions
rec	hobbies, leisure, and the arts	rec.sport.paintball rec.arts.startrek.info
sci	the sciences (other than computer science)	sci.med.physics sci.psychology.digest
soc	sociology and social issues	soc.culture.latin-america soc.college.gradinfo
talk	controversial stuff (proceed with caution)	talk.politics.guns talk.religion.newage

As the newsgroups scroll ...

For first timers, sitting down to read the news can be a daunting experience. That's because you're automatically "subscribed" to each and every one of the newsgroups carried on your Internet provider. Starting up the news reader program on your host invokes all their names at once, like Faust having a bad hair day.

Depending on the provider, this can run into the thousands of newsgroups. Scrollin', scrollin', scrollin' down your screen, for minutes. Type Ctrl-C or Ctrl-Z to get it to stop. Whew!

Figure 12-11 shows a huge listing of about half of all the new newsgroups that have flourished since I last logged on to CERFnet's news reader program. This is only a fraction of what I'd see if this session were my first one ever, though.

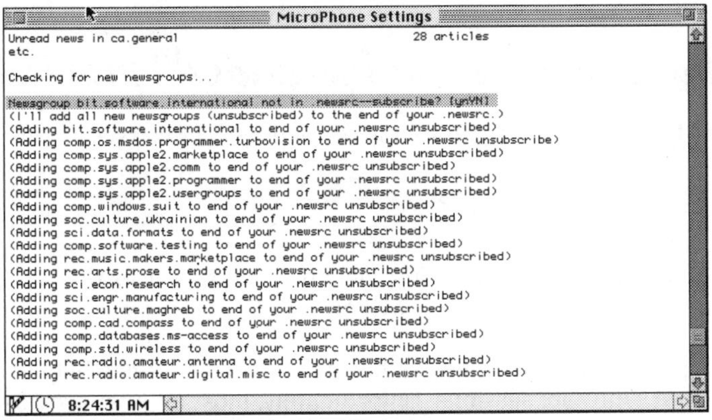

```
                         MicroPhone Settings
Unread news in ca.general                    28 articles
etc.

Checking for new newsgroups...

Newsgroup bit.software.international not in .newsrc--subscribe? [ynYN]
(I'll add all new newsgroups (unsubscribed) to the end of your .newsrc.)
(Adding bit.software.international to end of your .newsrc unsubscribed)
(Adding comp.os.msdos.programmer.turbovision to end of your .newsrc unsubscribed)
(Adding comp.sys.apple2.marketplace to end of your .newsrc unsubscribed)
(Adding comp.sys.apple2.comm to end of your .newsrc unsubscribed)
(Adding comp.sys.apple2.programmer to end of your .newsrc unsubscribed)
(Adding comp.sys.apple2.usergroups to end of your .newsrc unsubscribed)
(Adding comp.windows.suit to end of your .newsrc unsubscribed)
(Adding soc.culture.ukrainian to end of your .newsrc unsubscribed)
(Adding sci.data.formats to end of your .newsrc unsubscribed)
(Adding comp.software.testing to end of your .newsrc unsubscribed)
(Adding rec.music.makers.marketplace to end of your .newsrc unsubscribed)
(Adding rec.arts.prose to end of your .newsrc unsubscribed)
(Adding sci.econ.research to end of your .newsrc unsubscribed)
(Adding sci.engr.manufacturing to end of your .newsrc unsubscribed)
(Adding soc.culture.maghreb to end of your .newsrc unsubscribed)
(Adding comp.cad.compass to end of your .newsrc unsubscribed)
(Adding comp.databases.ms-access to end of your .newsrc unsubscribed)
(Adding comp.std.wireless to end of your .newsrc unsubscribed)
(Adding rec.radio.amateur.antenna to end of your .newsrc unsubscribed)
(Adding rec.radio.amateur.digital.misc to end of your .newsrc unsubscribed)

  8:24:31 AM
```

Figure 12-11: USENET's been busy since I last read the news.

✔ Here's how to automatically "unsubscribe" from all the new groups at once. Wait 'til you see the news reader asking you whether you want to subscribe, as in the shaded line of text in Figure 12-11. Leap for your keyboard and plant a big, uppercase **N** where it pauses. That's the massive unsubscribe option, shoving all those pesky newsgroups to the end of your newsgroup list for the present.

✔ Typing a lowercase **n** would have caused the reader program to go down the huge list of new groups, asking me to decide anew at each one.

✔ Typing **Y** adds all these new groups, at once, to my daily newsfeed. And typing a small **y** causes it to ask before adding each new newsgroup.

✔ *Umpteenth disclaimer:* Your news reader may vary, so if these commands don't seem to do the trick, scream for help and ask your site administrator how to effect this crucial stoppage.

✔ Until you learn how to unsubscribe on your particular account, pressing Ctrl-C or Ctrl-Z should get it to stop scrolling newsgroups.

✔ Most, though not all, Internet providers offer USENET newsgroups. To confuse matters, many local BBSs offer newsgroups as well. Ask BBS pals about local sources for USENET news.

✔ Because there are so many ways to access USENET newsgroups, the examples for Internet provider CERFnet's news program may be completely different from the program you use.

Ten Internet Tidbits

What Internet tour could end before making one last stop for Ten (more or less) Internet tips, facts, and findings?

Going through a "value-added" provider

This is a typical display of Internet goods and services you see at a host prompt:

```
%
```

Figure 12-12 shows a display of Internet goods and services that's possible when you go through the right provider.

This chapter is full of examples on how front-end menus can help you through your first few Internet sessions. Unfortunately, most Internet providers don't make much of an effort to be helpful.

Figure 12-12: Front-end programs like CERFnet's Internet Compass shield novices from first-time UNIX pangs.

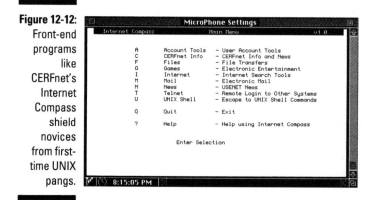

Table 12-6 shows a partial list of Internet providers across the nation. Some offer user-friendly menus and other services, some don't; many plan to offer such niceties. Be sure to ask what's included when you give one of these providers a call. Rate information was left out on purpose: It changes too fast!

Table 12-6	**Ten Internet Providers**		
Region	**Provider**	**Phone**	**E-Mail**
Nationwide	DELPHI	800-695-4005	info@delphi.com
Nationwide (toll-free access)	CERFnet	619-455-3900 800-876-2373	help@cerf.net
Major cities nationwide	NETCOM	800-488-2558 408-554-8649	info@netcom.com
West Coast	—	—	—
West Coast/ nationwide	The WELL	415-332-4335	info@well.sf.ca.us
West Coast Northwest	—	—	—
Rocky Mountains	Colorado Supernet	303-273-3471	info@csn.org
Capitol/VA	PSInet, UUNET Technologies, Inc.	703-620-6651 703-204-8000	info@psi.com info@uunet.uu.net
East Coast	ANS PANIX The World Holonet	313-663-2482 212-787-6160 617-739-0202	info@ans.net info@panix.com office@world.std.com

Using Netfind to find an Internet user

Netfind is discussed in the section "Finding People." Here's how to find Netfind and use it yourself to find people.

Netfind is a program that lives on a computer in Colorado. I found that out by looking up "finding people" in one of the few printed guides to the Internet, O'Reilly & Associates' *The Whole Internet*, by Ed Krol. (It's a classic.) So, in order to use Netfind, you have to use the telnet command to establish a remote connection with its computer. You do this by typing the following from your Internet host computer:

```
telnet bruno.cs.colorado.edu
```

Prune: a plum that's seen better days.

Then you press Enter. That's the telnet command, and then the name (_Bruno?_) of the other computer. Bruno greets you by saying that if you're here to use Netfind, you need to type **netfind** as your _login_. Happy searching.

Code/Decode rot13

The USENET newsgroups preserve free speech by requesting that anyone who is posting a potentially offensive message encode it first. The code is called _rot13_, a simple 13-letter alphabet rotation where _a_ becomes _n_, and consecutive letters take the same "turn."

Messages encoded in rot13 look way cool. Unfortunately, they're mostly just dorky, off-color jokes.

Sadly, new Internet accounts don't come with nifty secret decoder rings. Instead, most news reader programs contain a way to encode and decode messages with rot13.

Cyberpunk, w metal, seeks android, into silicon, long gawks

Long-time Internet users have been known to lose touch with reality. "Net-surfing" can become as addicting as any other pursuit taken too far. The Internet even has its own branch of science fiction, called _cyberpunk_.

Cyberpunk writers like William Gibson and Bruce Sterling characterize the near future as a vast "matrix" of disconnected humans interacting in disjointed bits and spurts while "net-running" a virtual grid, an electronic landscape. (This is supposed to be _fiction?_)

Get the Internet Resource Guide

A huge guide that contains pointers to specialized databases or archives of information, anonymous ftp sites, and other Internet resources can appear right in your electronic mailbox. Send e-mail to resource-guide-request@nnsc.nsf.net.

Earth: a solid substance, much desired by the seasick.

Meet the "Geek of the Week" on Internet Talk Radio

"Geek of the Week" interviews of computer industry luminaries, discussions of how all this technology is affecting our American scene, and computer geek restaurant reviews are only some of the refreshing stories you'll hear from Internet Talk Radio.

Of course, you need an advanced, sound-ready computer called a *workstation* to be able to "play" (and hear) the actual voices on one of the resource-hogging, hour-long shows. (You can try to run a segment by using a PC and Sound Blaster sound card, but you'll still need lots of hard disk space and memory. Plus, if you succeed, you'll be punted forth into the teeming ranks of computer nerd-dom.)

To obtain an Internet Talk Radio file for reading (or listening), send e-mail to info@radio.com. You can tap into a USENET newsgroup, alt.internet.talk.radio, that dishes up info on upcoming guests, schedules, and other tidbits.

Part IV
Modem on the Blink (and Not Just Its Little Lights)

The 5th Wave — By Rich Tennant

"THE TROUBLE-SHOOTERS FROM THE VENDOR AND THE PHONE COMPANY ARE HERE."

In this part...

Your modem's on the blink, but you've already called Pocket-Protector Man to your office too many times this week. Besides, you need to know what to do for those late-night sessions at home when there's no one to help.

Here's the book's Q&A section: the part to turn to when you're modem's acting up. Suggestions for talking nice to tech-support staffers fall into this part, as well as things you can do to prevent troubles from happening in the first place.

Chapter 13

Common Modem Mysteries

*"H*elp! It's broken!" Almost every modem user has uttered these anguished cries — under their breath, if not aloud. Before you yell it to tech support, here are some tips you can try.

Getting Hooked Up

Troubles getting your modem, computer, and software working together? Don't wait for the next Harmonic Convergence. Here are some of the most common glitches encountered during Phase I: the "Getting hooked up" phase.

My internal modem won't fit into my computer's slot!

A gentle back-and-forth motion should work 99 percent of the time to set the internal modem's connectors into the computer's slot. If this doesn't seem to be working, don't force it. Your modem doesn't fit your slot type.

- ✔ Look on the modem box. Does it say "MCA," or "Micro Channel"? If so, and you don't have an IBM PS/2 computer, you probably bought a modem for the "other" type of slot (these are rare, but it happens).

- ✔ Or, if you *do* have an IBM PS/2 computer that uses MCA slots, you might *need* an MCA-style internal modem. If you have a real IBM computer (the kind that says "IBM" on the front), you'd better check the manual to see whether it uses MCA or *ISA* cards.

- ✔ Head back to the store and exchange the MCA modem for a regular, ISA-slot modem.

- ✔ Internal modems come on two sizes of cards: short and long. The vast majority of internal modems are small cards. If yours has bulked up on special chips or other heft-inducers and comes on the larger size card, and your PC doesn't have room for it, go back to the store and exchange the modem for one on a smaller card. (It will still have the features you need.)

My serial cable fits in the modem's socket— but not in my PC's port

External modems attach to your computer with a serial cable. With most modems, the cable comes separately. As with anything else having to do with modems, this seemingly simple purchase can be a source of much confusion. The cable and port connectors share something called the "RS-232 standard," which should have been named the "RS-232 Nonstandard."

Sometimes the cables and ports don't match up. When things don't fit, head back to the computer store for one of the following fixes. (For best results, draw a picture of your modem's and computer's serial connectors and bring it to the store along with your current cable.)

- ✔ Computer serial ports come in two sizes: small, with 9 pins; or large, 25-pin connectors. If your cable doesn't fit, you may need to head back to the computer store for a special 9- to 25-pin adapter.

- ✔ If you're seeing prongs where there should be holes, you need a weird-sounding gizmo called a "gender bender."

- ✔ If possible, avoid adapters and gender benders altogether, and buy a cable that fits right in the first place. The more connectors and gizmos you string together, the more vulnerable your connections will be to line noise and other interference.

- ✔ Be sure that it's a modem cable. There are lots of RS-232 cables.

My external modem's lights don't light up at all

This is a power supply problem.

- ✔ Your modem comes with a special power adapter, which plugs into an electrical outlet or power strip. Make sure that the adapter is plugged in all the way, at both ends.

- ✔ If the modem still doesn't light up, try plugging a lamp into the same outlet. If the lamp works, your modem or power adapter may be defective.

- ✔ The lamp doesn't work either? Do you have an outlet that's controlled by a wall switch? Make sure that the switch is on, and try again.

- ✔ Can't get the modem to light up, no matter what? You might try a call to tech support; or just take the modem and its power supply back to the store to exchange them for one that works.

My modem software says "The CTS signal was not detected," "Device not present," or something similar

Your software is not "talking" to your modem.

- ✔ Perhaps your external modem's power switch is turned off. Turn it on.

- ✔ Make sure that the external modem is plugged into the computer's serial port with the proper serial cable.

- ✔ Also, make sure that you've told your modem software which COM port you've plugged the modem into. Head to Chapter 4 for more on talking to your software.

- ✔ If your software still can't detect your modem's presence, you may have a defective serial cable or modem. If you installed an internal modem, it may not be seated down all the way in the slot.

I tried typing AT to my modem from my software's terminal screen, but my modem didn't say "OK" back

It's probably sulking. Have you hugged your modem today? If it still doesn't work, try these measures, which appear in order of simplest to most desperate.

- Turn on your modem's power switch. (I know; this sounds obvious, but you'd be surprised how many beginners and experts alike forget to do this.) Internal modems draw power from your PC, so these turn themselves on at the same time you turn on your computer.

- Make sure that the serial cable is plugged in completely to the back of your external modem and your computer's serial port.

- Make sure that you typed **AT** in all uppercase or all lowercase letters — not a combination. Remember this for when you type any AT commands directly to the modem. And don't forget to press Enter at the end of a command.

- Perhaps your modem's capability to display result codes (like OK) has been disabled. You can make the modem reply to your commands by typing the following AT command on a line by itself from your software's terminal window:

`ATQ`

Press Enter.

- Seeing a 0 (zero) instead of the OK result code? Your modem is talking to you in numbers instead of words. To toggle from numeric result codes to "verbal" ones, type the following AT command from the terminal window:

`ATV1`

Press Enter.

- Still seeing nothing, not even a 0? Make sure that you told your software which COM port the modem is plugged into. Try changing the software COM port setting from COM1 to COM2. If these don't work, try COM3, then COM4. You may need to consult your software's manual to find out how to change this setting.

- Still no luck? You may be experiencing the dastardly IRQ conflict, which happens if the modem is demanding to share the exact same computer "interrupt" resource used by another device. Head to "Now I Gotta Learn About COM Ports" (it's next).

Now I Gotta Learn About COM Ports

Do you really need to know all about COM ports to use a modem? The answer is a resounding, "No!" Unless things aren't working right — or you regularly walk miles out of your way to hear the sound of fingernails scraping down a chalkboard — you'll probably never have to worry about COM port conflicts.

But what gives? You've tried changing your software's COM port settings to COM1, COM2, COM3, and COM4, but the software still can't find your modem. Nothing has worked. Even so, you may be able to escape this aspect of installing your modem.

✔ The "COM-Port Conflict Cry of Anguish" should get any self-respecting modem guru out to look at your computer in a hurry. The clever guru types love a good challenge (and this certainly qualifies). Bring on the usual arsenal of munchie bribes: Nacho Cheese Flavored Doritos or M&Ms (or top-flight Chinese food for the gourmands) will do perfectly.

✔ Go ahead; fork over a few extra bucks and have your computer dealer install the modem and configure it to work correctly with your software. This is the recommended way to go if you can't find a qualified modem pal.

✔ After you've bribed (or paid) somebody to get your modem up and running, ask them what COM port the modem is using, and write that information down on the little COM port table on this book's Quick Reference Card, right inside the front cover. It will come in handy later.

Why can't my software find my modem?

Normally, you can plug a new printer or mouse into your computer with no worries. Everyone welcomes the newcomer and things work the same as before.

Once in a blue moon, however, adding a new device results in it competing with one of the old devices for the computer's attention. Computer nerds call this a "conflict" — and that's what you're experiencing.

Device conflicts are slightly more common with serial port residents, like modems and serial mice. These conflicts seem to crop up most often when you add an internal modem, a scanner, or some other serial device inside your computer. Gizmos don't like to share serial ports on DOS computers.

Most PCs come with two serial ports (and a parallel port). These two serial ports usually come set up as COM1 and COM2. Choosing a COM port for your external modem is simply a matter of plugging its serial cable into the serial socket for COM1 or COM2.

`Two rights don't make a wrong; they make an airplane.`

When you add an internal modem, it provides yet another serial port for your PC to worry about. Happily, it's up to you which COM port your new modem will be assigned to. You check to see what devices are already using the PC's preinstalled COM ports by looking at what's plugged into the back of your PC. Then you choose a COM port that's not already being used by one of your PC's gizmos.

✔ Most DOS computers let you have up to four COM ports. They're COM1, COM2, COM3, and COM4. One device, one port.

✔ This sounds simple enough, but there's more to it: You can't use devices assigned to COM1 and COM3 at the same time. Furthermore, devices on COM2 and COM4 can't be used at the same time, either.

✔ The odd-numbered COM ports share the same "interrupt request" (IRQ 4) to get your computer's attention. The even-numbered ports also share an interrupt request (IRQ 3). Two devices trying to "talk" to your computer through the same interrupt request causes a conflict: Your PC doesn't know who to "call on" first.

How do I change my internal modem to use a different COM port?

Most internal modems let you change port assignments by moving little gizmos around on the circuit card. Depending on your modem, these can be either *jumpers* or *DIP* switches.

Your manual should show little drawings of the gizmos and offer a chart telling which position enables which COM port. When you're wiggling jumpers or flicking DIP switches, keep a magnifying glass handy — those things are *tiny*.

Some kinder, gentler internal modems let you change ports through a software set-up program. Easy setup is something to keep in mind when you're buying your modem.

✔ Most internal modems come preset to COM2. But a serial mouse may already be using COM2. Because your software can't seem to find your internal modem at COM2, try moving the jumpers, DIP switches, or software settings that make the modem look for COM1. (Don't try COM4, however; COM2 and COM4 share that same interrupt-request thing.)

✔ Using Windows 3.0? You should upgrade to Windows 3.1. Older versions have some trouble recognizing COM ports above COM1 and COM2.

✔ It's rare to find a serial printer that uses the COM port's interrupt request. So, if you have a serial printer plugged into COM1, try assigning your modem to COM3. Even though these two ports share interrupt request number 4 (IRQ4), only the modem will use it to seek the computer's attention. This won't work with other types of serial devices, however — especially a serial mouse.

TIP

✔ When assigning your modem to a COM port other than COM1 or COM2, make sure that your modem software supports that port. Most good programs support COM1 through COM4.

Trying to Connect

You're almost there but not quite? Read this section for answers to connection problems.

My modem dials, but it sounds extra loud and I don't hear a dial tone. Nothing happens after it finishes dialing.

Your phone line is the culprit here.

✔ Check to make sure that the phone cord is plugged into both the modem and the phone jack in the wall.

✔ Most modems have two phone jacks: you want the one that says *Telco* or *Line*. Not labeled? Because the modem is not working, try dialing with the cord plugged into the other modem jack.

✔ After you know which of the unlabeled jacks works with your modem, label it with a colored sticky-dot or some other visual aid.

✔ Everything's plugged in okay but it's still not dialing? Make sure that the phone line is working by plugging a regular telephone in at the wall and making a voice call.

✔ Make sure that your modem software is set for the type of phone service you have: most use *Touch-Tone* dialing, although a few older ones use *pulse* dialing.

Life is so uncertain; eat dessert first.

It makes that screeching noise and seems to connect, but then garbage flows across my screen!

Garbage characters stem from a number of causes. Try the following fixes:

- ✔ Make sure that your software's "communications settings" or "data parameters" (whatever the program calls this) match the BBS or on-line service you're trying to call. These settings almost always are **8,N,1**. If you're calling CompuServe or some very obscure place on-line, press your software's Hang Up command. Then redial with your settings at **7,E,1**.

- ✔ Make sure that your software's Auto Baud Detect setting is turned on if you have a 2400 bps modem and you're calling a BBS or service with a 1200 bps (or slower) modem on its end.

- ✔ If your software doesn't seem to have Auto Baud Detect, try calling back at a lower speed.

My modem connects okay, but the BBS's menus fill up with weird characters and garbage gets inside the message I'm trying to type.

Ignore the alien hieroglyphs; your computer is not possessed (well, no more than usual). You're just experiencing *line noise*, the visible by-product of a particularly bad phone connection.

- ✔ Keep typing, but try to wrap up whatever you're doing so that you can log off the BBS.

- ✔ Whatever you do, avoid backspacing to try to "correct" the garbage. Those backspaces are wiping out the characters you've typed, *not* the garbage characters you see on the screen.

- ✔ You can't do anything about line noise except hang up and try calling back. With any luck, you'll get a cleaner line.

I connect okay, but the text on-line is missing some letters and characters.

You're having a problem with flow control, where data from the modem is piling up faster than the computer can handle it.

- ✔ If you have a 2400 bps or slower modem, make sure that your software's *software flow control* setting is turned on. This is often seen as *Xon/Xoff*.

- ✔ Have a 9600 bps or faster modem? Then you can't rely on software flow control (Xon/Xoff). Instead, turn on your software's *hardware-handshaking* setting, often called *RTS/CTS*.

- ✔ You can't benefit from hardware flow control without buying one of the better hardware handshaking cables. (Never skimp when it comes to serial cables.) Head to Chapter 2 for the full story on serial cables.

- ✔ High-speed modems benefit most if you lock the software's DTE setting (modem to computer) to a higher speed than your modem's data-transfer "sticker" speed.

- ✔ For best results, make sure that you have a 16550AFN UART chip on your serial port, which helps with data flow. (Internal modems often have that chip onboard already — ask the dealer.)

I see weird characters in brackets where the BBS's menus should be.

Terminal emulation is messing around here.

Set your modem software's terminal emulation to PC/ANSI or whatever sounds closest to that in your software.

I can't hear the modem dialing, but the lights blink softly.

- ✔ If your modem has a volume switch, turn it up. It's usually a little knob along the modem's side.

- ✔ For modems with no external volume controls, you can make the speaker louder with an AT command, usually

   ATM1

   Press Enter. This turns on the speaker until you connect.

- ✔ Some modems don't have speakers; you won't be able to hear anything no matter what.

I don't want to hear the modem dial or do anything else!

You can turn off your speaker altogether with a variation on the preceding AT command. (This is handy for any late-night modeming sessions you may be experiencing). Type

```
ATM0
```

Press Enter. That should hush it up.

My internal modem doesn't show me any lights! Someone on a BBS told me about a program that provides software lights. I want lights like the external modem guys!

You may hear about handy *utility* programs with file names like LITES.COM and others. These programs throw a bank of status lights somewhere on your screen that make you look like a regular modem pro. (Actually, status lights prove quite useful for "watching" what's going on in your modem connections.)

LITES and other such programs aren't recommended, however, because they belong to a breed of "memory-resident" software that can interfere with your other software. If you want little lights, you'll just have to buy an external modem.

The modem seems to be connecting, but then that noise just blares constantly and nothing else happens!

Your modem can't relate to the modem on the other end of the connection. Try calling somewhere else and see whether the same thing happens there.

✔ Make sure that you're calling with the right communications settings. Most places expect **8,N,1**.

✔ If you have a modem with error correction, your modem may be taking too long to negotiate with the BBS's modem, which in turn feels snubbed and hangs up. Try looking up and typing the AT command that turns off error correction, and then call the BBS again. You need to look in your modem's manual for the specific AT command; better yet, get your modem pal to look it up and dog-ear the page for you. (Or, call tech support.)

✔ Try calling back at a lower speed, and see whether that helps.

My 9600 bps [your speed here] modem won't call anywhere faster than 2400 bps [your actual speed here].

Check to make sure that your dialing directory is set for your modem's highest speed. If that looks okay, perhaps your software's serial port setting (DTE rate) is set for a lower speed. Find your software's commands to set the serial port for a higher speed. Fast modems work best with the modem-to-computer rate set much higher than the modem's actual speed, and with the *Auto Baud Detect* setting set to Off.

When I dial someplace that's busy, my modem sits there listening to the busy signal instead of redialing.

Some modems just can't detect a busy signal. Most can, however.

✔ Go to the software's Modem Initialization String setting and type the AT command that returns the verbal result code Busy for your modem. With my brand of modem, I'd add this command to my software's initialization string:

```
ATV1X4
```

If AT is already there, you don't need to type it again; just type what comes after (**V1X4**).

Make sure that you press the keys that save the changes you've made in your particular modem software.

✔ If this doesn't seem to work, you need to dig through your modem's manual to find the appropriate AT command. No luck? Try a call to tech support. It could be that your modem doesn't detect a busy signal (some don't). In this case, just press Enter and redial the number.

✔ Finally, some programs, like Windows Terminal, won't recognize a busy signal, even if the modem flashes the word BUSY across the screen.

Troubles On-Line

Even after everything's hooked up and you're on-line, there's still a chance things can go wrong.

I'm seeing two of everything I type!

It's not your contacts acting up again; you just need to toggle your software's *Local Echo* or *Duplex* setting (programs call this one by either name).

- ✔ Does yours have *Duplex*? Set it to *Full*.
- ✔ Is your software's setting called *Echo*, instead? Or *Echo On/Off*? Set it to *Off*.
- ✔ It's okay to toggle the Echo/Duplex setting while you're on-line, by the way.

I don't see anything I type.

Your modem software's Echo or Duplex settings need to be set to Echo On, or Half Duplex.

The on-line service's text just froze on my screen!

You may have accidentally pressed your keyboard's Ctrl-S keys, which sent a signal to pause the display. Typing Ctrl-Q usually gets things rolling again. Take a look at your software status line to make sure that you're still on-line.

The download (or upload) just stopped!

The modem at the other end may have terminated the file transfer due to a bad phone connection. Take a look at your software status line to see whether you're still on-line.

The download (or upload) never started!

Your modem software is probably trying to use a different file-transfer protocol than the one used by the other modem. Unless the two modems agree on the same protocol, nothing happens.

- ✔ Press Ctrl-C or Esc or whatever key stops the transfer at your end. This command usually appears on the software screen: `To cancel, press Esc,` or whatever. Cancel it on the host computer, as well.

- ✔ This time, pick a file transfer that's available on *both* the host computer and your modem software. And remember which one you've selected.

- ✔ Some variants of XModem and other popular protocols sound remarkably alike. Be careful to pick the exact same protocol at both ends.

I'm using a fast modem, but the download/ upload is going very, very slowly.

Older computers have older components; sometimes these don't get along with the whiz-bang new stuff you buy for your PC. That's especially true of the UART chip that controls data flow through the computer's serial port.

- ✔ To get the most out of your speedy modem, you need a late-model UART: a 16550AFN chip provides the best data flow.

- ✔ Programs that tell you what type of UART you have can be downloaded from BBSs and on-line services. You might ask your modem pals about these "diagnostic" programs.

- ✔ Need the new UART? The simplest way is to buy a new I/O card, which provides your computer's serial and parallel ports. Make sure that each serial port on it has the 16550AFN UART chip. (There may be a newer "chip on the block" by the time you read this; consult your friendly local modem gurus and get the current "best" UART for the job.)

Whenever Mom calls and my call-waiting "beep" kicks in, my modem freaks out!

You need to tell your call-waiting to chill out until you're finished with your modem session.

- ✔ If you're using a Touch-Tone phone, disable call-waiting by adding the prefix ***70** in front of a phone number. For example, tell the modem to dial ***70555-1212**.

- ✔ If your phone uses pulse dialing, put the numbers **1170** in front of the phone number.

- ✔ If you can find your modem software's *dialing prefix* string, stick the ***70** or **1170** command in there. That way, your modem software automatically turns off call waiting before dialing any number.

- ✔ The preceding tips work in most situations, but not all areas. If they're not working, call your phone company and find out what commands work best in your area.

Everything was going fine on-line, when all of a sudden my software's terminal window said NO CARRIER and I was dumped off-line!

The other computer's modem just hung up on you. Hmmm.

- ✔ Were all the menu choices magically being made for you on your way out the door? Perhaps the sysop (system operator) was watching your every move and simply decided to dump you off-line. (Maybe some emergency maintenance or something had to be performed on the BBS computer just then.)

- ✔ Some BBS computers shut down for regular maintenance and mail swapping and other stuff. This "down" time usually occurs at a regularly scheduled time in the wee hours. Most BBSs give you ample warning that this is to take place, however; you could have missed the messages, maybe. Examine your software's scroll-back buffer for any `I'm about to log you off` messages from the BBS.

- ✔ Most callers are allowed a set number of minutes on-line per call, or per day; again, the BBS normally gives you several warnings that `your time on-line is about to expire`. Again, perhaps you missed the warning messages.

- ✔ Someone in your home may have picked up an extension phone on your modem line.

- ✔ You may have waited too long without pressing a key or typing a command. BBSs and other on-line systems hang up if they don't detect your presence after a preset interval, typically 5-10 minutes without any input from you.

✔ If your phone's call-waiting beep came through while you were on-line, that'll dump you off-line pretty quick. See the call waiting section in this chapter.

I tried connecting two computers in my office with my serial cable to exchange files, but it didn't work.

To make a direct connection between two computers (and no modems), you need to buy a special cable or adapter called a *null modem*. Its sending and receiving data pins are switched, so that all the data don't butt heads.

I tried to make it dial a number in Windows Cardfile, but it didn't work!

Cardfile uses your modem to dial, just like any other terminal program. That means it needs to know where your modem lives.

To introduce them, choose **S**etup from Cardfile's **A**utodial menu. Then fill out the little form, being sure to tell Cardfile what COM port your modem is connected to.

How can I set up my modem so that my friend can call my computer?

Most people use their modems to call someplace else. They work both ways, however, and getting a friend to call your computer can be fun and a convenient way to transfer files. When you can get everything working right, that is.

My host mode won't host!

Host mode is a way that most modem programs let you set up your modem and computer to take calls. Coordinating the efforts of your software, modem, and caller can be tricky, however.

- ✔ Make sure that your callers know what data parameters to use. And always make these **8,N,1**.

- ✔ Have you specified that a password must be used? If so, your caller must be "clued-in" in advance, too.

- ✔ Set your software's Auto-Baud Detect to On.

- ✔ Make sure that you and your caller are using the same terminal emulation setting. Use PC/ANSI if your software (and the caller's) supports it. Otherwise, try VT-102 or VT-100.

- ✔ If nothing you try works, you can always go to your software's terminal window and type the following, on a line by itself, when you see the words Ring Ring Ring on your screen:

> **ATA**

Don't forget to press Enter.

Now you can type **Hello** to your friend.

- ✔ If you decide to transfer a file, you must both agree on the protocol you'll use.

My computer won't let me into my own hard disk from host mode!

If you're getting an Access Denied message from your software's host mode, this mode won't do you much good on this trip. But here's how to set it up for future trips:

- ✔ Go into your software's host mode setup area, and tell it you want callers (you) to have unrestricted access anywhere on the hard drive. (Here's a good opportunity to get together for a Cheetos session with your long-lost modem guru pal.)

- ✔ Then choose a password you can remember and tell host mode what it'll be.

My modem keeps trying to answer the phone before me!

Turn off the setting for auto-answer. (This can be done through a setting in your modem software, an AT command, or by flipping a DIP switch on the modem, if your modem has DIP switches.)

The AT command is typically

```
ATS0=0
```

You then press Enter.

(Almost) Ten Tricks to Try Before Panicking

Exit from any programs you're running, turn off the computer, count to 30 slowly, and then turn the computer on again. This works more often than you'd believe; to fix all sorts of problems. (Pause for a decent interval before turning it on so that you can avoid jolting the computer or messing with the spinning hard disk.)

Make sure that the modem is plugged in.

Make sure that the modem is turned on.

Make sure that the modem's power cord is plugged in.

Plug a lamp or something else into the electrical outlet.

Make sure that you've set the right COM port setting in your modem software.

Go down to the next lowest speed and try dialing again.

Try using different communications software. **_Tip:_** Get the same one your modem guru uses.

If you're having trouble connecting, try the number as a voice phone call. If you hear a screech, you know you're probably dialing the right number.

Okay, now you can reread this chapter. Then you can call tech support!

The 5th Wave   By Rich Tennant

"THE PHONE COMPANY BLAMES THE MANUFACTURER, WHO SAYS IT'S THE SOFTWARE COMPANY'S FAULT, WHO BLAMES IT ON OUR MOON BEING IN VENUS WITH SCORPIO RISING."

Chapter 14

Modem Woes (and Where to Find Relief)

● ●

In This Chapter

▶ Your software's Help screen

▶ Finding customer support

▶ Discovering helpful folks on local BBSs

▶ Calling in a modem guru

▶ Visiting a users' group

▶ Other modem resources

▶ Ten (More or Less) Common Beginner Mistakes

● ●

*E*veryone knows where to find help with a new breadmaker or belt sander. You pry the manual out from under the Styrofoam packing. Voilà! All the know-how magically leaps from the pages of the little pamphlet.

Modems come with manuals, too. True, most of these are written in the pre-Colombian dialect reserved for the full-moon corn dance, and even the most helpful modem manual won't be much help for troubles you're having with your modem *software* or America Online.

Here are some of the people and places to turn to (okay, *run screaming to*) for help when you're having troubles. Plus, you'll find some general guidelines on getting help effectively.

Golden Rules for Getting Helped

This chapter points out several different ways you can get help. Whichever route you take to customer satisfaction, you can get the most prompt, efficient help by preparing your "material" in advance.

✔ Find out your computer's brand, model, chip type, speed in megahertz, and operating system version. Write the information on a piece of paper and keep it handy.

✔ What model of modem (or whatever gizmo's giving you grief) do you own? Consult Appendix A to find out the meaning of any those weird little **V.XX** (vee dot) numbers you see on the box or in the manual.

✔ Ditto for your software. What's its name and version number? (If this isn't clearly marked in the manual, and you can get it to start up, check the opening screen. Windows programs generally offer an About the Program window you can access from the **H**elp item on the top menu bar (see Figure 14-1). Mac users should be able to find the program's version number under the Apple menu or under Help.

✔ If your question has to do with any additional components — if your printer won't work with your modem software, for example — be ready with details about that, too.

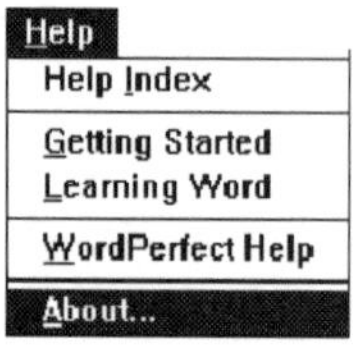

Figure 14-1:
Check
About menu
item in
Windows
programs.

Help
Help Index
Getting Started
Learning Word
WordPerfect Help
About...

Questions technical people will ask you

Before you pen that e-mail cry for help or phone tech support, sit down and ask yourself the following questions (and be prepared to answer them). When they get back to you, the tech support staff or guru will undoubtedly ask them too.

✔ Have you been able to get it to work before? If yes, has anything changed since then?

✔ What were you doing when the problem happened?

✔ Can you make the problem happen again?

✔ Did you get an error message? What did it say?

Phrasing the question

Whether you're on the phone or just connected to one, describe your problem in clear, measured tones. Refrain from getting emotional or hopped up in any way. It's just a computer, after all. Be friendly, and remember the old saying about being able to catch more flies with honey (blecch)....

✔ Have your computer up and running.

✔ Describe your problem in the most excruciatingly detailed terms possible. Why not take the time to jot down a "script" first, or at least a list of key points to cover in your conversation?

✔ When setting down your question in an e-mail message, make it as complete as you can muster. Phrase it with courtesy and professionalism. And send it off with a little prayer.

✔ Take notes when listening to the guru or tech support staffer's recommendations. In your own words, tell the guru what you think you heard to make sure that you get it right.

✔ Getting the troubleshooting in writing — in an e-mail message — is far superior to hearing it on the phone.

✔ When someone offers help, say thanks, and then try to solve the problem with the new information they gave you. Don't repeat the same questions — any parents of teenagers can vouch that it's frustrating to give advice when you think no one's listening.

Your Modem's Manual

Manuals aren't all bad; some are well written and offer general information about *telecommunications* (whatever *that* is) and other lofty subjects. Either way, it's all you've got.

✔ Before you lose your modem manual, look up any tech-support numbers, both for "voice" phone numbers and BBSs. Write them down somewhere prominent, or highlight them in neon green. (It's a good idea not to lose the manual altogether, because it may contain warranties or other policies.)

✔ Before you install an internal modem, get the serial number and write it down.

✔ Take a moment to add the modem company's BBS number to your modem software's dialing directory or settings. Refer to Chapter 4 for help with adding phone numbers to your software. Head to "Finding Customer Support for Any Hardware or Software Stuff You Own (Not Just Modem-Related Stuff)" to learn more about computer company BBSs.

✔ If you have an internal modem, you may need your manual to look up what COM port, or *serial port*, the modem is assigned to — and how to change this *default* setting if necessary. (Refer to Chapter 13 to find out more about changing an internal modem's COM port.)

✔ Another helpful section in your manual is the part that tells you what specific *AT commands* talk directly to your modem and get it to do stuff.

- ✔ Your manual should list the basic AT Command Set. Look here to find out what to type in your software's terminal mode to get the modem to redial a number, for example, or to perform a self-test. (Refer to Chapter 4 for more on the AT Command Set.)

- ✔ Your modem software shields you from reading over pages of AT commands and other gruesome spectacles.

- ✔ Some modems possess their own, unique command subsets that access cool features of that particular modem; the manual should list these, too. You may want to mark this page with a sticky note for your modem guru to peruse when he or she is trying some picky advanced stuff and typing the usual AT commands doesn't seem to be working.

- ✔ Does your external modem sport DIP switches? These control some of the more frequently used settings (also controlled through AT commands). The manual lists each switch and tells what switching it on or off does.

- ✔ If your modem does have DIP switches and something seems horribly wrong and nothing you try seems to be working, probably one of the switches is set wrong. Grab the manual and try changing them, one by one, in order of most logical to most outlandish (and write down everything you do).

Your Software's Help Screen

Staring at YAS (*yet another setting*)? If you can't possibly decipher what it does, some sort of help screen within your modem software can probably help you.

Depending on your software's make and model, Help springs to life when you press the F1 key, Ctrl-H, ⌘-H, or some other keyboard combo.

- ✔ Your software's manual or someone else may refer to the software's built-in help as *on-line help*. (With modem software, this can be doubly confusing, because modems are known as being *on-line* when they've connected to other modems ... not to mention the *on-line* services you can call....) Just remember that most computer programs — whether they're spreadsheets, word processors, or pet vaccination trackers — offer on-line help (even when it doesn't necessarily help *you* go on-line).

- ✔ It's important to find out, *before you connect with another modem*, what key brings up help in your software. I've found one program's help summoner (Alt-H) to be another's hang up command! If you never look in your software manual again, at least look up the command to get help (and possibly to hang up).

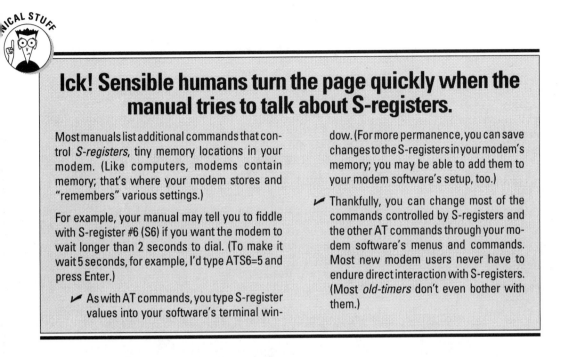

Ick! Sensible humans turn the page quickly when the manual tries to talk about S-registers.

Most manuals list additional commands that control *S-registers*, tiny memory locations in your modem. (Like computers, modems contain memory; that's where your modem stores and "remembers" various settings.)

For example, your manual may tell you to fiddle with S-register #6 (S6) if you want the modem to wait longer than 2 seconds to dial. (To make it wait 5 seconds, for example, I'd type ATS6=5 and press Enter.)

✔ As with AT commands, you type S-register values into your software's terminal win-

dow. (For more permanence, you can save changes to the S-registers in your modem's memory; you may be able to add them to your modem software's setup, too.)

✔ Thankfully, you can change most of the commands controlled by S-registers and the other AT commands through your modem software's menus and commands. Most new modem users never have to endure direct interaction with S-registers. (Most *old-timers* don't even bother with them.)

✔ The software's built-in help is usually much better than reading the manual. As with modem manuals, however, the software's manual probably contains tech-support phone numbers, any relevant BBS numbers, and other useful information. Try to keep it handy.

✔ *Context-sensitive help* means you can press the help key from a particular menu to get help for that specific command or menu item. Figure 14-2 shows a help screen for the Snapshot feature in Crosstalk for Windows. Most modem programs offer context-sensitive help.

✔ Windows communications programs (and *all* Windows programs, for that matter) share something called the Windows Help engine. Most menus offer a Help button, as in Figure 14-2. Another way of accessing Help is from the Help pull-down menu atop each program, like the one for Procomm Plus for Windows (see Figure 14-3). Select and click any green, underlined words you see to bring up more information (provided that you have a color monitor).

✔ It's helpful to print any help screens you find especially rewarding. Keep a folder near your modem to keep the printouts handy.

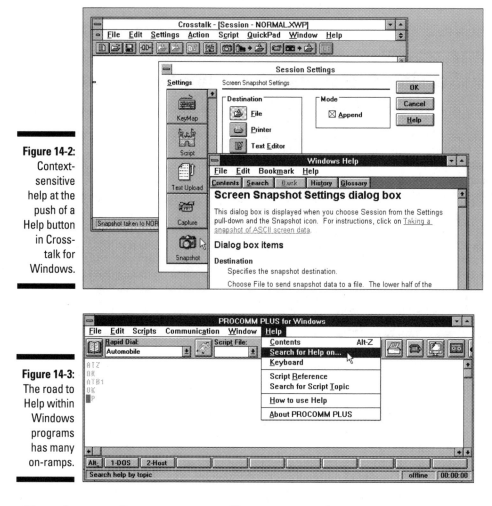

Figure 14-2:
Context-
sensitive
help at the
push of a
Help button
in Cross-
talk for
Windows.

Figure 14-3:
The road to
Help within
Windows
programs
has many
on-ramps.

Finding Customer Support for Any Hardware or Software Stuff You Own (Not Just Modem-Related Stuff)

The company that made your computer offers customer support. So does your modem's manufacturer, as well as your modem software's publisher, as well as the service you're trying to call ... with today's products, customer support is pretty much a given. The trouble comes when you try to reap any of this support and end up on hold for interminable periods, listening to doo-bee-dah renditions of "Hey Jude."

Having a modem is your ticket to quicker, more responsive technical support, provided that the company runs a BBS (and most of them do).

Of course, if your modem's acting up and you can't get on-line to modem tech support in the first place, you'll have to wait on hold (or call your favorite modem guru over) until the problem's fixed.

✔ Keep your product registration number handy when first signing on to the company's BBS. (You should be able to find this in the manual or on the registration card.)

✔ The first route to getting help or an answer to a nagging question is by sending an e-mail message to tech support. Generally, the Help!! message areas are clearly marked. The BBS may offer other message topics that focus on the company's other products, customer "wish lists," or general chatter.

✔ Often, computer and software companies offer text files that contain tips and troubleshooting "articles." You can also find program upgrades, shareware "add-ons" such as spreadsheet templates to figure a mortgage, and other files to enhance your experiences with the product. These are all free for the cost of the phone call.

✔ Most of the major hardware and software companies sponsor forums or other special interest groups on CompuServe, America Online, and several other on-line services. If you're already a member, this can be one of the best ways to stay in touch with the product upgrades, news, and opinions — even find users' groups in your area. If you're not already a member, this can be an excellent incentive for joining.

✔ Local BBSs may offer *networked* conferences (where a message you post is seen nationwide, as are responses you read) on specific brands of modems and modem software. By nature, BBSs are especially resource-rich when it comes to modems and modeming — for *some* reason.

✔ Networked conferences and USENET newsgroups abound on all aspects of computing.

✔ Many users' groups offer one or more BBSs; these are excellent places where you can post questions, learn about others' experiences, or simply cry for help. Refer to the following section for more on users' groups.

Visiting a Users' Group

Groups of computer users form around computer types, software, or sometimes just for the heck of it. These sociable assemblies are known as *computer users' groups*. Members meet on a regular basis and hold computer swap meets and fairs, exchange shareware disks, eat Cheetos, and host guest speakers from leading computer hardware or software companies.

✔ Many users' groups charge an annual fee for membership. This usually gets you a newsletter and a higher level of access to the users' group BBS (if they have one).

✔ Users' groups run some of the best BBSs in the country. Although they're usually open to all callers, your users' group membership fee may grant you a less-busy number to get through on, access to more file downloads, or other privileges.

✔ It's a good idea to look into your local users' group. Even the big, general Mac or PC groups generally splinter into subgroups covering special interests: There may be one devoted to modems, your modem software, or another aspect of computers you're forced to endure.

A quick glance at the San Diego Computer Society's newsletter shows a group devoted to Windows, a Macintosh User Group, a Home-Based Business group, a DIGSIG (Disabled Interest Group), and a Data Communications (modeming) group, among many others.

✔ You'll be well-received at the next meeting of your local group if you come laden with Mrs. Field's cookies and other goodwill gestures.

Calling in a Modem Guru

Maybe it's someone you met at a users' group meeting or on a local BBS. It could be the computer guru from work. Wherever you find him or her, when someone who's knowledgeable about modems comes forth, treasure and cultivate that person: It's your Guardian Modem Guru.

✔ After you find your guru, find out what modem software, off-line mail reader, and on-line services he or she prefers. If at all possible, obtain those and use them yourself. At the most, your guru may drop by to help set up everything. Barring that, at least he or she can help when things go wrong.

✔ Along the same lines, try to buy the same brand of modem as your modem guru — within your financial means, of course. (Modem gurus tend toward owning the fastest and most advanced — read *most expensive* — modems.)

✔ Don't be a one-way user. Offer your guru a sail out on the bay, a round of Frisbee golf, or a home-cooked meal. (And always spring for the pizza for house calls.) Develop some outside interests to share with your guru and you won't seem like such a pest the next time you phone.

Top Ten Computer Company BBSs

Having a modem opens up a whole new line of communication between you and the companies that make and sell computer stuff. (Just think: Most of the whiners sitting on phone-support's hold can't even access the company's BBS.) Figure 14-4 shows the many avenues for support awaiting a caller to WordPerfect's BBS.

As you can see by the following Top Ten, most of these BBSs will be long-distance calls for you. Still, it's a good idea to check in with a BBS — even when you're not experiencing a problem. The following tip should help you cut costs.

Save your calls for after-hours; that'll save on long-distance charges. Better yet, before you call, practice on a few local BBSs (the ones that don't charge membership fees) so that you'll have a few BBS tricks under your belt: knowing how to find and capture messages, finding and downloading the BBS's file listing, leaving an e-mail message, and other stuff.

More BBSs are offering off-line mail readers. If you find yourself calling a customer support BBS often, why not suggest to the sysop (system operator) that he or she set up a handy mail-reader program?

After an off-line reader is in place, a quick call can grab all the new messages in the conferences you specify and alert you to new files and other announcements — automatically — then log you off. That's a lot of support for relatively little time spent on-line. Another advantage: After the messages are on your hard disk, you can save them and go back to a topic's beginning, for example, when composing your own replies.

Sure, phone numbers change. And companies come and go, too. But at least a few of these numbers should help you find the Good Stuff.

General support

Tech-support program updates and fellowship with others who own and use the same products as you — all these are available on the company BBSs that follow.

Microsoft Download BBS (no tech support — just files)

206-637-9009

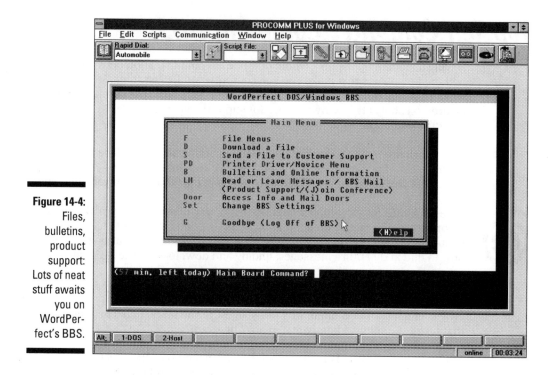

Figure 14-4:
Files,
bulletins,
product
support:
Lots of neat
stuff awaits
you on
WordPer-
fect's BBS.

Lotus Automated Support Dial-in
617-693-7000

617-693-7001 (9600 bps and 14,400 bps modems)

WordPerfect Corp. BBS
801-225-4414

Borland (Support for dBASE, Quattro Pro, and other programs)
408-439-9096

Orchid Technology (graphic card drivers, GIF viewers)
510-683-0327

Apogee Software (Commander Keen and other shareware)
508-365-2359

Creative Labs (SoundBlaster sound card support)
405-742-6660

Three kinds of people: those who can count and those who can't.

Media Vision (sound cards and other multimedia support)

510-770-0968

Modem software companies

Here's where to call for modem "drivers," programs and utilities that make modeming easier — even tech support BBS lists and e-mail swapping with other users.

Mustang Software (QModem Pro, OLX, Wildcat! Software)

805-395-0650

DCA (Crosstalk modem software, files, and other stuff)

404-740-8428

Software Ventures (MicroPhone II "technoids and techbase," other support)

510-849-1912

deltaComm Online BBS (Telix support)

919-481-9399

Procomm BBS (Procomm family of modem software)

314-875-0503

Hayes Microcomputer BBSs

800-US-HAYES (tech support for Hayes modems and Smartcom software)

404-HI-MODEM (shareware, user e-mail, special interest groups)

Modem companies

The BBS run by the people who made your modem contains tech-support BBS ads and drivers to work with various modem software — among other things.

Intel PC and LAN Enhancements Support BBS

503-645-6275

Boca Research (modems, video cards, memory cards)

407-241-1601

Hayes Microcomputer Products (Hayes modems)

800-US-HAYES (tech support for Hayes modems and Smartcom software)

404-HI-MODEM (shareware, user e-mail, special interest groups)

Complete PC

408-434-9703

Supra Corp.

503-967-2444

US Robotics

708-982-5274

Practical Peripherals

805-496-4445

ZyXEL

714-693-0762

Ten (More or Less) Common Beginner Mistakes

Here, in no particular order of dire consequences (except the first one), are the most common mistakes made by new denizens of the on-line world.

Failing to scan newly downloaded files for viruses

Don't download files until you buy (or download) an antivirus program. It's not media hype; viruses are real and they can destroy the stuff on your computer.

Assuming it's your fault

Don't always assume that you're the bonehead: Modems are temperamental.
Software can have bugs. The trouble you're having may not be your fault. Try
dialing a few times; turn your modem off and on between tries. This has worked
for me more times than I can count.

Choosing an obvious password

Everyone loves your new Kitty; won't it be cute to use Kitty's name as your
password for logging on to new on-line services and BBSs? *No!* Never use the
name of a pet or family member for your password; it's too easy for unauthorized jerks to figure out. In fact, don't even use a real *word*. (Bad people have
teamed fast dialer programs with dictionaries, trying *every word* to "crack" a
password.) All-number passwords are the worst idea of all.

Instead, choose an unlikely number, a few letters (capital and lowercase), and
some characters atop your number keys. Make it six characters long, at least.
And choose a different one for each place you call.

How to remember these off-the-wall passwords? Why, jot them down on a
sticky note and affix them to your monitor, like everyone else, of course. ***Just
kidding!!*** ***Never*** do that. Instead, try to memorize your passwords. It's much
more convenient that way.

Falling for the get-well card to Craig Shergold legend that won't die

Eventually, somewhere, somehow, when you least expect it ... you'll see a plea
on-line to send a get-well card or one of your business cards or something else
to the hospital housing a poor young boy dying of a brain tumor and trying to
make the *Guinness Book of World Records* for the most get-well cards ever
received....

Before you start rustling through the stationery for the card, know that Craig
did get well (yay) and made it into Guinness (swell), but the hospital has been
flooded with cards (more than 30 million) ever since (boo). Rumor has it that
Guinness has cancelled the "most cards ever" category forever (probably at the
request of the poor postal carrier serving that route). Say, maybe if you sent the
postal carrier a get-well card ... nahhh.

Getting all worked up about Prodigy's STAGE.DAT file

Prodigy can't hurt your computer or peek inside your data. Period. (For more about the legendary STAGE.DAT file, head on-line to Prodigy's About Prodigy section.)

Downloading too many files at once

Depending on the size of your hard drive, this won't exactly hurt anything. But take it easy with new files (and all new software, whether downloaded or store-bought).

Yes, there are a zillion files waiting for you on-line. But give yourself time to scan, unpack, and evaluate one program at a time. View the cute puppy in that GIF file. Make Windows wallpaper out of it. Play around with a new program's commands. Decide whether you like it. Software is confusing enough without that "So many files, so little time" feeling.

Changing too many settings at once

Don't tweak too many DIP switches or change too many settings in your modem software all at once. This is a sure way to confuse the Dickens out of yourself. Trust me; I've tried (and succeeded — at confusing myself, I mean) more than once.

Joining too many BBSs or on-line services at once

There are lots of interesting people to communicate with on-line. The trouble is, when you join too many places at once, you tend to lose track of what you said to whom where. So, as with files and settings, take things easy with joining new BBSs and on-line services. You wouldn't start corresponding with 50 new pen pals, would you?

Not buying a quality serial cable

You should buy a good, hardware-handshaking cable — someday you will want to go fast, if you don't already — and this cable works best with high-speed modems.

Is MultiMate the word processor for bigamists?

Part V

Bonus Part
of Even More Tens

The 5th Wave   By Rich Tennant

PORTRAIT OF A
CYBERHOLIC

SO, HOW'S DAD?

CYBERHOLICS DON'T WRITE LETTERS.
IF YOU DON'T HAVE A MODEM THEY
DON'T WANT TO TALK TO YOU.

In this part...

*I*nformation that's presented in a list just seems to catch the eye. And after you've been playing with a modem and its manuals, your eyes are probably tired enough to appreciate a good list.

So, that's what you'll find in this section — bunches of lists, all explaining ten things you'll want to know about modems.

However, even the most tired eyes will probably notice that there aren't *really* ten items in each list. Some have more than ten, others have fewer.

But who cares? Remember, this is the packaged Hot Dog Generation: Everybody's already grown accustomed to buns coming in a pack of 8 and hot dogs coming in a pack of 10. Then, when the usual two hot dogs get burned up in the barbecue, everything comes out even.

Each chapter in this book offers a Top Ten This and a Top Ten That ... so consider this a Bonus Part of Tens.

Chapter 15

Ten E-Mail Acronyms and Other Handy Shortcuts

In This Chapter

▶ Deciphering all those *other* weird acronyms you see in e-mail messages and forum postings

"*T*hat's a big 10-4, good buddy." Back in the heyday of citizen's band radio, you didn't need to be a trucker to know this was CBer talk for something like "Okay." Crowing "Breaker, Breaker" into microphones was as much a part of the seventies as platform shoes and mood rings. (Say, platforms are back; could CB be next? Naah. My mood ring says, "No way.")

The on-line world has its own, albeit quieter, form of lingo. As with the CBers of yore, modemers save time by using these shortcut phrases. Unlike the radio set, modem lingo saves on typing, too. And, when connected to a commercial on-line service that charges an hourly rate, these good buddies also save you connect charges.

When little strings of letters and numbers show up in chat or e-mail messages (and you're sure users aren't passing out their modem init strings), check here to find out what it all means.

AOL

What it means: "America Online"

Example: "I dropped by AOL to check out the new baked bean recipe."

BTW

What it means: "By the way"

Example: "BTW, do you remember the command to exit ProComm?"

CIS or CI$

What it means: "CompuServe Information Service"

Example: "Wow! My CI$ bill sure rocketed last month."

CU or CUL8R

What it means: "See you" or "See you later"

Example: "Well, I guess I'll CUL8R."

FAQ

What it means: "frequently asked questions." (New users are referred to such a list before asking a lot of, well, frequently asked questions.)

Example: "Download the FAQ file; you'll find the answer in there."

FWIW

What it means: "for what it's worth"

Example: "FWIW, the baked bean recipe on AOL smells awful."

FYA

What it means: "for your amusement"

Example: "Here, FYA, is the latest batch of Bill Gates jokes."

<g> or <vbg>

What it means: "grin" or "very big grin"

Example: "Of course, they hadn't read the FAQ first. <g>"

HHOS

What it means: "ha ha, only serious"

A more sardonic version of "ha, ha, only kidding."

Example: "The way your computer's whackin' out, I'd guess you've contracted a virus. HHOS."

IMHO

What it means: "in my humble opinion"

Example: "IMHO, that policy's doomed to failure. But what do I know?"

(When you stick this in front of any opinions or predictions you share, you won't be flamed as hard when you turn out to be wrong.)

Honest people tend to use IMO (leaving out the "humble").

IOW

What it means: "in other words"

Example: "Microsoft's stock rose last quarter due to significant new releases. IOW, Bill Gates is a rich man."

LOL

What it means: "laughing out loud"

Example: "Dutch, that computer repair horror story of yours had me LOL!"

OTOH

What it means: "on the other hand"

Example: "OTOH, that guy's been right before."

p

What it means: "Prodigy"

Example: "Yeah, I used to use *P* too."

PITA

What it means: "pain in the ***"

Example: "IMHO, people who don't read the FAQ are a real PITA."

PMJI

What it means: "pardon me for jumping in"

Example: "PMJI, but what baked bean recipe are you all talking about?"

ROFL or ROTFL

What it means: "rolling on the floor, laughing"

Example: "ROFL!"

RTFM

What it means: "read the ******* manual"

Examples: "If they don't read the FAQ, at least they could RTFM."

"How can I RTFM when I can't FTFM?" (*Find* the manual...)

TTFN

What it means: "ta ta for now"

Example: "Well, it's almost time to get up and get ready for work; TTFN."

Chapter 16
Ten (More!) Don'ts and Nevers

. . . .

In This Chapter

▶ Don't leave the computer turned on when adding a new gizmo

▶ Never hang up a modem connection by turning off your modem or computer (if you can avoid it)

▶ Don't use your external modem as a coffee mug holder

▶ Never pile any stuff around your external modem

▶ Don't use any other gizmo's power adapter with your external modem

▶ Never try to use two communications programs at once

▶ Don't use downloaded software without doing a virus scan

▶ Never give out personal financial/account information to unauthorized persons on-line

▶ Don't give out your password on-line

▶ Don't let strangers call your computer in host mode

▶ Never upload commercial software to anyone, anywhere

▶ Don't take part in "free long-distance calling" schemes

▶ Never change settings or switches without writing down what you did

▶ Don't try to pry open your old modem to "fix" it

▶ Don't string your external modem cables across the floor where someone usually you can trip over them

. . . .

*B*ooks about teaching always tell you to put things in positive terms. For example, you should never say "Never"; instead, you're supposed to emphasize the positive by saying things like: "Always wear heavy-duty shoes when walking through old minefields"; or, "Always string up pleasant curtains in your bomb shelter."

The warnings in this chapter are too serious for namby-pamby "Always" phrasing. So: *Never* do anything rash without *Always* turning here first.

Bring on the Don'ts and Nevers:

Don't Leave the Computer Turned On When Adding a New Gizmo

Exit properly from any programs that are running. Then turn off and unplug every component attached to your computer before opening its case or plugging new things into the ports in back. Touch a metal surface first to discharge pesky static.

This is especially true for adding an internal modem to your computer. But it's a good idea to make sure everything's turned off even before plugging in an external modem. Remember that sparing the expense of a computer technician may result in your calling an electrician! (Or a paramedic.)

Never Hang Up a Modem Connection by Turning Off Your Modem or Computer

Ending a connection in this manner is just plain rude. Besides, abruptly hanging up may "hang" the BBS or host computer, or even cause a system freeze on your own PC.

Tell the host computer you would like to hang up now, thank you. Then perform the commands to tell your software to hang up. When you see the NO CARRIER message, it's safe to exit your modem software and turn off your modem.

If your system has frozen or crashed and pressing Ctrl-C or Esc or ⌘-C doesn't help, then go ahead and press your reset button. But try to track down the problem's cause before you go modeming again.

Don't Use Your External Modem as a Coffee Mug Holder

Why risk a spill into the modem's sensitive circuitry? Especially as crowded as most computer desks get. Does your cat play the knock-over-your-water-glass-to-get-a-sip trick? Keep those liquid-loving kitties far away.

Never Pile Any Stuff Around Your External Modem

Notice the little vents on your modem's case? They're there to help the modem stay cool and let air circulate around its little parts. Don't thwart a beautiful job of engineering; keep stuff from piling under, around, or atop your modem.

Don't Use Any Other Gizmo's Power Adapter with Your External Modem

Lost your modem's power cord? Don't try to skimp by trying to use the one to your electric heater or sewing machine instead; they may use a different voltage (just because it fits the plug does not mean it's the same voltage), and it could fry your modem. Write the manufacturer if you must; wait for the replacement cord to come in the mail if you must; but use only the modem's recommended power adapter.

Never Try to Use Two Communications Programs at Once

Programs like Windows and DESQview can run more than one program at a time. Same with some Macintosh models. (Nerdy folk call this "multitasking," a word that invariably reminds me of that "I Love Lucy" episode with Lucy and Ethel trying to pack chocolate bon bons on an assembly line.)

Whatever the capabilities of your computer, don't make the mistake of trying to use more than one modem or fax program at a time. They're competing for the attention of the same COM port, and it won't work.

Of course, if you have two modems, each hooked up to its own COM port and phone line, go ahead. And congratulations for figuring it all out so well! Now go eat a bon bon.

Don't Use Downloaded Software without Doing a Virus Scan

Don't mess around with any unknown program without scanning it first for viruses. Most on-line services and BBSs tell you they've already screened all their files. That's nice. Scan anyway.

Whatever antivirus program you use, keep it current by calling the software's BBS and downloading updates. (Now that's putting your modem to good use!)

Never Give Out Personal Financial/ Account Information to Unauthorized Persons On-Line

Modems are a great way to automate and speed up financial dealings. You can bank by modem. Or maintain accounts with stockbrokers on-line. Be careful when giving out your personal information, however.

Sometimes, bad people go to great lengths to impersonate officials. (Remember the "unplugged" automated teller machine in that midwestern shopping mall? For weeks, it took unsuspecting cash-seeker's ATM cards, memorized their passwords and PINs, then spit the cards out with a "Sorry, I'm Out of Order" message. Cute — especially when the ringleaders were caught trying to draw money on those accounts.)

Crime doesn't pay. But until the criminals figure that out, keep an eye on your pocketbook and any numbers that can leave it open to others.

Don't Give Out Your Password On-Line

No on-line service employee will ever ask you for passwords or other confidential information. In the rare chance that someone does inquire about your password, ignore the scofflaw. If you're harassed again, note the user ID and report the offender.

Don't Let Strangers Call Your Computer in Host Mode

Letting a stranger call your computer through your software's host mode can leave your system files and private data open to exploration ... and maybe worse. Have your modem pal come over to help you set up host mode. Have your pal restrict the directories that a caller can access. Use a password to screen your callers. And only use host mode with friends you trust.

Never Upload Commercial Software to Anyone, Anywhere

Software is either commercial, shareware, or freeware. It's okay to pass along the latter two types to friends and BBSs. But if it's a program you bought and it doesn't say "Shareware" or "Registration Fee" anywhere, it's the first type: commercial software.

Sharing commercial software is a violation of copyright laws. Despite the many screensful of heated debate you'll see on-line, "electronic file freedom" or "software for all" are just that: topics for debate. Currently, it remains illegal for users to distribute their commercial software. You can be prosecuted if you do. You can get into trouble even by being associated with a *pirate board*, a BBS that specializes in illicit "warez."

Don't Take Part in "Free Long-Distance Calling" Schemes

It's unlikely that you'll ever be offered free long-distance account numbers, but remember: You can't get something for nothing in this world. Ignore anyone offering free phone calls. There's probably a stolen account number lurking behind the offer.

Never Change Settings or Switches without Writing Down What You Did

Don't change any modem or software settings "just to see what happens" unless you write down what you've changed. That way, you can change it back if something really weird happens.

Don't Try to Pry Open Your Old Modem to "Fix" It

You can't fix your old modem. Don't try. Unfortunately, it's rarely worth the expense of taking it to the repair shop, either. You may need to spring for a new modem, but don't injure yourself trying to patch up an old one. Refer to the first Don't at the beginning of this chapter.

Don't String Your External Modem Cables across the Floor Where Someone (Usually You) Can Trip Over Them

Don't fall for the old "I set up all the cables this way, so I know they're there" trick. You'll be the first one found face-down in your plush carpeting, smelling all the pet stains close-up and wishing you'd done a better job of cleaning them (and of stringing your modem cables).

Chapter 17

Ten External Modem Lights and What They Mean

In This Chapter

▶ Discovering what your modem is saying with these lights — and why you should care

| HS | AA | CD | OH | RD | SD | TR | MR | RS | CS | SYN | ARQ/FAX |

M all sorts of fascinating information before, during, and after a connection. Mostly, they can help in troubleshooting if things foul up.

Unfortunately, internal modems aren't as capable in the visuals department. (The blinking and flashing would only be seen by the computer's other innards: a waste.) Despite the absence of little lights, however, most people get along fine with their internal modems.

Here's a guide to the lights on your external modem. Remember, all brands differ slightly in the number and names of indicator lights.

The Usual Status Lights

Here are the status lights found on most external modems. As with everything else in the modem world, "your mileage may vary."

HS (High Speed)

Means: The High Speed light goes on when you connect at whatever rate your modem considers a high speed. All the brands vary; in fact, some modems flick on their HS light for several of their fastest speeds.

You see it: Most brands' HS lights stay on from the time you connect at high speed until you make your next modem call. Consult your manual for an exact translation of the HS light.

AA (Auto Answer/Answer)

Means: The Auto Answer light shows that your modem is ready to answer an incoming call.

You see it: Just before your modem answers the call; some brands turn it off during the ringing phase; others leave it on during the entire connection.

CD (Carrier Detect)

Means: The Carrier Detect light indicates that the modem is in touch with the remote modem; it reads the other modem's carrier signal, and the two modems start talking to each other.

You see it: When the modems have connected over the phone lines and are starting to talk to each other. The modems may still be speaking different languages, but at least they're aware of each other.

OH (Off Hook)

Means: The Off Hook light indicates that the modem has taken the phone line "off the hook" and is ready to start dialing and connect to another modem. Just like when you pick up the phone's receiver — you'll also hear a dial tone.

You see it: Both when the modem is trying to call and after it has made a successful connection and is on-line with the other modem. Everybody who calls you will hear a busy signal. If you think you're on-line but see that this light is off, think again.

RD (Receive Data)

Means: The Receive Data light blinks once with each bit of data your modem is receiving.

You see it: After you're on-line, and the remote computer is sending you menus and any information. Also, the RD light really starts blinking wildly after you've engaged your modem to receive a file you're downloading ... so if it's taking a long time to download, take a look at this light to confirm that things are moving forward.

SD (Send Data)

Means: The Send Data light blinks with each bit of data you send.

You see it: During a session with another modem, this light blinks when you type information to the other computer. During file transfers, it blinks regularly or looks like it's constantly on as you send the file you're uploading. As with RD, if you think you're sending data but it's taking forever, check whether this light is blinking to be sure that the send is going okay.

TR (Data Terminal Ready)

Means: The Terminal Ready light shows that the modem is on and talking with your computer and software. The computer has sent the modem a *DTR signal* through the serial cable.

You see it: When you load your modem software, the modem's power is on, and everything's working okay (so far).

MR (Modem Ready)

Means: The Modem Ready light comes on when the modem is turned on and plugged into the power source.

You see it: When you turn on your modem. It may flash when the modem is testing itself.

Less Commonly Seen Status Lights

The following status lights may or may not appear on your modem brand.

RS (Request to Send)

Means: The Request to Send light may come on when your computer sends your modem the RTS signal (depending on how the AT commands are configured).

You see it: During a modem connection, when your computer is asking your modem whether it's okay to start sending data.

CS (Clear to Send)

Means: The Clear to Send light stays on when receiving data — again, depending on the AT commands configured.

You see it: During a modem connection, when your modem is telling your computer that it's okay to start sending data.

SYN (Synchronous Mode)

Means: The modem is operating in synchronous mode. This mode is rarely required in personal computer communications; instead, mainframes and dedicated machines (like automated tellers) may use synchronous mode.

You see it: During a modem connection that's taking place in advanced, synchronous mode.

The ARQ/Error Control and Fax/Fax Mode Light

Means: This combination light says the modem has detected an error control signal on the other end and is working in that mode; it also flashes during a fax transmission.

You see ARQ/FAX remain on: During a connection in which the modem has achieved an error-controlled connection under specific AT command settings. It may flash briefly if the modem is resending data.

You see ARQ/FAX flashing: During a fax connection when the modem is sending or receiving a fax.

LB (Low Battery)

Means: A portable modem's battery is low.

You see it: The Low Battery light comes on when the battery is running low. (Or, just when you're about to finish transferring a large file ....)

Other, more specialized lights

You may encounter these lights on some brands of modems.

TXD (Transmit Data)

You see it: When the modem is sending data out the serial port.

RXD (Receive Data)

You see it: When the modem is receiving data into the serial port.

EC (Error Control)

You see it: When the modem is operating in Error Control or Data Compression mode.

SQ (Signal Quality)

You see it: When the connection is good; flashes when it's not so great.

TST (Test)

You see it: When the modem is performing a self-test.

Part VI
Appendixes

In this part...

This section contains all the stuff that you might need but that doesn't belong anywhere else in this book. Appendixes A and B give you all the ins and outs for splurging on a new modem and modem software. Appendix C tells you what to do with a downloaded file if it's compressed.

Finally, the Glossary defines all the computer lingo and jargon that you're likely to come across as you begin telecommunicating. Good luck!

Appendix A

Buying a Modem

You can't get started on any of this modem stuff until you buy a modem. And you shouldn't do *that* before giving these sections a once-over.

A Field Guide to Modems

The boxes filling the computer store's modem aisle are plastered with initials and acronyms. When the dust settles, you'll notice little *V-dot-this* and *MNP-that* numbers everywhere. The numbers after the letters *V* or *MNP* are code words for the performance *standards* the modem meets.

- ✔ *Protocol* is another word for standard; a modem's packaging may refer to the "V.XX protocol" instead of the "V.XX standard," for example. They both mean the same thing: the fussy rules a modem must follow to claim that it can do the things spelled out by that standard.

- ✔ Standards set rules for modem powers like *modulation*, or signaling *speed*; *error correction*; and *data compression*. These terms are dissected under their own headings elsewhere in this appendix.

- ✔ In a modem connection, the two modems try to talk to each other by using the highest standard they can both agree on. This negotiation, a noisy ritual known as *handshaking*, takes place at the beginning of the connection. (For those folks who've heard it before, those rushing/shrieking sounds are the two modems exchanging little modem résumés.)

- ✔ Expensive, cutting-edge modems simply meet more standards than do their bargain-basement brethren. No matter how many standards your modem meets, however, the modem on the other end must meet these standards too before they'll be put to any use. For example, you can't take advantage of your modem's error-correction standard if the other modem doesn't "understand" it.

- ✔ Broadly speaking, a modem that meets a certain standard embraces all the standards that came before it. (The computer industry calls this *downward compatibility*.) So a modem that conforms to the V.32, 9600 bps modulation standard can talk with *another* V.32 modem at speeds of up to 9600 bps. Plus, it also meets the earlier, V.22*bis* standard, which spells out how a modem talks at 2400 bps. (And the standard for 1200 bps, and 300 bps, and so on.) This rule makes it possible for a faster modem to drop down in speed when it realizes it's negotiating with a slower modem.

✔ Modem packaging usually lists several standards the modem can meet. So a V.22*bis* (2400 bps) modem may also include V.42 (error correction) and V.42*bis* (data compression) standards and boast about these on the packaging.

✔ The little *bis* word after a standard means it was the second version of that standard, which became the final, er, *standard* standard. (*Bis* means *second* in French, the language of *amour* ... and standards.)

✔ Modem standards are set by committees with firm ideas on how modems should behave. A committee from AT&T set the first *Bell* standards. (These date back to before AT&T set all the Baby Bell phone companies free to crawl around the floor on their own.)

✔ A company named Microcom named its *MNP (Microcom Networking Protocol)* standards after itself.

✔ A third committee outdid all the previous committees: its members agreed to meet in Switzerland so that they could eat stinky cheeses and gaze at chamois goats while thinking about standards. This is the *CCITT*, for *Comité Consultatif International Télégraphique et Téléphonique*. No one else could pronounce this, so the CCITT gave everyone a break, naming its standards vee-dot-this and vee-dot-that. To show its gratitude, and in recognition of the CCITT's superior pronunciation skills, the rest of the world awarded the CCITT jurisdiction over international modem standards. (Just to keep everyone awake, the CCITT has recently changed acronyms to be the ITU-T, which probably stands for something even less pronounceable.)

✔ Modem standards committees are composed of engineers, not marketers. That's why the standards are named after letters, dots, and numbers instead of moods, colors, or action words. These guys must have felt they were running low on numbers or something, because they certainly skimped when they chose numbers for the modem standards. This has resulted in the more popular standards having pretty darned similar numbers. In short, don't feel alone if you're confused about which modem standard does what. Everyone else is, too.

Modem Standards and How They Grew

Table A-1 gives the Berlitz Guide to Modem Packaging. The best way to use Table A-1 is to look at the modem you're thinking of buying and see what standards its box advertises. When you see an MNP number or vee-dot word, look at this handy table to see what the standard does. Then you can head to the section that deciphers the jargon word.

For example, if your modem box says "V.42" and Table A-1 shows that V.42 stands for something called _data compression_, look in the "Data Compression" section to see what that means. For a brief refresher so that you'll be ready when you're in the store and the modem salesperson is breathing down your neck, you'll find the terms recapped in this chapter's "Ten Modem-Buying Abbreviations" section.

Table A-1   Top Ten Most Common Modem Standards

This Standard Controls	_Standard Name_	_Capability the Standard Gives Modem_
Modulation (speed, basically)	Bell 103	Talks at 300 bps (bits per second)
	Bell 212	Talks at speeds up to 1200 bps
	V.21	Talks at speeds up to 300 bps
	V.22	Talks at speeds up to 1200 bps
	V.22_bis_	Talks at speeds up to 2400 bps
	V.32	Talks at speeds up to 9600 bps
	V.32_bis_	Talks at speeds up to 14,400 bps
	V.32terbo	Talks at speeds up to 19,200 bps, an extension of V.32_bis_ until V.Fast comes along
	HST	A proprietary (modem-brand specific) high-speed standard from US Robotics
(_near future_)	V.Fast	Will talk at speeds up to 28,800 bps
Error Correction	MNP 1, 2, and 3	Detects and controls errors from phone line noise during modem "conversation"
	MNP 4	Error correction that adapts data packet sizes to line conditions
	LAPM	Error correction
	V.42	Error correction; incorporates MNP 2-4 and LAPM
	ARQ	Error correction, for some specific modem brands

(continued)

Table A-1 *(continued)*

This Standard Controls	Standard Name	Capability the Standard Gives Modem
Data Compression	MNP 5, MNP 7	Compresses data during on-line sessions and in file transfers (up to 2:1).
	V.42*bis*	Compresses data even tighter than MNP 5 (up to 4:1).
Fax Standards	Group I	Ancient, slow standard for fax machines.
	Group II	Faster fax transmissions than Group I.
	Group III	Compatible with Group II; today's standard for 2400 bps (or faster) faxing by machines or fax modems; handles finer print quality (*resolution*).
	Group IV	Fax machine standard that includes Group III; standard for fax machines at very high speeds (19,200 bps); data compression.
	Class 1	Fax modem standard; supported by most software.
	Class 2	Faster standard in which the fax modem takes more of the work from the computer; also widely supported by fax software.
	Class 2.0	Upcoming fax modem standard; will allow fax modems to send/receive *data files* (wow).
	CAS	Fax modem standard by Intel Corporation that decides how PC/fax modem work together to send/receive; widely supported in fax software.
	SendFax	Send-only fax standard.
	V.17	Group III faxing at up to 14,400 bps (between two fax modems); 9600 bps to fax machines.

Examining Modem Features

Modems follow two basic camps. Generally, it's safe to say that the cheaper ones offer few or none of the data-compression, fax, or error-correction schemes. The more expensive ones offer most or all of them.

The rule for buying a modem is the same one you would follow when buying a new sport utility vehicle: get the best one you can afford at the time. While you own it, it'll take you pretty much wherever you want to go.

> ✔ A modem's sticker price isn't the only cost to take into account. Its initial price may zap your checkbook, but a high-speed modem greatly reduces the hourly connect charges levied by on-line services and pay BBSs, as well as any long-distance rates. (And your time is worth something, too, as you'll quickly discover during your first big download at 1200 bps.)

Here's a modem-buying lexicon to aid you in peeking under the hood.

Modem speed

Speed numbers like V.22*bis* or 14,400 bps tell you how many bits of information per second (bps) the modem can squirt back and forth across the phone lines. (You may see the term *baud* used interchangeably — and incorrectly — with *bps* to indicate modem speed.)

Whatever name it goes by, a modem's speed — more than anything else — determines how well it performs and how much you'll pay.

> ✔ A modem's top bps number isn't the only speed gauge that counts. Features like *error correction* and *data compression* are common on the high-speed models; these pitch in to boost a modem's *throughput*, or achievable speed, over its factory, or *raw speed*, limit. (The difference between the MPG on the sticker of your new car and what happens in real life.)

> ✔ I know a few folks who get by just fine using the older, slow 300 bps or 1200 bps modems. People who limit their modem use to calling one or two local BBSs and swapping e-mail messages with pals in various special-interest areas may be okay with these older models; there are times when speed's not much of a concern. (The average person can't read much faster than 300 bps anyway.) Even so, avoid the slowest speeds for your first modem purchase. (You would have to buy a used modem to even find these speeds for sale.)

> ✔ High-speed modems not only save you time and connect charges, they usually offer error correction and data compression, which are features that speed your data intact, even through noisy connections. When 9600 bps modems became common a few years ago, error correction and data

compression also became common: the faster the data "squirt," the more data will be affected by even a tiny line-noise glitch.

✔ Unfortunately, there have been cases where modem companies have falsely advertised their product's speed rates — touting enhanced rates achieved with error correction/data compression as the modem's raw speed. Remember, you won't always meet a modem capable of using these features on the other end of the connection. Look for the "V.XX" speed-standard on the box to guarantee that the modem is truly capable of the speeds spelled out by that standard.

Table A-2 wraps up the speeds you'll encounter in ads, on-line debates, and modem packaging.

Table A-2	Modem Speeds
Modem Speed	*When to Buy It*
300 bps	Accept this only if someone gives it to you, and then only for short-term practice sessions on local BBSs. You may quickly find that many BBSs and most on-line services don't even *allow* callers on at such a slow speed, because it ties up the phone lines for other users. You're better off shunning these slow dinosaurs.
1200 bps	Four times as fast as the 300 bps modems but still slow and creaky by today's standards.
2400 bps	This is the slowest speed commonly sold today; if you don't expect to do many file transfers or much long-distance dialing, you're okay with 2400 bps. Look for error-correction and data-compression standards that can boost this guy's speed and performance.
9600 bps	Today's workhorse standard, these 9600 bps speedsters are quickly being passed over in favor of their 14,400 comrades. High-speed modems (9600 bps and up) should feature error correction and data compression. Few on-line services top 9600 bps, the Internet excepted (but it's not really classified as an on-line service). Many BBSs zoom at much higher speeds, however.
14,400 bps	Prices are already falling on these top achievers. Make sure that the modem supports the true V.32*bis* standard by looking for "V.32*bis*" written on the box. Some modems advertise high speeds that are only possible when you connect to the exact same brand/model.
16,800 bps	Modems currently achieve this speed only when connected to the exact same brand/model of modem. Many BBSs delight in offering their callers a range of high-speed modem brands, however.
19,200 bps	As with the 16.8 Kbps just mentioned, the only modems that talk this fast use special, *proprietary* schemes to talk to the exact same modem brand and model.

Modem Speed	When to Buy It
28,800 bps (near future)	Many modems now advertise their "upgradability" to the future high-speed standard, informally known as V.Fast. They claim that you'll be able to torque your modem by opening up the case and inserting a circuit card, replacing a chip, or tweaking it in some other way. Buyer beware: To successfully transmit data at 28,800 bps, even with error correction a modem will need an almost perfect phone connection — difficult to find — plus another 28,800 bps modem on the other end — even more difficult to find.

Bps numbers and vee-dot words don't really say that much. Table A-3 shows you how modem speeds affect what really counts: your checkbook. The table shows what it costs at various speeds to download, read a text file, and *conference* (type back and forth in "real time") with other users on CompuServe, assuming Standard Plan membership rates current at the time this was written.

Table A-3	Modem Speeds: Real-life Version				
Task	Speed	Connect Charge	Time to Perform Task	Cents/ Minute	Total Cost
Downloading a 2M file* (*2 megabytes is a fairly sizable file)	300 bps	$6/hr.	18+ hrs ...forget it!	$.10	$108
	1200 bps	$8/hr.	4 hours, 32 min.	$.13	$36.26
	2400 bps	$8/hr.	2 hours, 16 min.	$.13	$18.13
	9600 bps	$16/hr.	34 min.	$.27	$9.07
Reading a 25K text file on-line	300 bps	$6/hr.	14 min.	$.10	$1.40
	1200/2400 bps	$8/hr.	4 min.	$.13	$.52
	9600 bps	$16/hr.	4 min.	$.27	$1.08
Typing to other users on-line or conferencing	300 bps	$6/hr.	1 hr.	$.10	$6
	1200/ 2400 bps	$8/hr.	1 hr.	$.13	$8
	9600 bps	$16/hr.	1 hr.	$.27	$16

As you can see, faster speeds are good for downloading. When you'll be conferencing and reading stuff on-line, call at 2400 bps instead.

✔ Buy a faster modem (9600+ bps) if you need to download and upload files of any real size. Do you need to connect with another modem long-distance? Speed pays off here, too.

✔ Along the same lines, if you're considering joining up with an on-line service or BBS that charges an hourly rate, consider a high-speed modem. As you can see in Table A-3, however, the savings depends on what you're doing on-line.

I know UART but what am I?

Skip this section unless you're considering a high-speed external modem. Even after you have the modem, don't bother reading this unless you're losing characters during file transfers or on-line sessions. Even then, make your modem guru read this, instead of you.

A special *UART* (pronounced YOU-art) chip on your computer's serial port directs data flow between the PC and the port. (In short, the UART converts 8-bit bytes from your PC into single bits to send out over the serial port — and bits into bytes in the other direction.)

With faster external modems transmitting at higher speeds, the information may stream faster than the serial port can handle. This bottleneck can cause dropped characters, seen as chunks of data missing from a transferred file, for example. This is particularly likely to happen when running a DOS modem program "under" a multitasking program like Microsoft Windows or DESQview, when many things called interrupts are competing for your PC's attention.

To take full advantage of your high-speed modem, make sure that your serial port has an updated UART, which provides a holding tank (nerds call it a *buffer*) for the speedy bits until your busy PC can get to them. Look for a 16550AFN UART or better. (Internal modems supply the UART, so any modem you buy now should have the later, more capable chip.)

If you have an older computer, your computer's serial port I/O card probably has an older UART

chip soldered onto it. Although diagnosing your UART is beyond the scope of this book, Microsoft Windows and DOS 6 users can use the MSD (Microsoft Diagnostics) program to tell what type of UART they have. Also, many BBSs and on-line services offer easy-to-use UART diagnostics in their communications areas. Ask your favorite BBS's sysop or your on-line guru pal to recommend one.

If you have the older UART, consider replacing your PC's I/O card, which supplies your computer's serial and parallel ports. If you buy a card with two serial ports (recommended), make sure that each port is equipped with the beefy 16550AFN UART chip.

Hayes makes a speedy serial port card called the ESP Communications Accelerator that goes inside a PC's case. Besides the updated UARTs, it offers a special chip called a *coprocessor*, which shoulders some of the communications tasks. The drawback? You'll still need a printer (parallel) port, which comes on a separate card — filling up yet another expansion slot inside your PC.

Replacing an I/O card is similar to installing an internal modem or any expansion card inside your PC. Just the same, I'd recommend teaming up with your modem pal when shopping for and installing a new I/O card. If you're on your own here, be sure to buy *Upgrading and Repairing PCs For Dummies* to help you out.

Error correction

Everyone has heard the clicks and hisses of a bad phone connection. Modems can hear them, too, and they earnestly try to make some sense of them, trusting that the errors are part of the data you're trying to send. The result: garbage characters on your screen and potential loss of data.

Error correction provides extra signals that let the two modems double-check the data and resend it if the check marks don't add up.

For error correction to take place, the modem at each end of the connection must support the standard.

Count on error correction to be included with almost any 9600 bps or faster modem. That's because the modem companies know that a line noise bleep that causes one bit to be lost at 2400 bps may well wreak major havoc at 9600 bps — and even more damage at higher speeds — requiring the modems to resend the afflicted parts and taking longer in the process.

Look for the error correction standard called V.42. When V.42 senses a modem using an older error-correction standard like MNP Levels 4 through 2, it can "fall back" to these standards. In short, the V.42 modem can talk error-free to a wider range of modems, old and new.

Data compression

The checks and balances of modem error correction pave the way for data compression. In fact, no modem can offer compression without first offering error correction.

Tedium to ignore about streaming protocols

If you achieve an error-correcting connection, an external modem usually has a special light that tells you so.

This comes in handy for the times you want to upload or download a file and you want to see your data rate really zoom. When the light ensures that error correction is taking place on the modem hardware end, you can opt for a non-error-correcting file-transfer protocol (like YModem-G or XModem-G), which sends the data in fast streams (and leaves the little checks and balances up to the modems).

To be safe, however, save the Whoopee-fast -_G_ protocols for recreational (and small, in case you need to transfer them over again) files. Use ZModem for the crucial file transfers; it has the capability to sense an error-correcting modem connection and turn off its own error-checking scheme. (But you're still safe if error correction isn't working for some reason.)

When two modems establish a link using data compression, they agree to pack the data down into shorthand — sending notes about the shorthand they're using along with the squished packets. Compressing the data in this manner allows more data to squeeze through — or *enables higher throughput*, as nerds phrase it. The data is unpacked by the receiving modem and sent through your computer's serial port to the PC.

For data compression to take place, the modem at each end of the connection must support it.

Most of the better high-speed modems offer two forms of data compression: the CCITT standard called V.42*bis*, along with an older (and inferior, some say) one called MNP 5. Unlike the error-correction standards discussed previously, MNP 5 isn't automatically included in V.42*bis*. You'll want MNP 5 for any encounters you may have with modems using this older standard, however. Again, most good modems offer both.

You'll hear the salesperson say that V.42*bis* compresses data at a ratio of 4 to 1, while MNP 5 only squeezes stuff down 2 to 1. Simply put, V.42*bis* is twice as efficient and fast as MNP 5.

Many modem companies advertise their V.42*bis* modem as capable of speeds greatly exceeding the modem's usual "sticker" rate. Beware: these speeds are more theoretical than truth. They assume the connection is taking place with error correction and under optimal conditions.

Still, the modem savants recommend that with data compression, you should set your modem software's *computer to modem rate* (your software setting may say *DTE rate*) to four times the speed of your modem if you want to achieve the fastest rates possible. That's because the data streaming over the phone lines is compressed, whereas the data streaming from your modem into your computer's serial port is not compressed. You want your computer to be able to handle all the extra data on its end. Refer to Chapter 4, where a table does the math for you.

If you buy a modem with MNP 5 data compression, ask your modem guru for help in finding the software setting or AT command that turns it off. You want to disable this standard to achieve the best rates when transferring a file that's already been compressed with PKZIP, LHA, Stuffit, or one of the other "packer" software products discussed in Appendix C. (With MNP 5 data compression enabled, it usually takes *longer* to send or receive precompressed data.)

Hayes compatibility

The term *Hayes compatibility* is confusing: there are many Hayes standards, and most modems offer at least some of them. Even so, you're best off buying a modem that proclaims Hayes compatibility. To be safe, test it with your modem software as soon as you get it home so that you can return it for a refund if need be.

Voice capability

Many of today's modems can serve as fax machines. Now, a new breed of modems can even serve as answering machines. A few advanced modems, like ZyXEL's U-1496 series, come with *digitized voice capability*. This lets your modem serve as a voice answering machine when used with the proper software. The incoming voice message is stored as a file on your computer; you then tell the software to "play back" your message. As with a normal answering machine, you actually hear the caller's voice message.

Because sound files are large and take up lots of hard disk space, the ZyXEL and other voice-capable modems use a scheme called *speech compression* to cut down on hard disk hogging.

Voice/data switch

The US Robotics Courier, the ZyXEL U-1496 series, and many other brands come with a voice/data switch that lets you alternate between voice and data communications during a call. Most of these modems let you program the switch for self-testing and other functions.

Price

You can pay anywhere from $25 to $1,250 for a modem. The lowest-priced modems are the stores' *loss leaders*: items you'll spot in the ad and come running down to check out. After you're in the store, don't be surprised if you end up looking at better models. In fact, I'd recommend you avoid the sub-bargain basement (sub $30) modems altogether.

Expect to pay a bit more for data compression and error correction. External modems cost a little more, too, from $10 to $50 more, depending on the lights and other *bells and whistles* included on the case. The big price hikes come with features you may not need, like synchronous transmission (used in specialty applications like ATM machines) or upgradability to future standards.

If it's possible, try to get error correction and data compression. For about 50 bucks extra, you can get a modem that also has fax capabilities. Because I like having a handy speaker volume-control knob and the diagnostic lights, I prefer external modems to the internal variety.

Not all modems offer a speaker, but be sure to get one that does.

Other features

Here are some other features to look for:

- ✔ Tone or pulse dialing to meet all dialing environments
- ✔ Automatic sensing of incoming calls
- ✔ Built-in test and diagnostics
- ✔ Full- and half-duplex transmission
- ✔ An aluminum or sturdy case
- ✔ A decent manual, warranty, and support — preferably a BBS

Internal Versus External

Two main species compete for food and territory in the modem world: *internal* and *external.*

- ✔ Internal and external modems work essentially the same, but a plastic or metal case houses the external kind.

- ✔ Internal modems come in the form of expansion cards; you install them inside your computer's case, where they conspire against you with all the other expansion cards.

- ✔ Mac owners rarely buy and install internal modems. It's hard even *finding* a Mac internal modem in the stores.

  One big exception to this rule is the crop of internal modems springing up around the PowerBook, those Mac laptop models. Let the authorized Mac store guy install this one, though.

- ✔ External modems sit outside your computer on your desk. Some people put them under their telephone. Their cases are adorned with little lights along the front. Modem experts derive great satisfaction from tracking their modem sessions by watching the blinking of these little lights. Actually, the lights are very helpful for being able to tell whether you're still connected to another modem. (If you end up getting an external modem and you want to see the other things the little lights say, page to Chapter 17, "Ten External Modem Lights and What They Mean.")

✔ External modems come with a power cord, which means you need a place to plug it in. Internal modems feed off the computer's power supply and "come on" when you turn on your computer.

✔ External modems also require a cable that plugs into your computer at one end and into the modem at the other end. You buy this *serial* cable separately; it runs anywhere from $10 to $25 in computer stores.

✔ All modems, regardless of affiliation, come with a phone line that plugs into your wall phone jack.

Considering an external modem? You'll need a serial cable for it. Serial cables come in various connector types and grades (but they're all the same boring gray color). The connector type you buy depends on what the serial port in back of your computer needs. As far as grade goes, buy the best quality hardware-handshaking cable with the full complement of 25 pins. (Be sure to go to Chapter 2 and look over Table 2-1, "What Serial Cable Should I Buy?" before you head to the store.)

Unless you have a desktop Macintosh, buying an internal or external modem really depends on which advantages or disadvantages count most to you. For example, an external modem can easily be moved from one computer to another. To see how the two stack up, turn to the end of Chapter 2 to read "Ten Reasons to be Glad You Installed an Internal/External Modem."

The important thing about a modem isn't really whether it's internal or external. Rather, you need to ask yourself what speed and other capabilities you'll need. And that depends on what you'll be *doing* with your modem. After you decide that, review this Appendix's sections "Modem Speed," "Error Correction," "Data Compression," and "Hayes Compatibility" — and the *real* deciding factor, "Price."

Buying an Internal Modem?

Here are some things to keep in mind as you look for an internal modem:

✔ Make sure that you have a spare slot inside your PC.

✔ Make sure that you buy the correct type of card to match your computer type: very few cards are designed for an off-the-wall slot called MCA, or Micro Channel.

✔ Be sure that the internal modem fits in the spare slot you have.

Which Modem Should I Buy?

No problem ... get the BahZoomBah 257,600-zillion bps vee-dot model with combined oven-cleaning and hard-disk optimizing features! It's on sale for a low, low $1,200 down at CompuWarp.

Fortunately, most people don't need the latest, greatest state-of-the-art modem. In fact, the most high-powered modems bring a raft of troubles trailing in their wake. First of all, decide what tasks your modem will be called upon to do.

Calling GEnie, Prodigy, and America Online

You can't connect with these services at speeds higher than 9600 bps (after America Online begins to offer 9600 bps service). If this will be the main extent of your modeming, you won't need a faster modem. Prodigy doesn't charge extra for calling at this fast rate, incidentally; rumor has it that AOL won't charge extra, either. CompuServe just started offering 14,400 bps services in some lucky "test" cities; look for 14,400 service nationwide by the end of 1994.

Look for a reliable brand that complies with the V.32 standard. V.42 and V.42*bis* come in handy here, too. US Robotics, Hayes, Boca, ZyXEL, AT&T, and Intel are a few of the top brands to investigate.

Calling other on-line services and BBSs

You need general modem software to call most places. *Make sure that your modem software supports the modem brand.* This can save you hours of fiddling with software settings to get your modem to work at its optimal rate. Unfortunately, you can't really tell whether your modem is supported without installing the software and heading for its "modem brand choosing" setting. Happily, the software or modem company's BBS is likely to contain a modem driver program to optimize the modem/software relationship.

Transferring files from computer to computer

If you plan to transfer files often between specific modems, you'll save time and long-distance charges if you buy modems that match in speed and other capabilities (like error correction/data compression).

Downloading and uploading files

Consider buying a faster modem if you'll be moving great big piles of data over the phone lines. Look at features like error correction and data compression to make sure that you can use the fastest file-transfer protocols.

Reading text on-line

You won't need a super fast modem if you're just planning to call a few services and read messages on-line. (You can't read as fast as the fastest ones go without getting whiplash.)

Conferencing and chatting on-line

These two words mean the same thing: typing messages back and forth with people in actual time. You can imagine that you won't need such a fast modem for this.

If you live in an area with noisy phone lines and conferencing is something you're serious about, look for a modem with error correction.

Do I Need to Fax with My Modem?

Most of the modems on the market include the capability to send and receive faxes. If you need a quick way to transmit letters and other computer-generated documents to people with fax machines, buy a modem with fax capabilities.

Whereas a modem sends a sound-wave "breakdown" of a digital document, a fax modem sends a *picture* of that document. When you receive a fax on your computer, you can view it on your screen — or you can print it out. You can't work with the text in any way, however, because it's only a picture of text.

Naturally, there's an exception to this rule. Some fax modems come with special *optical character recognition* (OCR) software that translates your fax picture into your computer as data. Early OCR software didn't work well; luckily, it's getting better.

WinFax Pro fax software includes OCR software; this works pretty well.

Fax Standards

Fax machines have been around for years. Today, the fax modem offers a cost-effective and convenient alternative for people who need to fax documents that originate in their computers.

Look for a Class 2 fax modem to support the Group III send/receive fax capabilities: both at 9600 bps, if possible. The modem should also operate at that speed. The modem should support the proprietary CAS standard, but avoid fax modems that can only receive faxes.

Laptop Talk

Modems are the shrunken heads of the computer world. They get smaller each year until they're hardly recognizable as modems. You can buy one of the stylish, cubist portable modems, for example, that will fit inside your laptop carrying case.

Small *pocket modems* work great on either your laptop or regular-sized desktop computer. Two modems in one!

An even smaller breed of modems comes on something called a PCMCIA card. (The acronym-artists were feeling emboldened the night they came up with that one.) These come on credit-card sized plastic rectangles that fit into similarly-sized slots inside many portable computers.

Today, even the high-speed, 14,400 bps, fax-capable blazers can be found on PCMCIA cards. By the time you read this, the standard may be a bit more developed; right now, I've found that everyone is still sorting out the details.

Before you rush out for one of these modems, be sure to read current reviews in *PC World*, *PC Magazine*, and other monthlies. One of the benefits of an on-line service is that you can search for and read — even download — reviews from magazines that maintain a presence on-line. Some of these even offer you e-mail access to the editors who wrote the review — for in-depth discussions.

Ten Modem-Buying Abbreviations

Most of those strange "vee-dot" words you see stand for three things: how fast a modem transmits data, or its *modulation rate*; its capability to spot errors speeding by, or *error correction*; and whether it can squeeze down data into a smaller packet, or *data compression* (and unsqueeze).

Here's a quick blow-by-blow to the buzzwords you're likely to see on packages in the modem aisle and how to decipher them.

AT: The modem's box usually advertises the modem's "Hayes AT Command Set" compatibility. This means that the modem complies with standards set by the Hayes Microcomputer Products company for modem commands. What this really means is that the modem can "get along" with most modem software and other modems.

Important: Do not consider buying a non-Hayes-compatible modem, no matter how much money you think you're saving.

bis: French, for *second*; a standard tagged with *bis* is the second version of that standard.

bps: Bits per second; the number directly preceding this tells you the maximum speed at which the modem can transfer data over the phone lines. Today's cutting-edge modems' maximum bps? About 57,600 bps, under fantastically ideal conditions at both ends of the transmission.

Don't go by the "bps word" when trying to tell how fast a given modem can go. Instead, look for the vee-dot standard and any MNP words. Then see what they mean in Table A-1.

CCITT: This group gathers periodically to set the standards for modem speed and data compression and other stuff that you see as the "vee-dot" words on the modem's package. Also, this committee recently changed its initials to ITU-T.

When buying a modem, consider whether you'll be communicating overseas. If so, be sure to buy a modem that adheres to the CCITT standards, which are accepted internationally.

HAYES: A modem company that played a pioneer role in the industry; Hayes set some basic standards for modem commands, including the "AT Command Set," back in the early days of modems. Look for the modem's package to say, "Hayes Compatible." See the warning icon near the beginning of this section.

MNP: A company that set standards for error correction and data compression. The better modems manage to agree both with MNP *and* CCITT standards.

MNP4: An older standard for error correction; it includes MNP standards 2 and 3.

MNP5: An older standard for data compression; it's best to disable this standard when transferring files compressed by means of software packers like PKZIP, LHA, Stuffit, and others.

V.22: A CCITT standard for modems that communicate at 2400 bps.

V.32: The standard for modem communication at 9600 bps. V.32 modems can drop down to lower speeds in case they hear line noise that can cause errors to infiltrate the data being transmitted. Brought to you by the CCITT.

V.32*bis*: The standard for modem communication at 14,400 bps. These modems can drop down to lower speeds as they sense the speed the modem on the other end is capable of.

V.42: A CCITT standard for error correction; includes MNP 2-4 standards for "downward compatibility" with older error-correcting modems.

V.42*bis*: The CCITT's data-compression standard, capable of squeezing a file down to a quarter of its size. V.42 is for error correction; V.42*bis* is for data compression. The only difference in the name is the *bis*.

RS-232: A quasi-standard for serial cables and interfaces. Look for this to be displayed somewhere on the serial cable's packaging if you're opting for an external modem.

VISA: Also **MC**, **AE**; credit card standards; something you'll probably put down on the table when paying for your new toy.

Appendix B

Buying Modem Software

*W*hich modem program is the best? That's easy: the one that meets your needs the best. With dozens of packages on the market, often with very little difference among them, it's more important than ever to decide what *you* want from a modem program.

Some people, for example, merely want to check their e-mail once a day. Other people want to download six files automatically at 3 a.m., when the rates are low and the phone lines aren't busy.

One of the best ways to decide what you need in a modem program is to jump right in: start using one and keep track of what you don't like about it. And with shareware modem programs, you don't even have to pay until you find one you like.

Where to find these shareware programs? Your best bet is to seek out and attend a meeting of your local computer users' group. Most computer dealers can recommend a group in your area and can let you know about meeting times and locations.

During a users' group meeting, a shareware *librarian* usually sits in the back corner, selling programs for a moderate fee — usually just enough to cover the cost of the floppy disk it's on. Ask the librarian to recommend a simple program for getting on-line right away.

When you're on-line, you'll find plenty of enthusiastic advocates for each of the commercial and shareware programs. So which modem program do you choose? That's easy: the one your modem guru uses.

Seriously, consider buying something that your on-line pals like and recommend. The learning curve is a lot shorter when you have company. And you can always invite everyone over to your place for a "script macro recording" session (when you really want help just installing the program and setting it up).

Best of all, you may find yourself peeking over your friends' shoulders and getting some of those pesky questions answered without having to moan anything out loud.

Although you may pay a few dollars for a shareware program on a disk, remember you're only paying for the disk. You'll need to register the program with the shareware author when you find yourself using and enjoying it.

Check the coupons in back of this book for some good leads, too. When you've decided that you need a full-featured program and you've identified a winner on-line, you can take advantage of some pretty great discounts offered by these software publishers.

In the meantime, here are some features to consider when shopping for a modem program.

Top Ten Modem Software Considerations

Context-sensitive help

A modem program should bring help for any command you're puzzled by — right from that part of the program — at the touch of a single key. *Context-sensitive help* is much more convenient than traipsing over to a Help menu, reading about a command, and then finding your way back to where you were, especially if you happen to be on-line at the time.

Marking entries for mass dialing

Some programs call this feature *queue* dialing — a way to make your modem call several different places by moving ahead to the next number if one's busy.

Without queue dialing, a busy signal will keep you stuck for a long time as the modem dials the same place over and over until it finally gets through.

Support for high-speed modems

If you have a 9600 bps modem or one that's even faster, buy a modem program that can keep up with you. It should allow both hardware and software flow control, the ZModem file-transfer protocol, and modem-to-computer (DTE) speeds that are at least four times faster than your modem. All these zonko terms are explained in Chapter 4.

Built-in ZModem

If you'll be sending and receiving many files, buy a program that offers ZModem, a powerful file-transfer protocol. Most programs offer ZModem, thank goodness, but a few skimp.

Ease of use

Are the menus clear in terms of what they do? What keys do you press to make various boxes pop up? Can you change a setting quickly with a mouse click or minimal typing?

Look for a program that's easy to understand, with settings that can be changed without much hassle.

Easy script recording

This feature is not for everyone. But it's a joy to find a program that can *watch* all the words you type while you're logging on to an on-line service, record and remember them, and then type them back for you automatically the next time you call.

This feature is called *script recording*, like a movie script actors read. You can use scripts to do much more powerful things, but using them to log on to places automatically can be a real time-saver.

Alternate phone numbers

Many BBSs or on-line systems offer more than one phone number to call, so look for a program that recognizes this fact. The program should let you enter three or four different numbers for a service and then automatically shift among them, dialing the next number if the first is busy.

Too many programs make users fill out long forms for every new phone number they enter — even if those numbers are for the same on-line service.

Support for your modem brand

Modems talk the same basic language, but each brand has its own dialect. Look for a modem program that can talk specifically to *your* brand of modem. Otherwise, you'll be tossed upon the shores of Lake "Modem Init Strings" (and other unmentionables) and left to perish at the hands of the cruel natives.

Attached notes

A few programs offer a built-in editor for jotting down and storing notes about the places you're calling. You can record stuff about folks you meet on-line or files you want to download the next time you're on. (Avoid storing notes about passwords, on-line nicknames, and other information you don't want getting out.)

Easy mail processing

Not all modem programs have a built-in off-line mail reader (sigh). But it's a wonderful feature to look for because it allows you to automate the vast e-mail and messaging that comes as a natural part of being on-line.

Mouse capability

Most modem programs offer menus you can access with your mouse (all Mac programs do, of course). If you don't have a mouse yet, you'll enjoy using one to get on-line. (Get your modem guru to come over and install it, though, to avoid dealing with COM port conflicts and other miseries.)

Note that few BBSs and on-line services let you navigate using a mouse — but most modem programs are mouse-conscious (and much easier to use as a result).

Built-in fax sending

If you have a fax modem, look for a program that lets you dash off a quick fax. (Increasing numbers of programs offer both send and receive fax capabilities.) High-traffic fax users should opt for Delrina's line of dedicated fax software or another fax-specialty program.

Custom-written software

If you're only going to be calling one place, look for software designed with that place in mind.

For example, Prodigy and America Online come with their own built-in software — you can't call them with any other modem program.

CompuServe, however, can be reached from any modem program, so some people have written modem programs designed specifically for CompuServe. Programs like TapCIS, Navigator, and OZCIS can call CompuServe to grab your mail and log off as quickly as possible. Or you can use the CompuServe Information Manager software designed to keep you on-line forever.

The other on-line services offer similar "make-easy" navigation programs. Try one out!

Appendix C
I've Downloaded a File: Now What?

Congratulations! You're the proud recipient of a newly downloaded file. But stifle the urge to check it out and play with it right away. To keep your hard disk safe and organized, you need to make the new file jump through a few hoops first.

This appendix hands you the whips and tells you when to crack 'em. Mac users can skip down to the Macintosh icons and start reading there. If you need a refresher course on downloading, head back to Chapter 5.

Who You Callin' "JUNK"?

The last thing you want is for new, untested files to be roaming around all over your computer. Make it a point to set up a new-file "holding tank," and use it, whether the file is something your friend hands you on a disk or your latest download. That way, your computer runs less risk of contracting a virus — and all your new programs stay handy in one place.

- PC users can make a JUNK directory by using the DOS Make Directory (MD) command. At the DOS prompt, type these two lines:

```
C:\> CD \
C:\> MD JUNK
```

Press Enter. Now you can use the DOS COPY command to get your file from the directory it downloaded to into your JUNK directory. At the DOS prompt, type:

```
C:\> COPY C:\WHEREVER\FILENAME.ZIP C:\JUNK
```

Press Enter. Substitute your download's location directory name for \WHEREVER in this example. You type your downloaded file's name instead of FILENAME.ZIP, too.

- From the Desktop's File menu, choose New Folder. Select the folder's meek *untitled folder* bar and type **JUNK**, instead. Voilà! Newly downloaded files can be dragged into the JUNK tank until you get a chance to uncompress them and check for viruses. (These inspections are covered in the next few sections.)

> ✔ After you sit the newcoming file down under the bright lights and cross-examine it, you can promote it to its own "real" subdirectory or folder. Keep reading to see how.

Shareware is the name for the typical try-before-you-buy programs and files you'll be downloading from BBSs and on-line services. For a refresher, flip back to Chapter 1.

Freeware and public domain software are different: you can pass it around and use it all you want; there's no guilt message reminding you to *register* (pay for) it, which you get with shareware.

Unpacking the New Arrival

You can't start having fun in Hawaii until you rip into your luggage and unpack your shorts and sandals. It's the same with most files you download: they come packed and can't run around in the sand until you run a program to unpack, or *uncompress*, them.

> ✔ You're glad to bear the slight inconvenience this causes: when files are squished down (to less than half their uncompressed size, usually), they take up less room on the BBS or on-line service — and less downloading time. This saves you precious on-line minutes (tick ... tick) and maybe some money, too. (Didn't Ben Franklin or someone who's always harping on these things say, "Time is money"?)

> ✔ Compression programs work both ways, incidentally — so if you were in an ultra-nerdy mood you could use its *compress* feature to pack down several files into one smaller file and save room on your hard disk or a floppy. Computer folk call this *archiving*, and even the nerdiest ones put it off until running out of hard disk space makes archiving absolutely necessary.

> ✔ Table C-1 displays file endings, called *extensions*, that you're likely to see on files you want to download, along with the file-compression program you'll need and what command leads to file liberation. Science project prizewinners call packers *data-compression utilities*, just in case you hear it bandied about outside the lab one day.

> ✔ Many shareware packers are available on BBSs and on-line services. Software stores are also starting to carry shareware. And there's always that old standby, mail order. Check the coupons in the back of this book for some great offers.

Table C-1	Compressed Files Have Even Weirder Names than Usual and Won't Do Anything Unless You Unpack Them	
File Ending	*This Packer Did it*	*Type This to Unpack*
DOSFILE.*ZIP*	PKZIP.EXE pronounced "P.K. Zip"	C:\> **PKUNZIP -o-d DOSFILE**
DOSFILE.*ARJ*	ARJ.EXE rhymes with "Marge"	C:\> **ARJ X DOSFILE**
DOSFILE.*ARC*	ARC.EXE older, say "ark"	C:\> **UNARC DOSFILE** (older)
DOSFILE.*PAK*	PAK.EXE older, say "pack"	C:\> **UNPAK DOSFILE** (older)
DOSFILE.*LZH* or .*LHA*	LHA213.EXE pronounced "L.H.A."	C:\> **LHA X DOSFILE**
MACFILE.*SIT*	StuffIt ... (Lite, Deluxe, and so on)	Use freeware Extractor or StuffIt Expander; follow Menu to expand
MACFILE.*CPT* Expander	CompactPro	Use freeware Extractor or StuffIt
MACFILE.*PIT* use	PackIt III	Obsolete packer: fire up StuffIt and its Unpack Translator
MACFILE.*SEA*	SEA stands for Self-Extracting Archive	Simple: just double-click the file icon (no extra program needed)
MACMORE.*DD*	Disk Doubler (commercial)	Disk Doubler works; so does freeware unpacker DDExpand

Who cares how packers work?

Packers are the kindly big sisters of the computer world. After watching you try to cram two pairs of boots, six trousers, and three shirts into your bulging suitcase, Sis steps in and tosses out the duplicates (and helps you sit on the suitcase to close it).

Like Sis, a packer watches for duplicates in data — cramming them down by using shorthand for repeated patterns it spots in the file.

✔ Packer programmers continuously update their wares, so the file names may look slightly different by the time you get around to downloading or sending away for them. For example, the current file name for the program that contains PKZIP and PKUNZIP is PK204G.EXE. The packers are packed in self-extracting files, meaning that you need only type the file name **ARJ**, for example, to get the program ARJ.EXE to unpack itself.

✔ Mac packers self-extract when you click their icons. StuffIt and CompactPro are shareware and start at $25, so if you just want to unpack files and have no need to pack 'em, get the freeware programs Extractor or StuffIt Expander instead; these are one-way programs.

✔ In the weird event that you need to unpack DOS files in the .ZIP format, StuffIt's Deluxe (commercial) version (is it just me, or are there twenty thousand versions of StuffIt?!) or a shareware program called MacZip will do the job.

On the path to righteousness and unpacking

Stick your packer in a directory that's on your DOS path so that it can work on a packed file lurking anywhere on your hard disk.

✔ Here's how: make a \UTIL subdirectory and COPY the packer into UTIL. (For a blow-by-blow account, refer in this chapter to the instructions for making a JUNK directory.) Then have a nerdy pal put the UTIL directory in the PATH statement that's part of your computer's AUTOEXEC.BAT start-up file and that looks something like this:

```
PATH C:\DOS;C:\UTIL
```

✔ Utility directories and folders are great places for storing your virus scanner (keep reading to see what this does and why you need it) plus all the other utilities you want to keep handy.

✔ If you shun this advice, you'll be forced to an eternity of copying the unpacker along with each downloaded file to your JUNK directory (or typing some very long, unwieldy commands each time you want to unpack a file). Some people add their packer to their DOS directory (and add it to the PATH), but this is rather untidy.

Scanning for Viruses (and Other Pleasures of Modern-Day Living)

Don't even think of downloading files onto your computer before you get your hands on an antivirus program and set it up. *Viruses* are screwed-up programs that can screw up your computer; they're written and spread around by screwed-up malcontents who derive some sort of screwed-up satisfaction from ruining others' computers and data.

End of rant. (Rant, rant, grrr.)

- ✔ Viruses replicate by attaching to real-life programs like a file you download or a disk a friend hands you. Viruses have even been found in shrink-wrapped, commercial software boxes.

- ✔ Two types of antivirus programs safeguard against infection two different ways. You set up the first type to lurk in your computer's RAM, awaiting signs of suspicious activity. With the other type, you command it to scan your computer when you first install it (and at regular intervals thereafter). Then you make it scan each new disk and file you want to use. Commercial and shareware versions of each type abound. (The Mac even has a freeware virus scanner.)

- ✔ A virus can't go to work on your files until you trigger its evil machinations — usually by running the program it's attached to. That's why it's crucial not to run any new programs until they're scanned. (Some Mac viruses can spread just by sticking an infected floppy into your Mac, though.)

- ✔ You can't scan a packed file for viruses. It's too easy for one to hide there. Unpack the program first into your JUNK directory or folder and then be sure to scan all its files.

- ✔ Macs get lucky in the virus department. Not only do fewer viruses menace Macs, but their owners can download Disinfectant, a top-rated freeware program of the scanner variety, from CompuServe or America Online. With Macs, it's best to scan each new floppy even before unpacking the program files.

- ✔ PC users can get SCAN, a shareware program that works similarly to Disinfectant. Plenty of commercial antivirus programs compete on the shelves, but the shareware ones don't cost as much and updates are easier to find. Plus, they're every bit as good as their commercial counterparts.

As with human diseases, new computer-virus strains pop up all the time. Fortunately, new strains of virus scanners are just about as persistent. Updates to Disinfectant, SCAN, and the other antivirus programs show up regularly on BBSs and on-line services.

McAfee Associates, SCAN's publishers, have thought of a smart way to help new users locate and download protection for the latest viruses.

A program called The Software Exchange comes on each McAfee shareware disk. After you install it, Exchange notes your computer settings, modem type, and other icky details. Then it displays a menu of various shareware programs and offers to automatically download one or more for you.

You peruse the menus for details on any of the programs and if one sounds good, you can tell Exchange to go get it. After you select a file, Exchange scans your hard disk to make sure that you don't already have a copy of that version.

As Exchange prepares to dial the McAfee Exchange and download your program, it tells you how long the call should take based on your modem's speed and system settings. After it has finished downloading, Exchange makes a subdirectory, sticks the new program there, uncompresses the program, and takes you on a test drive.

Neat, huh? PC users can get SCAN and The Software Exchange with the coupon in back of this book.

You have lots of choices when it comes to antivirus programs. Just make sure that you choose one and use it before you try to run any of a new program's files.

README. 1ST, First (Using Your New File)

After you have uncompressed and scanned your new program, you can use the DOS DIR command to see all the names of the files that were hiding inside there.

One of the files you're likely to see will say README or something similar.

Although they appear to have been named by a programmer with an Alice in Wonderland fixation, it's a good idea to humor these bossy files and give them a once-over before trying to run the program.

✔ README files tell you all the last-minute additions and leftovers they stuck in the program. Reading this information will make you a better person.

✔ Sometimes the README file points to the manual's file name and tells you what the other files do. If you're lucky, it tells you what to type to get the program to run.

✔ Many README files run automatically when you type the file's name. Mac users generally can click the file's icon to get the file to pop up in a readable form. Clicking a Windows README file almost always brings up Windows' bare-bones word processor, Notepad.

✔ If you can't seem to get a DOS README file to work, try using the DOS TYPE command. First, make sure that you're in the same directory as the README file. Then type the following at the DOS prompt:

```
C:\> TYPE README.TXT | MORE
```

Press Enter. On most keyboards, the weird | character is usually down by the right Shift key. Substitute the file's real name for README.TXT in this example, of course.

✔ If you're using DOS 5 or a later version, you can use the *editor* program that comes bundled with it. (An editor is like a no-frills, nerdy word processor.) Type **EDIT** and the name of the README file and then press Enter.

✔ There's an easier way. Ask a friend to give you a copy of a shareware program called LIST. Running LIST is as simple as typing **LIST FILENAME.TXT** (substituting your README file's name for FILENAME.TXT, of course). This is another good one to stick in your utility directory, incidentally. Also, if LIST is in a directory on your path, you can use it anywhere on your hard disk.

See Program Run. Run, Program, Run.

Mac users have it so good. Getting a program to run is usually a matter of clicking the program icon. If you have a PC, however, you may need to do a bit of sleuthing before you can figure out what "secret word" gets the program to do its stuff.

✔ If you've forgotten what files were in the program, use the DOS DIR command again. DIR spits out a listing of all the program's files. Type the following at the DOS prompt:

```
DIR
```

Press Enter.

✔ If too many files scroll right off your screen, type **DIR /P** and press Enter.

✔ The file name to type will have one of the following three-letter *extensions* after the dot: BAT, EXE, or COM.

✔ If the file ends in PCX, EPS, TIF, GIF, it's a picture, not a program, and you need to download a special *graphics viewer* program before you can view the naked lady or the parrots.

Discovering How to Exit the Program

Some of the clingy types of programs don't want you to leave. "No, stay just a bit longer," they implore, plying you with salted plums and herbed walnuts. You know it's time to go when they start dragging out the photo album of their Niagara Falls vacation. But how to say good-bye politely?

✔ If you're stuck in a new program and there's no clear escape route, try pressing the Esc key.

✔ No luck? Press Alt-X or Alt-Q instead.

✔ Sometimes pressing the function key F1 brings up a help menu that points the way to the exit sign. Function keys are the ones with F and a number on them; they do varied things in software, depending on the program.

If all your endeavors to quit a program elude success, there's always the "three-fingered salute": pressing the Ctrl-Alt-Del keys simultaneously. Technical types call this *rebooting* your computer, a drastic measure best saved for the times when your computer balks, freezes up, and refuses to play nice.

Treat this tip as the last resort it is: rebooting your computer in the middle of a program causes you to lose any unsaved data or work you did. The reboot may be harmful to your PC as well. And, in any case, it lacks the graceful aura you're striving to maintain.

Giving a Home to a "Keeper"

You've downloaded it, unpacked it, scanned it, and tried it. And you like it. If the program's a keeper, or even a wait-and-see-er, it's time to give it a home.

✔ Use the DOS Make Directory (MD) command to provide your new program with its own place, called a subdirectory, on your hard disk. At the DOS prompt, type the following:

```
C:\> CD\
```

Press Enter. Now type:

```
C:\> MD NEWSTUFF
```

Now use the DOS COPY command to get the file from C:\JUNK over to C:\NEWSTUFF, or whatever you want to call it. At the DOS prompt, type the following, and don't forget to press Enter after each step:

```
C:\> CD JUNK
C:\> COPY *.* C:\NEWSTUFF
C:\> CD\
C:\> CD NEWSTUFF
```

Now you're in the Newstuff program's NEWSTUFF directory, ready to do stuff with Newstuff.

✔ You get to choose a name for your new directory, but it's best to give it a name that's short and yet somehow evokes a sense of what the heck's in there.

✔ Mac programs get their own private folders (and the names you give them can be as long and eloquent as you like). Moving the new programs is a matter of dragging their files over into the folder. Refer to *Macs For Dummies* for basic command pointers.

✔ After you've tried out the new file and everything is working okay, be sure to delete the packed file to keep your hard disk tidy (and to keep it from filling up with needless duplicates).

✔ Special packer programs called *shells* make this whole process easier. A great Windows shell is WINZIP. A DOS program worth checking out is Shez. Mac users will enjoy Stuffit Deluxe or one of its many variants.

Glossary

●●●

*70: To disable call-waiting on most Touch-Tone phones, place ***70** (star-7-zero) in front of the phone number in your software's dialing directory. (Or, if you can find it, change your modem software's Dialing Prefix command to **ATDT*70** instead of **ATDT**.)

14,400 bps: Modems that support the CCITT V.32*bis* standard can connect at speeds up to 14,400 bps. (Often abbreviated as 14.4 Kbps, the K is short for *thousands* of bps.) Fax modems that support the V.17 standard can achieve 14,400 bps when connected to another V.17 fax modem.

24, 24 lines: The number of lines of text most PCs can display on their monitors. When a BBS or on-line service asks, "How many lines can your monitor display?" answer with **24**.

2400 bps: Modems that support the V.22*bis* standard can connect at speeds up to 2400 bps.

7,E,1: A rarely seen standard, this data parameter is used for some on-line services. Before calling CompuServe, set your modem software's data word format, port setting, or communication parameters to **7,E,1**.

8,N,1: The setting used most often for calling on-line services and BBSs. Modem software usually refers to **8,N,1** as a data word format, port setting, or communications parameter.

80: The number of columns (characters per line) most computers can display on their monitors. When a BBS or on-line service asks, "How many columns can your monitor display?" say **80**.

9-pin DIN connector: A Mac serial cable has this type of connector at one end.

9- to 25-pin adapter: If your cable is too big to fit your serial port, buy this small connector at computer or hardware stores, which lets a serial cable plug into your PC's serial port connector.

9600 bps: Modems that support the CCITT V.32 standard can work at speeds up to 9600 bps.

access number: The number you dial to reach a BBS or on-line service.

America Online: One of the fastest-growing on-line services, it's a haven for Mac users, yet offers an interesting mixture of service for everyone else. America Online requires its own special software; your normal telecommunications software can't call it up.

ANSI: The acronym for American National Standards Institute, the ANSI standard lets terminals agree on how to transmit and display graphics. If a BBS asks whether you would like ANSI graphics, reply Yes. If the screen starts to look funny, head for the BBS's terminal settings area and change it to No.

ANSI BBS: A terminal setting used by many BBSs and modem software to display on-screen graphics. See **ANSI**.

archive: The modem world uses this in several ways. A file that's been compressed to save space is referred to as *archived*; common compression programs, like PKZIP, Stuffit, and Disk Doubler are called *archiving utilities*. On the Internet, an *anonymous ftp archive* is a file-laden computer that's open to callers who log on as "anonymous" — they don't need to set up an account on that machine to access certain files.

ASCII (American Standard Code for Information Interchange): Pronounced *Ask-eee*, this lengthy acronym stands for a plain text format understood by just about any brand or model of computer. On-line, an ASCII file often goes by the name "text file." (Tech alert: ASCII assigns each alphabetical letter, number, and other symbols one of 128 code numbers; the code number for the capital letter B is 66, for example.) See **text file**.

ASCII transfer: Computers can send ASCII files back and forth by swapping the actual words and letters in the ASCII file. Relatively simple, an ASCII transfer is supported by almost every type of modem software because it doesn't use the newer or fancier transfer protocols. That *also* means it doesn't offer any of the error-correction or speed those fancy protocols bring.

AT Command Set: Modems using this popular language are called *Hayes compatible*. Most modem software speaks this language, so your modem needs to speak it, too. For example, sending the AT command **ATDT555-5555** makes a modem dial the phone number 555-555. You needn't bother with AT commands because your modem software handles all that stuff. See **Hayes compatible**, **command mode**.

auto-answer: A common feature to make the modem answer the phone automatically — regardless of whether it's another modem

or your mom on the other end of the line. If it's your mom though, watch out!

baud: Technoid term for how quickly the modem's carrier signal changes state. Most other people use the term to describe their modem's speed (used inaccurately in all but the slowest modems) ... as in "I just bought a 9600-baud modem." Often abbreviated as *bps*.

BBS: Short for "bulletin board system." Some hobbyists or professionals put special BBS software on their computers, hook the computer to a phone-linked modem, and let other modem users call it up. BBS callers can commonly transfer files, read and leave messages, and type messages back and forth with other callers. When you buy a modem, ask the sales clerk for a list of BBSs to call. To run your own BBS, you'll need a computer, BBS software, a modem, a phone line, and a sense of humor for all the wacko callers you'll get.

binary, **binary transfer:** Binary files, unlike ASCII files, usually cannot be shared between incompatible computer types or different software. CompuServe, the Internet, and other hosts may ask you to specify whether you want to transfer an ASCII or binary file.

bit: A bit is the smallest unit of data a computer can understand. A bit has a value of either 1 or 0; like tiny switches, bit signals are either on (1) or off (0).

block: When sending a file, modem software breaks that file up into little packages called *blocks* and then sends the blocks across the phone lines, one at a time. The block's size depends on the file-transfer protocol being used. During a file transfer, your modem software often shows how many blocks have been transmitted.

bps: A modem's speed is measured in bps, or bits per second; the more bits it can transfer, the faster (and more expensive) the modem.

buffer: Buffers are places to store data temporarily, so that the computer can grab the data when it has some free time. Buffers in the newer UART chips can hold the last 16 characters that streamed in through the serial cable, for example. The result of using buffers is increased speed. Also, a modem program's *scroll-back buffer* stores the last few screens of text that have passed through your software's terminal window while you've been on-line.

byte: Most personal computers deal with *bits* in groups of eight, called *bytes*. Files, hard disks, and computer memory are measured in bytes. While on-line, you'll often see the byte size next to downloadable files. The more bytes, the longer the file will take to download. The letter K stands for kilobytes, or 1,024 bytes. The letter M, or MB, stands for megabytes, or 1,048,576 — about a million bytes. (After you're off-line with all your new files, you'll see your hard disk fill up in thousands and millions of bytes, too.)

call waiting: Many phone companies offer call-waiting: your phone beeps during a conversation to tell you another call is coming in. This beep can put a serious whammy on your modem sessions; most call-waiting services can be disabled with a simple command prefacing a phone number.

capture: Most modem software lets you store — *capture* — incoming text to a file on a disk or send it your printer. Some things worth capturing from a BBS or on-line service are file lists, welcome announcements, lists of message subject areas, and a helpful commands list.

carrier: The high-pitched squeal of joy expressed by two modems who've established a connection across the phone lines, which means they can start speaking to each other. When a modem hangs up or loses its connection, it "drops carrier" — often sending you the message NO CARRIER.

CCITT: An acronym for the Comité Consultatif International Télégraphic et Téléphonique, the international standards committee that nailed down V.32, V.32*bis*, V.42, and V.42*bis*, among other standards. The CCITT recently swapped acronyms with the International Unit of Transistor-Toting shareholders: now they're the IUT-T.

chat: To type "conversationally" with other callers who are on-line at the same time you are. On-line services and BBSs often offer chat rooms where several callers gather to exchange pleasantries. Conferencing is a more-focused form of chatting, usually devoted to a single topic or an on-line "interview."

Class 1: A fax modem standard governing the sending and receiving of fax information between the computer, the fax modem, and Group 3 fax machines.

Class 2: A later fax modem standard refining Class 1; the modem takes on more of the work formerly performed by the computer.

COM port: Short for "communications port," this refers to the serial port your modem uses to get stuff into and out of your computer. Most PCs offer at least one COM port; internal modems provide their own serial port. Your modem software must know what COM port your modem is using.

COM port settings: Some modem software makes you type the necessary COM port, data parameters, and modem speed for each place you call. These are often found in a "COM port settings" menu.

command: An order that forces your computer — or DOS or your modem— to do your bidding: print a file, copy a disk, or call another computer, for example. When you select an option from your modem software's menu, the software sends an AT command to your modem. (Some technoids type these AT commands directly to the modem through the software's terminal window. See **AT Command Set**.) While you're on-line and connected to another computer, the commands you select or type go to the _other_ computer — not to your software or modem. Weird, huh?

command mode: When a modem is in command mode, it's listening for _AT commands_. AT commands force the modem to do something — redial the last number you tried to call, for example. See **data mode**.

communications software: Another term for modem software, these programs let you talk to your computer's modem. That's how you talk to other computers. Often abbreviated to simply _comm software_, some folks call it _term software_ instead.

Communications Toolbox: A feature on Macintosh System 6 and later operating systems that eases the process of getting modems and modem software to talk to each other.

compression program: PKZIP, ARC, Stuffit, and similar programs make files smaller — _compress_ them — by spotting and removing redundancies. Most files available on-line are compressed. That means you'll have to _uncompress_ them before they'll do anything. Look for the uncompress portions of PKZIP, ARC, Stuffit, and others on popular BBSs' file utilities areas. You can download them for your own computer.

CompuServe Information Service: CompuServe, or CIS for short, is the foremost on-line service, sporting the largest number of special-interest forums, files, customer-support forums — and some of the highest connect-time charges.

conferencing: Various on-line services hold conferences where forum members meet at a scheduled time to discuss issues, chat with special guest "speakers," and type back and forth at each other. Anything you type during these live sessions is broadcast to everyone else in the conference.

cps: Short for "characters per second," it's commonly used in the modem world to measure data-transfer rates. A 2400 bps modem can transmit about 240 cps.

cursor: That little blinking square or line in a program or operating system that awaits your next typed character or action. A computer mouse is one way of moving the cursor; the keyboard's arrow keys move the cursor, too.

data: Information stored on a computer — anything from sounds that make your computer burp, to programs that make your modem call places, to a text file with recipes for chokecherry brandy.

data bits: One of the _data parameters_ your modem software asks you about. Modems communicate _serially_, meaning that the bytes that make up your data are broken down into single bits on their way to the phone lines. By agreeing on the number of data bits, the modems can keep better track of arriving data. The number of data bits is either 7 or 8.

data compression: A way to shrink a file before sending it so that it won't take so long to squirt it over the phone lines. Most high-speed modems support data-compression standards called MNP 5, V.42*bis*, or both. When the calling and answering modems agree on data compression, they work out a system for identifying and cutting redundancies in the data bits to be sent. Then the receiving modem restores everything back to its usual long-windedness on its end. Hardware modem data compression isn't the same thing as software file compression.

data mode: When the modem is in data mode, stuff you type goes through to the host computer — the one you're connected to.

data parameters: One of the many ways modems agree on how data will be transmitted and received while they're connected. Before you can connect to a host computer, you need to know what *data parameters* it requires and set your modem software's dialing directory or "settings" accordingly. (Your software may call this "communications settings" or "data word format.") Most on-line services and BBSs ask for **8,N,1**, meaning 8 data bits, No parity, and 1 stop bit. (CompuServe requires 7,E,1.)

database: On-line databases contain vast amounts of information for subscribers to search, sort, or download into their computers. Some special services — Nexis, Lexis, and Knowledge Base, among others — focus exclusively on providing database search and retrieval to subscribers.

default: This simply means the automatic choice or setup that something is set to: **8,N,1**; COM1 is the default communications setting on most modem software, for example.

dialing directory: A "phonebook" for BBSs and on-line services within your modem software.

DIP switch: A row of tiny switches used to change a gizmo's settings. Tiny DIP switches on internal modems can change the modem's COM port, for example. Some external modems have a bank of DIP switches that control auto-answer or other functions. See **jumper**.

download: To copy a file off another computer and onto your own computer. Computers can transfer files by using several different protocols, each having its own advantages and disadvantages. See **file-transfer protocol**.

duplex, full or half: A modem software setting to be changed if you don't see anything you type on-screen (set to *full*) or you see two of everything (set to *half*). See **echo**.

echo: On a BBS, an echo is a networked message base, or special interest topic, usually distributed to Fido-Net BBSs worldwide. Many BBSs share a group of messages and pass them on, like a round-robin story. For example, your local BBS may carry the Firearms Echo; you'd look here for discussions on which semiautomatic weapons the drug dealers are using now. On modem software, the *duplex* setting is often called echo; you would set your local echo to *on* if you weren't able to see what you're typing, and set it to *off* if you saw double characters. In a long tunnel, an echo is what you hear when you shout "Ho, ho" real loud.

editor: A simple, no-frills word processing program used to quickly compose a text file, like an e-mail message, README.TXT file, or other document. Most on-line services and BBSs offer built-in editors in the mail area for typing replies. Some modem software offers an editor, as well. Finally, *off-line readers* work in tandem with an editor to let you compose replies and new messages for uploading by the reader program. See **off-line reader**.

e-mail: Short for "electronic mail," it refers to messages sent or received through BBSs and on-line services, as well as office networks and the Internet. MCI Mail and other subscription services specialize in electronic mail.

emoticon: Smiling faces and other "emotional icons" created with punctuation marks to add humor, irony, and other sparks to an e-mail message. To view my California sunglass smiley, tilt your head to the left: 8^).

error correction: A way for modems to detect and correct any haphazard errors when sending or receiving files. A modem with MNP 2-4 or V.42 standards (or both) can detect and correct data errors when connected to another error-correcting modem. Most file-transfer protocols also check for errors. If you have a V.42 modem and you're *sure* you've achieved an error-correcting connection, try using YModem-G or one of the other streaming protocols on your next file transfer to achieve higher speeds (but only if you're *sure*).

expansion slot: A slot built inside a PC for sliding in internal modems, sound cards, and other gizmos. A typical PC comes with five to eight of these slots inside the case, ready for cards that will upgrade your PC's powers.

external protocol: Sometimes a new file-transfer protocol is *too* new: No modem software includes it. So, some programs add on that new protocol's capabilities for use with your existing modem software. Dubbed *external* protocols because they didn't come with the modem program, they're found on BBSs and on-line services for the most gotta-have-it power users. Most folks get by just fine without ever encountering or using an external protocol.

F1: Press this function key to see some helpful information while using most modem software.

fax/data modem: Most modems today conform to Group 3 fax standards; they're called fax/data modems.

file-transfer protocol: A set of wrestling rules modems must agree on before they can upload or download files to each other. ZModem, XModem, and Kermit are a few of the many protocols you'll see. Your modem software *and* the computer you're calling must offer the same protocol before you can use it.

flame: To flame someone on-line is to send unreasonably insulting, angry, or otherwise rude messages to him or her. A longtime-flame session between two or more people on-line is known as a flame war.

flow control: The way in which modems and computers receiving data can call a halt and process data that's already piled up. There's *software flow control* (Xon/Xoff) and *hardware flow control* (RTS/CTS).

forum: A place where people with similar interests leave messages for each other on-line to yak about their favorite topic, be it Tonkinese cats, internal chip architecture, or French merlots. Every on-line service and most BBSs offer forums or their equivalent. Other names for forums include special interest groups (SIGS), RoundTables, bulletin boards, networks, echoes, newsgroups, message topics, message boards, and clubs.

garbage: A nickname for random characters that stream across your screen when a connection is not going well. Phone line noise (crackling), mismatched data parameters, or calling at the wrong speed can all produce varying degrees of garbage.

Group 3 Fax: A fax standard; a Group 3 fax modem with this on the label should be able to communicate with most other fax machines.

GUI: Short for "graphical user interface." Electrical engineers got tired of putting "Don't Touch" in 15 languages on electrical wires. So, they posted a picture of a guy getting shocked and falling off. That's a GUI, similar to ones in Macintosh or Microsoft Windows. GUIs let people point at pictures with a mouse and click to boss their computers around; that's often easier than typing in commands — known as a "command-based interface."

handshaking: A friendly bunch, modems do handshaking when they're trying to settle on rules and protocols for the ensuing on-line session. Different from *hardware handshaking*. See **hardware flow control**.

hardware flow control: Also known as hardware handshaking, or RTS/CTS, this is a way for modems to stop the data flow while they catch up on stuff that's already been sent. Hardware-handshaking cables send little "stop" signals down two of the cable wires (RTS and CTS). You must set your communications software for this method of flow control, but only if you're sure you have this handshaking cable.

Hayes-compatible: A modem using a language used by 99 percent of the other modems out there. Buying a modem that's not Hayes compatible is like buying a car without a steering wheel: It will be difficult to drive.

host, host mode: Any computer you call is known as the host computer. In a slightly different vein, many modem programs offer a "host mode": That lets you set up your computer so a friend can call it and see

menus offering file downloads, messages, and chatting (practically like a real BBS).

installation program: A miniprogram offered by many modem programs that takes you step-by-step through placing it on your hard drive and making it work with your particular computer. A good modem program should be able to sniff out and diagnose a modem's COM port.

integrated software: All-in one software with several applications, usually including a simple modem program. Currently, they all seem to have names with *Works* or *Suite* in them.

Internet: A large network of even-bigger computer networks spanning the globe and offering access to information, people, and other resources.

jumper: Jumpers are tiny push-on/pull-off doohickeys that can set an internal modem's COM port to a different port.

Kermit: A relatively slow file-transfer protocol, Kermit comes with most modem programs. It's seldom chosen for any but the most specialized connections (usually between PCs and mainframes or other large computers).

line feeds: A setting on your modem software that scrolls material up the screen by one line.

line noise: Phone connections are sometimes less than perfect. We humans simply talk louder or call back later when we perceive a bad connection, but modems usually interpret phone line "noise" as data. That's when random characters (see **garbage**) come through on your screen, usually looking somewhat like this: }}}}}}}a{4$@{{{. Try calling back at a lower speed or simply calling back later.

log file, or **session log:** A feature in modem software that lets you record an entire on-line session, and then save it to a file. When you call up the file, you can review every command you typed, and all the information that passed through your terminal window.

log off: To hang up from a modem connection. See **carrier**.

log on: To type in your name, password, and any other required gobbledygook required by newly called computers.

lurkers: People who hang around on forums or in conferences on-line but never type any messages or contribute anything to the conversation. Lurking is okay until you become accustomed to the BBS or on-line service's atmosphere.

MNP 4: An MNP standard for error correction, it helps keep files from getting damaged as they bounce across the telephone poles from computer to computer.

MNP 5: The MNP standard for data compression, able to squeeze a file down to half its normal size. (Handy for long distance callers, MNP 5 reduces phone costs when you're downloading files.)

modem: A contraction of two old technical words — *mod*ulator and *dem*odulator — that's turned into a new word: *modem*. Modems receive digital data from your computer's serial port and *modulate* it — turn the data into sound waves — before sending it through a phone line. A modem on the other end of the phone lines *demodulates* these sounds — turns them back into data again — for the owner of the other computer to play with.

modem driver: A special program that configures your software to recognize and take advantage of your modem brand. To find modem drivers, look first on your modem's BBS; second, on your software's BBS.

modem initialization string: A place in your modem software to add AT commands that help your software work better with your particular modem; also, the string of commands. You may find yourself changing this, say, if you want to disable error correction and data compression for some reason.

moof!: The cry of the Dogcow. You're likely to encounter this on America Online and other places.

null modem adapter: A special gizmo that lets you connect two computers directly, through their serial ports.

OCR: Short for "optical character recognition" — software that translates a received fax from a picture on your computer screen into real text — words you can edit or otherwise work with.

off-line menu: Some programs give this name to the menus and other software commands you see when you're not in the software's terminal window.

off-line reader: Programs that automate all your on-line correspondence. Off-line readers can automatically call the other computer, grab your mail or messages, and hang up. You can reply to the messages at leisure, when the connect-time charges aren't adding up. Done? Then the off-line reader can call the other computer back up and send your responses to the appropriate people or message areas. The off-line reader you use must be compatible with the BBS or on-line service; it's a good idea to get the same one used by your modem guru.

on-line: On-line means *connected*. When your printer is on-line, it's hooked up and ready to roll, for example. And when you're on-line, you're connected to a host computer. Often expressed as "going on-line" (in hopeful tones).

on-line service: A place you can call with your modem that charges subscription fees or hourly rates for files, fellowship, and games, plus shopping, news, sports, weather, and other stuff. CompuServe, America Online, Prodigy, GEnie, DELPHI, BIX, and The WELL are the primary on-line services.

parity: One of the communications settings or data parameters; you almost always will choose N, for *None* (CompuServe requires E, for *Even*). Parity controls how the sending and receiving modems check for errors.

PCMCIA modem: A credit-card-sized card that holds a modem and fits into a portable computer's PCMCIA slot.

phone cable: The flat, plastic-coated (usually beige or gray) wire that runs between your phone's wall jack and your modem's *line* jack. New modems usually come with a phone cable. Also known as RJ-11 cable. See **RJ-11 cable**.

power adapter: A little black box that plugs into an AC outlet. A little black wire on the box plugs into your external modem to give it power.

protocol: The modem world is full of *rules*, or *protocols*, for how things should take place. There are file-transfer protocols, speed protocols, error-correction protocols, and many other rules. Another word for protocol is *standard*.

READ-ME file: A text file usually tossed onto the disk with a program, it often contains last-minute information or installation instructions, or warns you of mistakes printed in the manual.

"real" comm software: Modem software used by chatty modem gurus you'll meet on-line. For example, don't be surprised to hear, "You're still using Windows Terminal? Get some *real* comm software like Unicom."

remote control software: An advanced program that lets you connect to a computer with a modem and use it as if you were sitting there. For example, Jerry could install remote control software on his work computer and his laptop computer; next, he would call the work computer by using the modem in his laptop. From the office, Jerry's computer screen would look like an invisible Jerry was sitting there, working on spreadsheet figures. From home, Jerry would be in bed, working on spreadsheet figures on his laptop, with a smile on his face.

RJ-11 cable: Average normal phone wire. See **phone cable**.

RS-232 cable: The nerdy, official name for a serial cable, the thick wire that connects the butts of your computer and external modem.

RTS/CTS: See **hardware flow control**.

scanner: A serial device resembling a pocket vacuum cleaner that plugs into a serial port or internal slot (just like a modem does). By sliding the scanner over text or a picture, the scanner sends a picture of it into your computer.

script: A way to "record" keystrokes, save them, and play them back at a later date. For example, many people tire of typing the same letters each time they log on to the same computer. So, they record the keystrokes as a script and tell the modem software to play back that script whenever they call that same computer. Some software offers advanced "script programming languages" that excite "power users" — those people who have all the AT commands memorized.

scroll-back buffer, or **screen buffer**: A place where your modem software stores the past few screens of text that's flowed by your terminal window. A "scroll-back" command in most modem software lets you see the buffer, in case something went by too fast the first time. Most programs also let you send the buffer's contents to a file for later review.

second phone line: What the phone company will soon be installing at your home when all your friends — the people you haven't seen for a while — start complaining about how your phone is busy all the time.

self-extracting archive: A file that's been compressed using a program that self-uncompresses when you run the program — either by typing the file's name (for DOS programs) or clicking its icon with your mouse (for Mac programs).

serial cable: External modems need a serial cable, also called an RS-232 cable. One end plugs into the modem, the other into the PC's serial port. Make sure that you buy one with all the pins included, capable of "hardware handshaking." Special, "multiheaded" serial cables can adapt to all sizes and genders of serial port.

shareware: Software given away by programmers on a "try it, you'll like it" basis. You send the programmer a check — called a registration fee — only if you find yourself using (and presumably liking) the program. The majority of programs on the file areas of BBSs and on-line services are shareware.

Shell to DOS: A software command where you can temporarily exit the program and drop into the DOS operating system — usually to look for a file or to create a directory.

slot cover: The long, shiny metal shields that you unscrew to prepare your PC's expansion slot for an internal modem. Saving these in a drawer somewhere makes you feel virtuous (but if you're like most people, you'll never be able to find one when you need it).

stop bit: The ending "framing bit" that tells the receiving modem a byte has been sent. See **data parameters**.

streaming protocols: Most file-transfer protocols contain error-correction schemes; YModem-G and a few other protocols use a streaming technique that leaves any error correction to error-correction-capable modems. See **error correction**.

surge suppressor: Like ocean waves, electrical current is subject to surges ... Buy a surge suppressor to keep any surges from capsizing the delicate innards of your PC stuff. Special suppressors are available with phone-line surge suppression.

telecommunications: A fancy word for the stuff a modem lets you do, plus even more technical concepts like cellular and wireless communications.

terminal: The smaller, less-expensive computers used to access larger, more expensive host computers — usually mainframes. Terminals usually contain merely a terminal and a keyboard; they're wired directly to the big guy, so they're "dumb" — they can't do much computing on their own. This host/terminal relationship now describes *any* two computers in a given modem connection: The calling computer is the *terminal*, and the computer answering the phone is the *host* or *hostess*. : -)

terminal emulation: The capability of modem software to "pass off" your computer as a terminal type that's expected by (and compatible with) the host computer. A good program offers several types of terminal emulation; special, "dedicated terminal emulator" programs are required for people who need to emulate more obscure terminals.

terminal mode: When in terminal mode, your modem software puts most of its menus away and you see a big open "window" that soon fills up with data if you're talking to another computer.

text file: Another name for an ASCII file. A text file has no formatting codes — margins, line spacing, bold-face, and other bits of fanciness — so it can be shared across computer and software types.

throughput: The actual speed you achieve during a file transfer. May be higher than a modem's "sticker" speed, say, under error correction/data compression, a "streaming protocol," and an ideal phone connection.

UART: A chip on a computer's serial port that controls the way the serial port talks to modems and other devices.

Unix or **UNIX:** A more advanced (read, cryptic) operating system than DOS; often used on Internet-connected computers. See **religion**.

upload: To send a file to a host computer with your modem. (One wag wrote that she remembered the difference between upload and download by picturing the other, host computer up in the sky, where she could receive data *down* from it or send data *up* to it.)

users' group: A regularly scheduled, actual, live-human, face-to-face, drive-there-in-your-car meeting of people who all *use* the same particular software package, computer type, or computer application. Your city may have a desktop publishing users' group, for example, or one devoted to modems. Users' groups are great sources for shareware, news and opinion, nerdy snacks, and technical support.

V.17: The CCITT standard for fax-modems capable of faxing at 14,400 bps with other fax modems.

V.32: CCITT standard for modems that can achieve speeds up to 9600 bps.

V.32*bis*: CCITT standard for modems that can achieve speeds up to 14,400 bps (the CCITT was running short of numbers that day...)

V.32*terbo*: Talks at speeds up to 19,200 bps; an extension of V.32*bis*.

V.34: The name everyone expects the CCITT (now the IUT-T) to pick for the nearly settled V.Fast standard that will talk at 28,800 bps.

V.42: CCITT standard for modems that have error correction; incorporates the MNP 2-4 error-correction schemes.

V.42*bis*: CCITT standard for modems that can do data compression, squeezing a file down to a quarter of its byte-size during actual transmission and restoring it back to normal on the receiving end.

V.Fast: Near-future standard for modems capable of working at speeds up to 28,800 bps. See **V.34**.

virus: Malicious program that can harm your data; the reason we scan each new file before giving it a home on our computers.

VT100-102: Terminal emulation that acts like a terminal used with DEC VAX minicomputers. Choose this setting if in doubt about what to use.

WELL, The: Commercial on-line service that focuses on message topics and user interaction (as opposed to file downloads or shopping).

XModem: An older, slower file-transfer protocol that's widely found on-line and in modem software. This one's a good standby when neither ZModem nor YModem are available. _Caution:_ Many variants are out there; be sure that the XModem on the service and your modem software match each other, or your file transfer won't work.

Xon/Xoff: Software method of flow control that you specify through a setting in your modem software. This is slightly less efficient than **hardware flow control**.

YModem: The YModem file-transfer protocol improved on XModem's speed and efficiency, allowing several files to be transferred in a "batch" — if this is supported by the host computer you've called.

ZIP: A file ending in the letters ZIP — FILE.ZIP, for example — has been compressed with the PKZIP compression utility. You can't use a ZIP file without uncompressing it first. PKZIP and PKUNZIP are among the first files new modem users should download.

ZModem: Currently the favorite, this file-transfer protocol offers error correction, speedy transfer rates, and batch transfers. Best yet, it can resume _where it left off_ when a file transfer has been interrupted because of line noise or other problems.

Index

• G •

• H •

● *p* ●

Which On-Line Service Has Real People Playing All The Time?

The ImagiNation Network.

The ImagiNation Network is the world's first interactive computer entertainment network dedicated to fun and games. Your PC & modem are your passports to all the adventures and new friends just waiting for you on-line. Quest in a medieval dungeon fighting Trolls and Orcs. Shoot down your new friend's plane in a flight simulator. Play Bridge or Chess or join in one of our many conferences. There are bulletin boards, e-mail, and clubs. Literally hundreds of activities and thousands of people all waiting for you on-line 24 hours a day, 7 days a week.

Our membership kit is FREE and includes 7 megs of software containing over 20 games with graphically rich animation and sound card support. But best of all, it comes with 10 FREE hours for you to explore your new world.

If you have a 386 (or better) personal computer, VGA monitor, mouse, a 2400 baud modem (or faster), and $4.95 for shipping and handling we'll send you the complete software membership kit. Call 1-800-462-4461 to order the kit by phone, or just fill out and send in the coupon below — and we'll see you on-line!

ImagiNation™
THE IMAGINATION NETWORK

000448

TM indicates a registered trademark of The ImagiNation Network ©1993 The ImagiNation Network

YOU'VE READ THE BOOK...
NOW OWN THE MODEM!

Sportster is designed for beginning modem users -- at home or in the office!

Sportster is easy to set up and use. With a comprehensive, easy-to-read manual, Quick Reference Guide and a toll-free help line, you can be on-line just minutes after you open the box. And **Sportster** has all the performance, compatibility and reliability **U.S. Robotics** is famous for.

❋ **Rated #1 by Computer Shopper Magazine!**

❋ **PC/Computing calls Sportster "a perfect balance of performance and price."**

There's one word for the support you get on CompuServe:

Personal.

CompuServe, the world's most comprehensive network of people with personal computers, is also a worldwide network of people with answers to your hardware and software questions. You'll find quick solutions and information from thousands of members, including product developers.

All you need to get started is your computer, a modem, communications software, and a CompuServe membership. To get a free introductory membership, complete and mail the form on the back of this page. Or call **1-800-524-3388** and ask for Representative 370.

Act now to receive one month of free basic services plus a $15 usage credit for extended and premium CompuServe services.

Put the world at your fingertips.

Join the world's largest international network of people with personal computers. Whether it's computer support, communications, entertainment, or continually updated information, you'll find services that meet your needs.

Your introductory membership will include one month of free basic services plus a $15 usage credit for extended and premium CompuServe services.

To get connected, complete and mail the card below. Or call **1-800-524-3388** and ask for Representative 370.

CompuServe®

☐YES!

Send me my FREE CompuServe Introductory Membership including a $15 usage credit and CompuServe basic services membership free for one month.

Name:_____

Address:_____

City:_____State:_____ZIP:_____

Phone:_____

Clip and mail this form to: CompuServe
P.O. Box 20212
Dept. 370
Columbus, OH 43220

Explore the Internet-FREE!

DELPHI is the only major online service to offer you full access to the Internet. And now you can explore this incredible resource with no risk. You get 5 hours of evening or weekend access to try it out for free!

Use DELPHI's Internet mail gateway to exchange messages with over 10 million people at universities, companies, and other online services, such as CompuServe and MCI Mail.

Download programs and files, using **FTP**, or connect in real-time to other networks using **Telnet**. You can also meet people on Internet. **Internet Relay Chat** lets you "talk" with people all over the world, and **Usenet News** is the world's largest bullentin board with over 3500 topics!

To help you find the information you want, you'll have direct access to powerful search utilities such as "Gopher," "Hytelnet," "WAIS," and "the World-Wide Web." If you aren't familar with these terms, don't worry; DELPHI has expert online assistants and a large collection of help files, books, programs, and other resources to help get you started.

Over 600 local access numbers are available across country. Explore DELPHI and the Internet today. You'll be amazed by what you discover.

DELPHI
Internet Services Corp.

Questions? Call 1-800-695-4005.
Send e-mail to INFO@delphi.com

Put the power of GEnie® at your fingertips ...

Want to make the most of your new modem? Try out GEnie! You can download thousands of new programs, applications, utilities, and files ... talk to other computer owners and experts to learn more about your machine (and get help fast!) ... explore special interest RoundTables on everything from scuba to Food and Wine, Microsoft to Medicine, Internet to Investing ... play the most incredible multi-player games in the business ... access daily closing stock quotes ... and much more!

And GEnie has it all at a standard connect rate of just $3.00 an hour![1] That's the lowest hourly connect rate of all the major online companies! Plus – because you're a reader of Modems For Dummies – you get an even sweeter deal.[2] If you sign up before December 31, 1994, we'll waive your first monthly subscription fee (an $8.95 value) and include ten additional hours of standard connect time (another $30.00 in savings). That's **fourteen free hours** during your first month – a $38.95 value![3]

You can take advantage of this incredible offer immediately – just follow these simple steps:

1. Set your communications software for half-duplex (local echo) at 300, 1200, or 2400 baud.
2. Dial toll free in U.S. 1-800-638-8369 (or in Canada 1-800-387-8330). Upon connection, type **HHH** (Please note: Every time you sign onto GEnie, you need to enter the HHH upon connection).
3. At the U#= prompt, type **EXPLORE** and press <Return>.
4. Have a major credit card ready. In the U.S., you may also use your checking account number. In Canada, VISA and MasterCard only.

Or, if you need more information, contact GEnie Client Services at 1-800-638-9639 from 9am to midnight, Monday through Friday, and from noon to 8pm Saturday and Sunday (all times Eastern).

[1] U.S. prices. Standard connect time in non-prime time: 6pm to 8am local time Mon. - Fri., all day Sat. and Sun. and selected holidays.
[2] Offer available in the United States and Canada only.
[3] The offer for six additional hours applies to standard hourly connect charges only and must be used by the end of the billing period for your first month. Please see our GEnie brochure for more information on pricing and billing policies.

Effective date as of 7/1/93. Prices subject to change without notice. Offer limited to new subscribers only and one per customer.

What would it take . . .

MSI
MUSTANG
SOFTWARE
INC.

. . .to persuade you?

Mustang Software, Inc. would like to introduce you to the *Qmodem* family of professional communications software for your PC.

That's why we're making this special offer, exclusively for purchasers of this book.

Use the coupon on the other side of this page to order your own copy of *QmodemPro* for DOS, or *QmodemPro* for Windows, the finest communications software in the world, *for less than half price!*

This isn't a "special edition" with limited functionality, or an old, outdated version of the software. This is the latest, most up-to-date release, complete with printed manuals, extensive online help, and technical support when you need it.

This is exactly the same product sold in software stores, at half the usual retail price! And you risk nothing. If you're not satisfied, return the package to Mustang Software, Inc. within 30 days, for a full refund.

Simply tear out the coupon on the other side of this page, and mail it to Mustang Software, Inc. with your payment. Within a few days, you'll be using the most powerful communications software on the market today.

Need more convincing?

How would you like to try *Qmodem* for only $1.00?

Over a quarter of a million modem users world wide have already "test driven" *Qmodem*. It's one of the most popular communications software programs in the world — for ten years running!

One dollar buys you the Test Drive version of *Qmodem*. That's practically free. In fact, that's *less than our cost* for disks, packaging and postage.

This is a DOS version of *Qmodem*, with many of the same features as our retail versions, including a complete, powerful scripting language, host mode, automatic quick modem configuration, and more.

This is not a "slide show" demo, this is a fully functional communications software program with online help, and full documentation on disk. Use it, copy it, even give it to your friends — with a clear conscience.

Try it. You have nothing to lose!

Mustang Software, Inc., P.O. Box 2264, Bakersfield, CA 93303
805-395-0223 • FAX 805-395-0713 • BBS 805-395-0250

Okay, I'm convinced.

I'm ready to try QmodemPro for DOS or QmodemPro for Windows, at less than HALF PRICE!

☐ **Send me QmodemPro for DOS for $49.** List price $99

☐ **Send me QmodemPro for Windows for $69.** List price $139

Name: _____

Address: _____

City: _____ State: _____

Country: _____

Phone: _____

Payment method (do not send cash):

☐ Check/Money Order ☐ Visa/MC/AMEX/Discover

Card No. _____

Signature _____ Exp._____

Include shipping:

☐ $10/US ☐ $15/Canada ☐ $35/Overseas
**CA residents include sales tax.

Place this coupon in an envelope and mail with your payment to:

Mustang Software, Inc.
QmodemPro Special Offer
P.O. Box 2264
Bakersfield, CA 93303

* Original coupon must be mailed to MSI, no FAXes or photocopies.

- -

☐ **YES!** **Send me Qmodem Test Drive (DOS only) for $1**
That's right, one dollar.

What's the catch?

We simply want you to try *Qmodem*.

To introduce you to Mustang Software, Inc.'s line of communication software products, we're prepared to *give* you a copy of our classic *Qmodem* Communications Software for a small handling fee of only $1.00

Use it for as long as you like, with our compliments. This is fully functional communications software, with features comparable to programs costing $100 or more.

Name: _____

Address: _____

City: _____ State: _____

Country: _____

Phone: _____

Place this coupon in an envelope and mail with your payment to:

Mustang Software, Inc.
Qmodem Test Drive Offer
P.O. Box 2264
Bakersfield, CA 93303

NETCOM INTERNET SERVICES

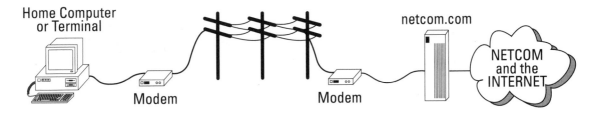

Connect to NETCOM and we will put you in touch with the world.

NETCOM On-Line Communication Services, Inc. provides full Internet access for a flat monthly fee with *NO CONNECT TIME CHARGES.* Our services include:

E-Mail	Stock Reports	USENET
World News	US News	Spreadsheets
Editors	Multi-user Chat	Menu Driven Interface
Telnet	WAIS	Compilers
ftp	Shell (all)	S/W Archivers

M*ention this coupon you found in* Modems For Dummies *with your account registration verification and*
receive a $10.00 discount
on your $20.00 setup fee. The monthly rate is $19.50.
Autobill to a credit card and receive an additional 10% off!

NETCOM is a full-service network provider committed to affordable and reliable Internet access.

Call 1-800-488-2558 for a list of our nationwide access numbers! NETCOM supports hundreds of high speed modems allowing access from 2400 bps to 14.4 kbps. To register, call your local access number and log in to our guest account. Set your communications parameters to 8-1-N and log in as "guest." Please remember to use lower case letters.

You may also contact our main office at 1-800-501-8649 or
408-554-8649 and ask for a personal account representative.

4000 Moorpark Ave
Suite 200
San Jose, CA 95117

**member of the
Commerical Internet eXchange
(CIX)**

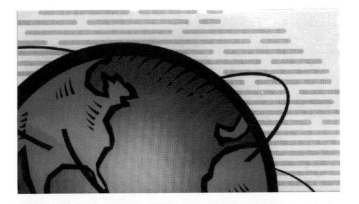

FREE PRODIGY® SOFTWARE OFFER

If you have a computer, you're probably familiar with the PRODIGY service, the online network that will let you get more out of your computer than you ever imagined possible. Now, with this offer, you can get **FREE** PRODIGY software ($4.95 S&H), so you can try Membership in the PRODIGY service for a month.

From financial information to investing services, to continually-updated news and sports, to learning via an encyclopedia, to bulletin board communications — with millions of other Members, PRODIGY brings you this and much more.

For details about this offer, and PRODIGY service pricing plans, or to order the PRODIGY service, call **1 800 PRODIGY, ext. 199.**

PRODIGY®
Service

Why ZyXEL?

INTELLIGENCE

ZyXEL U-Series modems offer an array of intelligent features that aren't found in other modems. These features ensure fast, reliable operation—and true "plug-and-play" operating ease.

❑ Digitized Voice Capability with Speech Compression
❑ Caller ID Decoding and Distinctive Ring Detection

❑ Fast Retrain With Auto Fall-Forward/Fall-Back
❑ Call-Back Security with Password Protection
❑ Decodes Touch Tone Response to Fax/Voice
❑ Remote Configuration Capability

SPEED

The U-Series offer the highest speeds available. And with auto fall-forward/fall-back plus fast retrain, throughput is maximized—even during adverse line conditions.

❑ Data Speed: ZyXEL 16.8Kbps (19.2Kbps for PLUS Series)
❑ DTE Speed: 57.6 Kbps (76.8Kbps for PLUS Series)
❑ G3 FAX, EIA Class 2 (send and receive):V.17-14.4Kbps

❑ V.32bis/V.32: 14.4Kbps/9.6Kbps
❑ V.22bis: 2.4Kbps
❑ V.22 Bell/212A: 1200bps
❑ V.21 Bell/103: 300bps
❑ V.23: 1200/75Kbps
❑ V.33, V.29, V.27ter., V.26 (with U-1496 PLUS)

COMPATIBILITY

ZyXEL modems operate with asynchronous and synchronous* systems—and in all environments, including DOS®, Windows®, OS/2®, Macintosh®*, UNIX®*, NeXT®*, and Amiga®*. They are also compatible with most other modems, and a wide variety of popular communications and fax software.

The U-Series support V.25bis* and the AT Command Set, and comply with CCITT standards V.42bis/V.42, and MNP® 5/4/3 data compression and error correction protocols.
*Not available with ZyXEL internal modems.

OUR WARRANTY

All ZyXEL modems come with a 5-year parts and labor warranty (2-year warranty for Rackmount models). And they're backed by our knowledgeable technical support team, and a 24-hour BBS Technical Support Line.

You also may rest assured that the U-Series modem you buy today will be the modem of tomorrow. Because the U-Series offers easy product upgrades through programmable EPROMs.

A WISE INVESTMENT

The ultra-high speed, universal compatibility, and intelligent features of the U-Series work together to save you time, effort, and money. And the U-Series come in Desktop, Plug-in-Card, and Rackmount models. So there's a version to fit every need—especially since a cellular option is available with the new PLUS Series.

Plus, ZyXEL modems come with Voice/Fax software—a powerful, easy-to-use, software package to meet your needs.
Which means you've made a wise investment by choosing a ZyXEL U-Series modem.

PRODUCT SELECTION GUIDE   NEW

ZyXEL U-Series	U-1496 PLUS	U-1496E PLUS	U-1496E	U-1496B PLUS	U-1496B	U-1496R*	U-1496P	ZyCellular Option
Model Type	external	external	external	internal (PC bus)	internal (PC bus)	rackmount	portable	available with all models
Display	LCD	LED lights	LED lights	N/A	N/A	LED lights	LED	N/A
Data Speed	ZyX 19.2/16.8Kbps	ZyX 19.2/16.8Kbps	ZyX 16.8Kbps	ZyX 19.2/16.8Kbps	ZyX 16.8Kbps	ZyX 19.2/16.8Kbps	ZyX 16.8Kbps	Cell 9.6Kbps (cellular mode)

*RS-1600 Rack System offers 16 U-1496R modem slots; 20 x 4 LCD display, single power supply.

*RS-1600N Rack System offers 16 U-1496R slots; 20 x 4 LCD display, redundant power supply, NMS control card & software.

The Intelligent Modem

4920 E. La Palma Avenue, Anaheim, CA 92807 (714) 693-0808 Fax:(714) 693-8811 24-hr. BBS Tech Support:(714) 693-0762
®ZyXEL and U-Series are registered trademarks of ZyXEL Communications Corp. Other trademarks listed are the properties of their respective owners. US FCC and Canadian DOC approved.

ZyXEL High Speed Modem/Fax

Now that you have become an expert on everything you ever wanted to know about modems, it's time to actually choose one. Let us introduce you to the ZyXEL U-Series. ZyXEL U-Series modems offer an array of intelligent features that just aren't found in other modems:

- •Data Speed: 19.2/16.8/14.4 Kbps
- •DTE Speed: Up to 76.8 Kbps
- •Digitized Voice Capability
- •MNP® 3-4-5 and V.42/V.42bis
- •Line Probing Techniques

- •Auto Fall-Forward/Fall-Back
- •Remote Configuration
- •24 Hour BBS Hotline
- •Firmware Upgradable
- •V.17/14.4Kbps Fax

- •EIA Class II G3 Fax (Send & Receive)
- •Caller ID & Distinctive Ring
- •Call-back Security/Password Protection
- •Decodes Touch Tone Response to Fax/Voice
- •5 Year Warranty

We know that the technical gurus are very impressed with these features, and we're sure that you will be too, once you see how much time, effort, and money ZyXEL's intelligent features will save you. You don't have to be a modem genius to operate a ZyXEL modem.

ZyXEL's fast. ZyXEL's friendly. ZyXEL's reliable. And now ZyXEL can be yours...**for $50 less**. Make this your intelligent choice and let ZyXEL be your wise investment.

✂ Please cut along the dotted line

Special Discount Coupon

$50 OFF!*

Have you ever used a BBS or CompuServe?

If so, you've probably encountered ZIP files.

Are you a Windows user?

If so, WinZip is *THE* way to handle these achived files.

WinZip brings the convenience of Windows to using ZIP files. It features an intuitive point-and-click, drag-and-drop interface for viewing, running, extracting, adding, deleting, and testing archives with optical virus scanning support.

Shareware *evaluation* versions of WinZip are available
on better bulletin boards everywhere.

 Special Offer— *Save $5 off the $29 price* (includes shipping)
with this coupon.

Offer good only when paying by check to
Nico Mak Computing, Inc.
P.O. Box 919
Bristol, CT 06011-0919

Connecticut residents, please add 6% sales tax.

Name: _____

Company: _____

Address: _____

City, State, ZIP: _____

Country: _____

Payments must be in US funds. Offer expires Christmas 1994. Original coupons only.
Price subject to change without notice. This coupon is not valid in France or French-speaking territories.

The Standard in Communications Software.

PROCOMM PLUS
$64⁰⁰*

PROCOMM PLUS
for Windows
$89⁰⁰*

The world's best-selling communications software is now available to you directly from DATASTORM TECHNOLOGIES, INC. Order today and you will receive PROCOMM PLUS for $64* or PROCOMM PLUS *for Windows* for $89*. Customer service representatives are available to take your call 8 a.m. - 5 p.m. (Central Time) Monday through Friday.

Call 314.443.3282
and place your order now!

DATASTORM TECHNOLOGIES, INC.

P.O. Box 1471 • Columbia, MO 65205 • 314.443.3282

How Many of These America Online Services Could You Use Right Now?

▲ Stock Quotes and Portfolio Management

▲ More than 70,000 Software Files and Programs

▲ Fast, Easy Multi-File Downloading

▲ Searchable Online Encyclopedia

▲ International E-Mail Gateway, Fax, and U.S. Mail Capabilities

▲ Multi-Player Graphic Games

▲ Microsoft Small Business Center

▲ Access to Computer Industry Experts

▲ Special Interest Groups and Clubs

▲ Graphic Windowing Interface

▲ Homework Help and Tutoring Sessions

See other side for detail about your FREE trial of America Online!

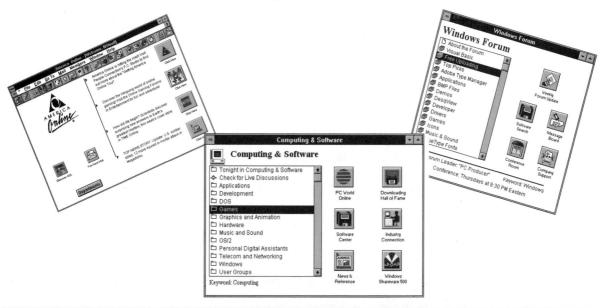

Own a Modem?

Try America Online for FREE

AVAILABLE FOR MAC, DOS, AND MICROSOFT WINDOWS™!

If you own a computer and a modem,
we invite you to take this opportunity to try
**the nation's most exciting online service, America Online®,
for FREE.**

▲ Build a software library by downloading selected files from a
library of thousands — productivity software, games, and more!

▲ Get computing support from industry experts at online conferences
and through easy-to-use message boards.

▲ Tap into expert advice on running a small business in the exclusive
Microsoft® Small Business Center.

▲ And much more! Order now and you'll receive FREE software
and a FREE trial membership to try America Online.

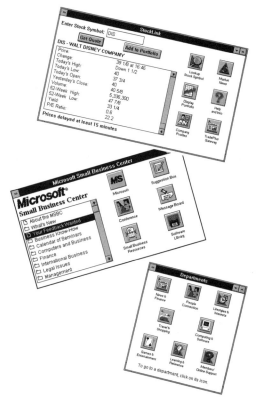

Join us in this exciting opportunity
Simply complete the form below,
clip and mail to:
America Online
Distribution Department
8619 Westwood Center Drive
Vienna, VA 22182
Or Call 1-800-827-6364, Ext. 8383

- -

[ ] **Yes**, I want to try America Online! Send me FREE software and a FREE trial membership to try the service.

Name: _____

Address:_____ Home Phone: (___)_____

City:_____ State:_____ ZIP:_____

Disk Type and Size:
IBM-compatible*:   [ ] 5.25   [ ] 3.5   [ ] High Density   [ ] Double Density   **Apple**: [ ] Macintosh®
Do you use Windows 3.1? [ ] Yes [ ] No
*The IBM-compatible version of America Online requires a PC/XT or higher; 512K memory or higher; a hard drive; a
mouse; and a Hurcules or EGA monitor or above.

Use of America Online requires a major credit card or checking account.
America Online is a registered service mark of America Online, Inc. Other product and service names are trademarks and service marks of their respective owners.

8383/W

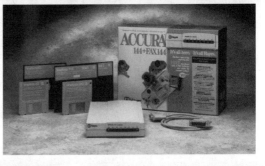

The Leader Of The Pack In Windows Comm Software.

Naturally, you'd expect *us* to say that Crosstalk® for Windows is the biggest, baddest comm software around. But the experts agree with us: For both ease-of-use and power, Crosstalk blows past the other Windows™ comm programs– including Procomm Plus® for Windows.

Easy-To-Use Rider.

Crosstalk has always been known for its horsepower. But now all that power is combined with tremendous ease of use. That's why *PC Magazine,* in naming us Editors'

"Even with all the added interface pizzazz, Crosstalk keeps an air of clean, no-nonsense functionality, unlike Procomm for Windows, which offers enough glitz to be distracting."
—*PC World,* October 1992

Choice for the second year in a row, said, "This package has changed a great deal since last year, when it earned our Editors' Choice Award. But it has changed only for the better."

Crosstalk Leaves The Competition In The Dust.

Fact is, if you're ready for a Windows comm program, there are two products you're most likely to consider: Crosstalk for Windows and Procomm Plus for Windows. And reviewers have been clear

> *Crosstalk covers all bases with its sophisticated scripting for high-end users and quick on-line access for novices. This rich package is suitable for any environment....*
> —PC Magazine Editors' Choice, April 1993

about their preference.

PC Magazine called our interface "elegantly designed" while noting that with Procomm Plus for Windows, "so many non-essential elements remain on-screen during terminal sessions." And *Windows Sources* commented that Crosstalk, "unlike Procomm for Windows, provides a full screen for sessions."

InfoWorld summed it all up: "Crosstalk is stronger than Procomm for Windows." Nuff said.

Ride With The Leader.

Through June 30, 1994, registered users of Procomm Plus can upgrade to Crosstalk for Windows 2.0 for only $49, by calling our toll-free number.

Give it a try, and you'll fall for the leader of the pack. And yes, we *are* talkin' to you.

1-800-348-3221, ext. 48X*

CROSSTALK
Another Communications Solution From DCA®

IDG BOOKS

Order Form

Order Center: (800) 762-2974 (8 a.m.-5 p.m., PST, weekdays) **or (415) 312-0650**

For Fastest Service: Photocopy This Order Form and FAX it to : (415) 358-1260

Quantity	ISBN	Title	Price	Total

Shipping & Handling Charges

Subtotal	U.S.	Canada & International	International Air Mail
Up to $20.00	Add $3.00	Add $4.00	Add $10.00
$20.01-40.00	$4.00	$5.00	$20.00
$40.01-60.00	$5.00	$6.00	$25.00
$60.01-80.00	$6.00	$8.00	$35.00
Over $80.00	$7.00	$10.00	$50.00

In U.S. and Canada, shipping is UPS ground or equivalent.
For Rush shipping call (800) 762-2974.

Subtotal _____

CA residents add
applicable sales tax _____

IN residents add
5% sales tax _____

Canadian residents
add 7% GST tax _____

Shipping _____

TOTAL _____

Ship to:

Name _____

Company_____

Address_____

City/State/Zip _____

Daytime Phone _____

Payment: ❏ Check to IDG Books (US Funds Only)   ❏ Visa   ❏ MasterCard   ❏ American Express

Card # _____   Exp. _____   Signature _____

Please send this order form to: IDG Books, 155 Bovet Road, Suite 310, San Mateo, CA 94402.
Allow up to 3 weeks for delivery. Thank you!

BOBFD

IDG BOOKS WORLDWIDE REGISTRATION CARD

RETURN THIS REGISTRATION CARD FOR FREE CATALOG

Title of this book: Modems for Dummies

My overall rating of this book: ❏ Very good [1] ❏ Good [2] ❏ Satisfactory [3] ❏ Fair [4] ❏ Poor [5]

How I first heard about this book:

❏ Found in bookstore; name: [6]   ❏ Book review: [7]

❏ Advertisement: [8]   ❏ Catalog: [9]

❏ Word of mouth; heard about book from friend, co-worker, etc.: [10]   ❏ Other: [11]

What I liked most about this book:

What I would change, add, delete, etc., in future editions of this book:

Other comments:

Number of computer books I purchase in a year: ❏ 1 [12] ❏ 2-5 [13] ❏ 6-10 [14] ❏ More than 10 [15]

I would characterize my computer skills as: ❏ Beginner [16] ❏ Intermediate [17] ❏ Advanced [18] ❏ Professional [19]

I use ❏ DOS [20] ❏ Windows [21] ❏ OS/2 [22] ❏ Unix [23] ❏ Macintosh [24] ❏ Other: [25]_____
(please specify)

I would be interested in new books on the following subjects:
(please check all that apply, and use the spaces provided to identify specific software)

❏ Word processing: [26]   ❏ Spreadsheets: [27]

❏ Data bases: [28]   ❏ Desktop publishing: [29]

❏ File Utilities: [30]   ❏ Money management: [31]

❏ Networking: [32]   ❏ Programming languages: [33]

❏ Other: [34]

I use a PC at (please check all that apply): ❏ home [35] ❏ work [36] ❏ school [37] ❏ other: [38] _____

The disks I prefer to use are ❏ 5.25 [39] ❏ 3.5 [40] ❏ other: [41]_____

I have a CD ROM: ❏ yes [42] ❏ no [43]

I plan to buy or upgrade computer hardware this year: ❏ yes [44] ❏ no [45]

I plan to buy or upgrade computer software this year: ❏ yes [46] ❏ no [47]

Name: Business title: [48]   Type of Business: [49]

Address (❏ home [50] ❏ work [51]/Company name:)

Street/Suite#

City [52]/State [53]/Zipcode [54]:   Country [55]

❏ **I liked this book!** You may quote me by name in future
IDG Books Worldwide promotional materials.

My daytime phone number is _____

IDG BOOKS

THE WORLD OF COMPUTER KNOWLEDGE

☐ YES!

Please keep me informed about IDG's World of Computer Knowledge.
Send me the latest IDG Books catalog.